Teacher Edition

Eureka Math Grade 3 Module 5

Special thanks go to the Gordon A. Cain Center and to the Department of Mathematics at Louisiana State University for their support in the development of *Eureka Math*.

For a free *Eureka Math* Teacher Resource Pack, Parent Tip Sheets, and more please visit www.Eureka.tools

Published by the non-profit Great Minds

Printed in the U.S.A.
This book may be purchased from the publisher at eureka-math.org
10 9

ISBN 978-1-63255-367-6

***Eureka Math: A Story of Units* Contributors**

Katrina Abdussalaam, Curriculum Writer
Tiah Alphonso, Program Manager—Curriculum Production
Kelly Alsup, Lead Writer / Editor, Grade 4
Catriona Anderson, Program Manager—Implementation Support
Debbie Andorka-Aceves, Curriculum Writer
Eric Angel, Curriculum Writer
Leslie Arceneaux, Lead Writer / Editor, Grade 5
Kate McGill Austin, Lead Writer / Editor, Grades PreK–K
Adam Baker, Lead Writer / Editor, Grade 5
Scott Baldridge, Lead Mathematician and Lead Curriculum Writer
Beth Barnes, Curriculum Writer
Bonnie Bergstresser, Math Auditor
Bill Davidson, Fluency Specialist
Jill Diniz, Program Director
Nancy Diorio, Curriculum Writer
Nancy Doorey, Assessment Advisor
Lacy Endo-Peery, Lead Writer / Editor, Grades PreK–K
Ana Estela, Curriculum Writer
Lessa Faltermann, Math Auditor
Janice Fan, Curriculum Writer
Ellen Fort, Math Auditor
Peggy Golden, Curriculum Writer
Maria Gomes, Pre-Kindergarten Practitioner
Pam Goodner, Curriculum Writer
Greg Gorman, Curriculum Writer
Melanie Gutierrez, Curriculum Writer
Bob Hollister, Math Auditor
Kelley Isinger, Curriculum Writer
Nuhad Jamal, Curriculum Writer
Mary Jones, Lead Writer / Editor, Grade 4
Halle Kananak, Curriculum Writer
Susan Lee, Lead Writer / Editor, Grade 3
Jennifer Loftin, Program Manager—Professional Development
Soo Jin Lu, Curriculum Writer
Nell McAnelly, Project Director

Ben McCarty, Lead Mathematician / Editor, PreK–5
Stacie McClintock, Document Production Manager
Cristina Metcalf, Lead Writer / Editor, Grade 3
Susan Midlarsky, Curriculum Writer
Pat Mohr, Curriculum Writer
Sarah Oyler, Document Coordinator
Victoria Peacock, Curriculum Writer
Jenny Petrosino, Curriculum Writer
Terrie Poehl, Math Auditor
Robin Ramos, Lead Curriculum Writer / Editor, PreK–5
Kristen Riedel, Math Audit Team Lead
Cecilia Rudzitis, Curriculum Writer
Tricia Salerno, Curriculum Writer
Chris Sarlo, Curriculum Writer
Ann Rose Sentoro, Curriculum Writer
Colleen Sheeron, Lead Writer / Editor, Grade 2
Gail Smith, Curriculum Writer
Shelley Snow, Curriculum Writer
Robyn Sorenson, Math Auditor
Kelly Spinks, Curriculum Writer
Marianne Strayton, Lead Writer / Editor, Grade 1
Theresa Streeter, Math Auditor
Lily Talcott, Curriculum Writer
Kevin Tougher, Curriculum Writer
Saffron VanGalder, Lead Writer / Editor, Grade 3
Lisa Watts-Lawton, Lead Writer / Editor, Grade 2
Erin Wheeler, Curriculum Writer
MaryJo Wieland, Curriculum Writer
Allison Witcraft, Math Auditor
Jessa Woods, Curriculum Writer
Hae Jung Yang, Lead Writer / Editor, Grade 1

Board of Trustees

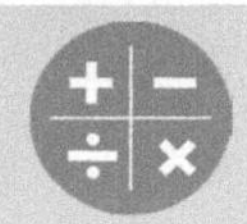

Table of Contents

GRADE 3 • MODULE 5

Fractions as Numbers on the Number Line

Grade 3 • Module 5

Fractions as Numbers on the Number Line

OVERVIEW

In this 35-day module, students extend and deepen Grade 2 practice with equal shares to understanding fractions as equal partitions of a whole (**2.G.3**). Their knowledge becomes more formal as they work with area models and the number line. Throughout the module, students have multiple experiences working with the Grade 3 specified fractional units of halves, thirds, fourths, sixths, and eighths. To build flexible thinking about fractions, students are exposed to additional fractional units such as fifths, ninths, and tenths.

Topic A opens Module 5 with students actively partitioning different models of wholes into equal parts (e.g., concrete models and drawn pictorial area models on paper). They identify and count unit fractions as *1 half, 1 fourth, 1 third, 1 sixth,* and *1 eighth* in unit form. In Topic B, students are introduced to the fraction form $\frac{1}{b}$ (**3.NF.1**) and understand that fractions are numbers. Just like any number, they can be written in different forms.

Students compare and make copies of unit fractions to build non-unit fractions. They understand unit fractions as the basic building blocks that compose other fractions (**3.NF.3d**), which parallels the understanding that the number 1 is the basic building block of whole numbers (e.g., 1 and 1 and 1 make 3 just as 1 third and 1 third and 1 third make 1). In Topic C, students practice comparing unit fractions using fraction strips. They specify the whole and label fractions in relation to the number of equal parts in that whole (**3.NF.3d**).

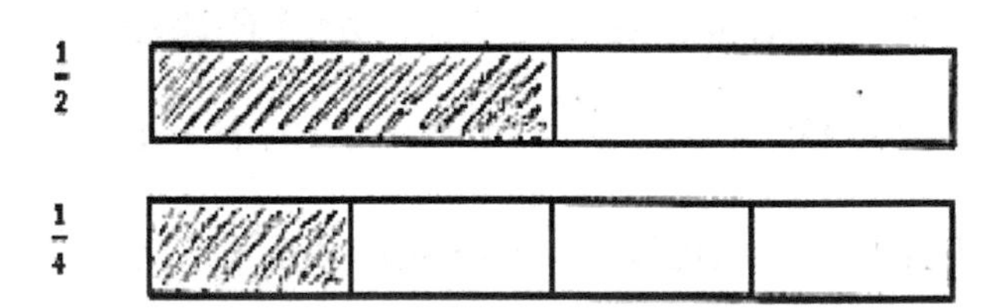

Compare unit fractions using fraction strips.

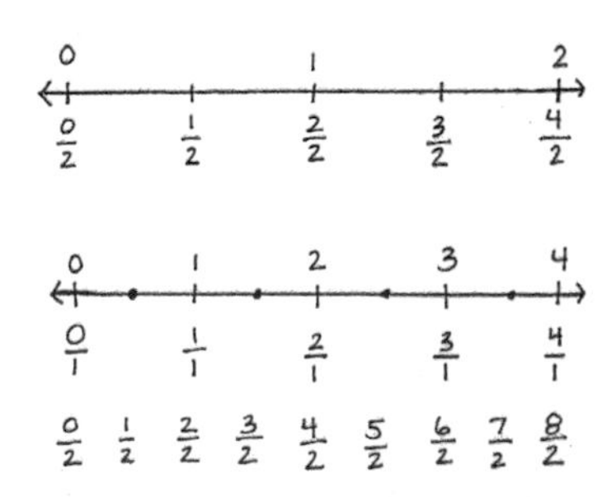

Students transfer their work to the number line in Topic D. They begin by using the interval from 0 to 1 as the whole. Continuing beyond the first interval, they partition, place, count, and compare fractions on the number line (**3.NF.2a**, **3.NF.2b**, **3.NF.3d**). In Topic E, they notice that some fractions with different units are placed at the exact same point on the number line, and therefore, are equal (**3.NF.3a**). For example, $\frac{1}{2}, \frac{2}{4}, \frac{3}{6}$, and $\frac{4}{8}$ are equivalent fractions (**3.NF.3b**); they are different ways of naming the same number. Students recognize that whole numbers can be written as fractions, as exemplified on the number lines to the left (**3.NF.3c**).

Topic F concludes the module with comparing fractions that have the same numerator. As students compare fractions by reasoning about their size, they understand that fractions with the same numerator and a larger denominator are actually smaller pieces of the whole (**3.NF.3d**). Topic F leaves students with a new method for precisely partitioning a number line into unit fractions of any size without using a ruler.

Notes on Pacing for Differentiation

If pacing is a challenge, consider the following modifications and omissions.

Omit Lesson 3. Lesson 3's objective is similar to Lesson 2's. The difference is a shift from concrete to pictorial. Students will have exposure to extensive pictorial practice throughout the module.

Omit Lesson 4. Although Lesson 4 is an exploratory lesson that affords students the opportunity to synthesize their learning, no new material is presented.

Consolidate Lessons 10 and 11, both of which have nearly identical objectives and provide practice comparing unit fractions pictorially. Within the lesson that results, incorporate a variety of models into practice.

Omit Lesson 13. Lesson 13 provides practice with concepts and skills taught in the three preceding lessons. Although this lesson deepens practice, no new material is presented.

Omit Lesson 19. Lesson 19, designated as an optional lesson in the teaching sequence, provides practice with concepts and skills taught in the five preceding lessons.

Omit Lesson 20, an equivalent fractions lesson. The seven subsequent lessons in Topic E provide practice that is more targeted toward specific understandings about equivalent fractions.

Consider omitting Lesson 25 since its content is embedded into the work of prior lessons. Ensure that students have practiced counting and labeling whole number fractions as part of their work with fractions on the number line.

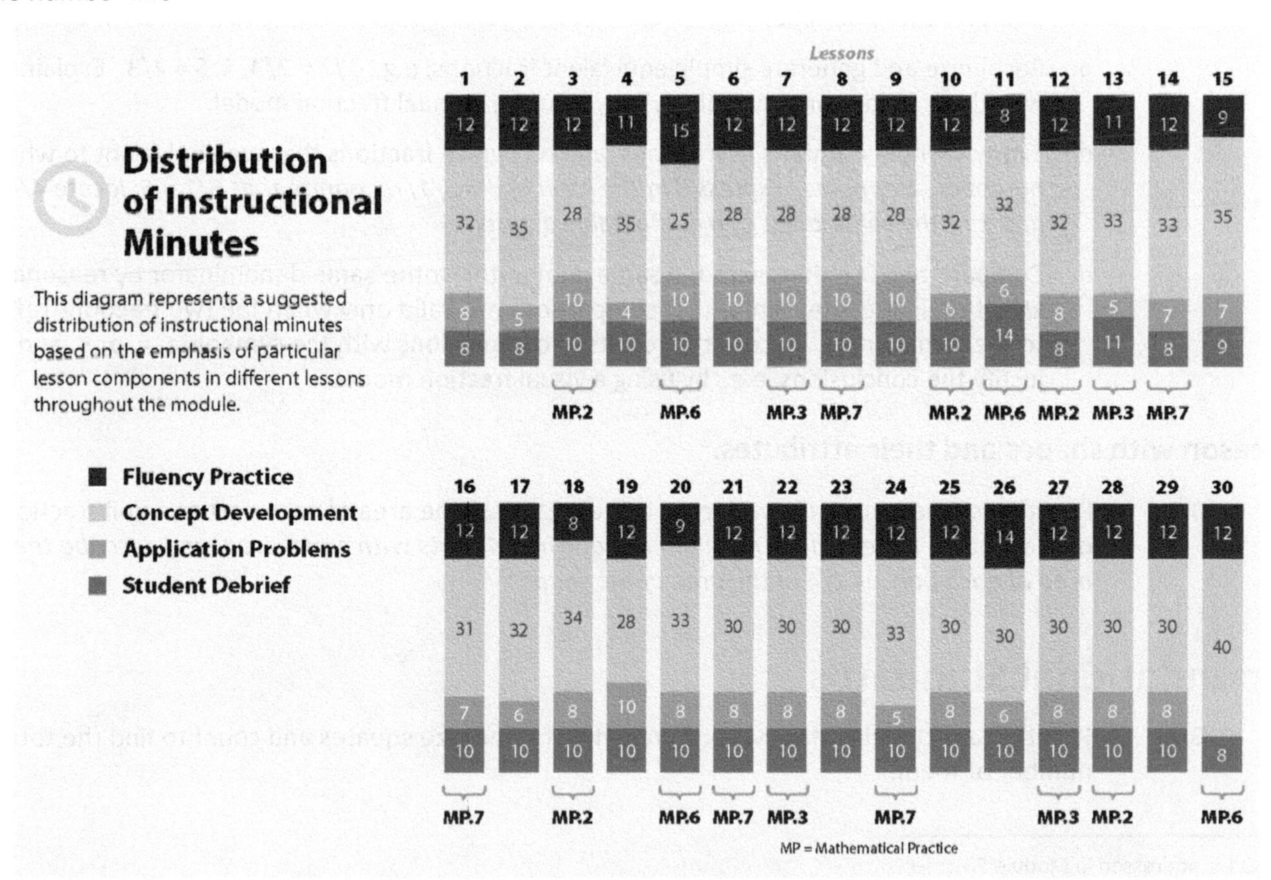

Lesson	1	2	3	4	5	6	7	8	9	10	11	12	13	14	15
Fluency Practice	12	12	12	11	15	12	12	12	12	12	8	12	11	12	9
Concept Development	32	35	28	35	25	28	28	28	28	32	32	32	33	33	35
Application Problems	8	5	10	4	10	10	10	10	10	6	6	8	5	7	7
Student Debrief	8	8	10	10	10	10	10	10	10	10	14	8	11	8	9
Mathematical Practice			MP.2		MP.6		MP.3	MP.7		MP.2	MP.6	MP.2	MP.3	MP.7	

Lesson	16	17	18	19	20	21	22	23	24	25	26	27	28	29	30
Fluency Practice	12	12	8	12	9	12	12	12	12	12	14	12	12	12	12
Concept Development	31	32	34	28	33	30	30	30	33	30	30	30	30	30	40
Application Problems	7	6	8	10	8	8	8	8	5	8	6	8	8	8	
Student Debrief	10	10	10	10	10	10	10	10	10	10	10	10	10	10	8
Mathematical Practice	MP.7		MP.2		MP.6	MP.7	MP.3		MP.7			MP.3	MP.2		MP.6

Focus Grade Level Standards

Develop understanding of fractions as numbers. (Grade 3 expectations in this domain are limited to fractions with denominators 2, 3, 4, 6, and 8.)

3.NF.1 Understand a fraction $1/b$ as the quantity formed by 1 part when a whole is partitioned into b equal parts; understand a fraction a/b as the quantity formed by a parts of size $1/b$.

3.NF.2 Understand a fraction as a number on the number line; represent fractions on a number line diagram.

a. Represent a fraction $1/b$ on a number line diagram by defining the interval from 0 to 1 as the whole and partitioning it into b equal parts. Recognize that each part has size $1/b$ and that the endpoint of the part based at 0 locates the number $1/b$ on the number line.

b. Represent a fraction a/b on a number line diagram by marking off a lengths $1/b$ from 0. Recognize that the resulting interval has size a/b and that its endpoint locates the number a/b on the number line.

3.NF.3 Explain equivalence of fractions in special cases, and compare fractions by reasoning about their size.

a. Understand two fractions as equivalent (equal) if they are the same size, or the same point on a number line.

b. Recognize and generate simple equivalent fractions, e.g., 1/2 = 2/4, 4/6 = 2/3. Explain why the fractions are equivalent, e.g., by using a visual fraction model.

c. Express whole numbers as fractions, and recognize fractions that are equivalent to whole numbers. *Examples: Express 3 in the form of 3 = 3/1; recognize that 6/1 = 6; locate 4/4 and 1 at the same point of a number line diagram.*

d. Compare two fractions with the same numerator or the same denominator by reasoning about their size. Recognize that comparisons are valid only when the two fractions refer to the same whole. Record the results of comparisons with the symbols >, =, or <, and justify the conclusions, e.g., by using a visual fraction model.

Reason with shapes and their attributes.[1]

3.G.2 Partition shapes into parts with equal areas. Express the area of each part as a unit fraction of the whole. *For example, partition a shape into 4 parts with equal area, and describe the area of each part as 1/4 of the area of the shape.*

Foundational Standards

2.G.2 Partition a rectangle into rows and columns of same-size squares and count to find the total number of them.

[1]3.G.1 is addressed in Module 7.

2.G.3 Partition circles and rectangles into two, three, or four equal shares, describe the shares using the words *halves*, *thirds*, *half of*, *a third of*, etc., and describe the whole as two halves, three thirds, four fourths. Recognize that equal shares of identical wholes need not have the same shape.

Focus Standards for Mathematical Practice

MP.2 **Reason abstractly and quantitatively.** Students represent fractions concretely, pictorially, and abstractly, as well as move between representations. Students also represent word problems involving fractions pictorially and then express the answer in the context of the problem.

MP.3 **Construct viable arguments and critique the reasoning of others.** Students reason about the area of a shaded region to determine what fraction of the whole it represents.

MP.6 **Attend to precision.** Students specify the whole amount when referring to a unit fraction and explain what is meant by *equal parts* in their own words.

MP.7 **Look for and make use of structure.** Students understand and use the unit fraction as the basic building block or structure of all fractions on the number line.

Overview of Module Topics and Lesson Objectives

Standards		Topics and Objectives		Days
3.G.2 3.NF.1	A	**Partitioning a Whole into Equal Parts**		4
		Lesson 1:	Specify and partition a whole into equal parts, identifying and counting unit fractions using concrete models.	
		Lesson 2:	Specify and partition a whole into equal parts, identifying and counting unit fractions by folding fraction strips.	
		Lesson 3:	Specify and partition a whole into equal parts, identifying and counting unit fractions by drawing pictorial area models.	
		Lesson 4:	Represent and identify fractional parts of different wholes.	
3.NF.1 3.NF.3c 3.G.2	B	**Unit Fractions and Their Relation to the Whole**		5
		Lesson 5:	Partition a whole into equal parts and define the equal parts to identify the unit fraction numerically.	
		Lesson 6:	Build non-unit fractions less than one whole from unit fractions.	
		Lesson 7:	Identify and represent shaded and non-shaded parts of one whole as fractions.	

Standards		Topics and Objectives	Days
		Lesson 8: Represent parts of one whole as fractions with number bonds. Lesson 9: Build and write fractions greater than one whole using unit fractions.	
3.NF.3d 3.NF.1 3.NF.3a–c 3.G.2	C	**Comparing Unit Fractions and Specifying the Whole** Lesson 10: Compare unit fractions by reasoning about their size using fraction strips. Lesson 11: Compare unit fractions with different-sized models representing the whole. Lesson 12: Specify the corresponding whole when presented with one equal part. Lesson 13: Identify a shaded fractional part in different ways depending on the designation of the whole.	4
		Mid-Module Assessment: Topics A–C (assessment 1 day, return 1 day, remediation or further applications 1 day)	3
3.NF.2ab **3.NF.3cd**	D	**Fractions on the Number Line** Lesson 14: Place fractions on a number line with endpoints 0 and 1. Lesson 15: Place any fraction on a number line with endpoints 0 and 1. Lesson 16: Place whole number fractions and fractions between whole numbers on the number line. Lesson 17: Practice placing various fractions on the number line. Lesson 18: Compare fractions and whole numbers on the number line by reasoning about their distance from 0. Lesson 19: Understand distance and position on the number line as strategies for comparing fractions. (Optional)	6
3.NF.3a–c	E	**Equivalent Fractions** Lesson 20: Recognize and show that equivalent fractions have the same size, though not necessarily the same shape. Lesson 21: Recognize and show that equivalent fractions refer to the same point on the number line. Lessons 22–23: Generate simple equivalent fractions by using visual fraction models and the number line. Lesson 24: Express whole numbers as fractions and recognize equivalence with different units.	8

Standards		Topics and Objectives		Days
		Lesson 25:	Express whole number fractions on the number line when the unit interval is 1.	
		Lesson 26:	Decompose whole number fractions greater than 1 using whole number equivalence with various models.	
		Lesson 27:	Explain equivalence by manipulating units and reasoning about their size.	
3.NF.3d	F	**Comparison, Order, and Size of Fractions**		3
		Lesson 28:	Compare fractions with the same numerator pictorially.	
		Lesson 29:	Compare fractions with the same numerator using <, >, or =, and use a model to reason about their size.	
		Lesson 30:	Partition various wholes precisely into equal parts using a number line method.	
		End-of-Module Assessment: Topics A–F (assessment 1 day, return ½ day, remediation or further applications ½ day)		2
Total Number of Instructional Days				**35**

Terminology

New or Recently Introduced Terms

- Copies (refers to the number of unit fractions in 1 whole)
- Equivalent fractions (fractions that name the same size or the same point on the number line)
- Fraction form (e.g., $\frac{1}{3}, \frac{2}{3}, \frac{3}{3}, \frac{4}{3}$)
- Fractional unit (half, third, fourth, etc.)
- Non-unit fraction (fraction with numerator other than 1)
- Unit form (in reference to fractions, e.g., 1 half, 2 thirds, 4 fifths)
- Unit fraction (fraction with numerator 1)
- Unit interval (the interval from 0 to 1, measured by length)

Familiar Terms and Symbols[2]

- =, <, > (equal, less than, greater than)
- Array (arrangement of objects in rows and columns)
- Equal parts (parts with equal measurements)

[2]These are terms and symbols students have used or seen previously.

- Equal shares (pieces of a whole that are the same size)
- Half of, one third of, one fourth of, etc. ($\frac{1}{2}, \frac{1}{3}, \frac{1}{4}, \frac{1}{6}, \frac{1}{8}$)
- Halves, thirds, fourths, sixths, eighths ($\frac{1}{2}, \frac{1}{3}, \frac{1}{4}, \frac{1}{6}, \frac{1}{8}$)
- Number line 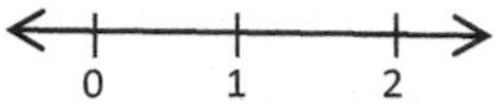

- Partition (divide a whole into equal parts)
- Whole (e.g., 2 halves, 3 thirds, etc.)

Suggested Tools and Representations

- 1-liter beaker (optional)
- 1 m length of yarn
- 12″ × 1″ strips of yellow construction paper
- 2″ × 6″ strips of brown construction paper
- 200 g ball of clay or play dough
- 4″ × 4″ orange squares
- $4\frac{1}{4}$″ × 1″ paper strips
- Arrays
- Clear plastic cups
- Concrete fraction models (e.g., water, string, clay)
- Food coloring (to color water)
- Fraction strips (made from paper, used to fold and model parts of a whole; see the example to the right.)
- Number line
- Pictorial fraction model (e.g., drawing of a circle or square)
- Rectangular- and circular-shaped paper
- Rulers
- Sets of <, >, = cards
- Shapes partitioned into fractional parts
- Tape diagram

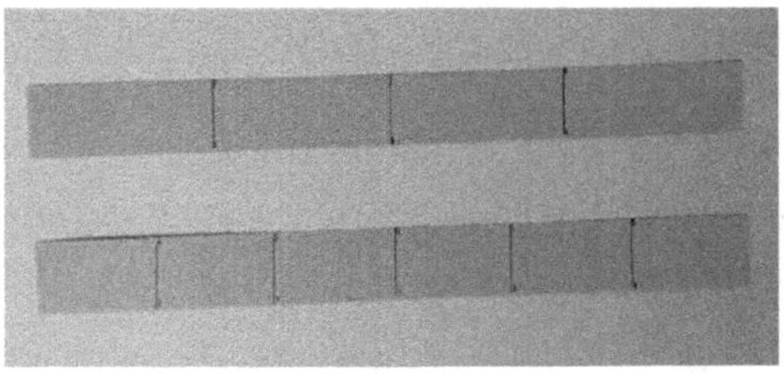

Fraction Strips

Scaffolds[3]

The scaffolds integrated into *A Story of Units* give alternatives for how students access information as well as express and demonstrate their learning. Strategically placed margin notes are provided within each lesson elaborating on the use of specific scaffolds at applicable times. They address many needs presented by English language learners, students with disabilities, students performing above grade level, and students performing below grade level. Many of the suggestions are organized by Universal Design for Learning (UDL) principles and are applicable to more than one population. To read more about the approach to differentiated instruction in *A Story of Units,* please refer to "How to Implement *A Story of Units*."

Assessment Summary

Type	Administered	Format	Standards Addressed
Mid-Module Assessment Task	After Topic C	Constructed response with rubric	3.G.2 3.NF.1 3.NF.3cd
End-of-Module Assessment Task	After Topic F	Constructed response with rubric	3.NF.2ab 3.NF.3a–d

[3]Students with disabilities may require Braille, large print, audio, or special digital files. Please visit the website www.p12.nysed.gov/specialed/aim for specific information on how to obtain student materials that satisfy the National Instructional Materials Accessibility Standard (NIMAS) format.

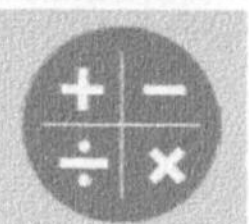

Topic A

Partitioning a Whole into Equal Parts

3.G.2, 3.NF.1

Focus Standard:		3.G.2	Partition shapes into parts with equal areas. Express the area of each part as a unit fraction of the whole. *For example, partition a shape into 4 parts with equal area, and describe the area of each part as 1/4 of the area of the shape.*
Instructional Days:		4	
Coherence	**-Links from:**	G2–M8	Time, Shapes, and Fractions as Equal Parts of Shapes
	-Links to:	G4–M5	Fraction Equivalence, Ordering, and Operations

In Topic A, students partition a whole using a ruler to precisely measure equal parts. They then see how cups can be used to measure equal parts of water. From there, students are invited to fold fraction strips and then estimate to draw pictorial models. The topic culminates in an exploration, wherein they model a designated fraction with a meter string, 12 ounces of water, 200 grams of clay, a 4″ × 4″ square, a 12″ × 1″ strip, and a 6″ × 2″ strip. Students then tour the fraction displays created by their peers and analyze their observations. They specify that the whole contains a certain number of equal parts.

A Teaching Sequence Toward Mastery of Partitioning a Whole into Equal Parts	
Objective 1:	**Specify and partition a whole into equal parts, identifying and counting unit fractions using concrete models.** **(Lesson 1)**
Objective 2:	**Specify and partition a whole into equal parts, identifying and counting unit fractions by folding fraction strips.** **(Lesson 2)**
Objective 3:	**Specify and partition a whole into equal parts, identifying and counting unit fractions by drawing pictorial area models.** **(Lesson 3)**
Objective 4:	**Represent and identify fractional parts of different wholes.** **(Lesson 4)**

Lesson 1

Objective: Specify and partition a whole into equal parts, identifying and counting unit fractions using concrete models.

Suggested Lesson Structure

■ Fluency Practice	(12 minutes)
■ Application Problem	(8 minutes)
■ Concept Development	(32 minutes)
■ Student Debrief	(8 minutes)
Total Time	**(60 minutes)**

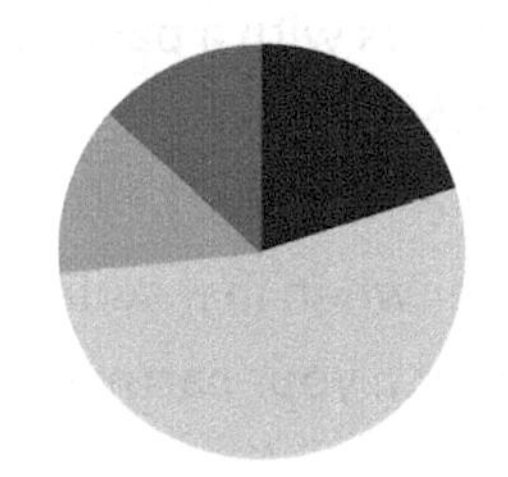

NOTES ON STANDARDS ALIGNMENT:

In this module, students work with a variety of fractional units. These fractional units include halves, thirds, fourths, sixths, and eighths, which are specified in the Grade 3 standards, as well as additional fractional units such as fifths, ninths, tenths, and twelfths. These additional fractional units are not part of the Grade 3 standards. Their inclusion in this module combats rigid thinking, encouraging students to see any number as a fractional unit. This bridges to content in Grades 4 and 5 (**4.NF.1-7 and 5.NF.1-7**). Module 5 assessments do not test the additional fractions.

Fluency Practice (12 minutes)

- Group Counting **3.OA.1** (6 minutes)
- Multiplication by Four and Eight **3.OA.4** (6 minutes)

Group Counting (6 minutes)

Materials: (S) Personal white board

Note: Group counting reviews interpreting multiplication as repeated addition. Count forward and backward by fours twice using personal white boards. Pause between each counting sequence so students see improvement on the second try. After doing the fours twice, have students underline multiples of 8 (e.g., <u>0</u>, 4, <u>8</u>, 12, <u>16</u>, 20, <u>24</u>, 28, <u>32</u>, 36, <u>40</u>, 36, <u>32</u>, 28, <u>24</u>, 20, <u>16</u>, 12, <u>8</u>, 4, <u>0</u>). Then, count forward and backward by eights twice, pausing between each counting sequence to analyze weak points.

Multiplication by Four and Eight (6 minutes)

Materials: (S) Personal white board (optional)

Note: Choose a mode of delivery (e.g., oral work, personal white boards). This activity reviews multiplication using units of four and eight.

Guide students to write and pair facts of 4 and 8 and uncover the doubling:

$2 \times 4 = 8$	$2 \times 8 = 16$
$3 \times 4 = 12$	$3 \times 8 = 24$
$4 \times 4 = 16$	$4 \times 8 = 32$

Application Problem (8 minutes)

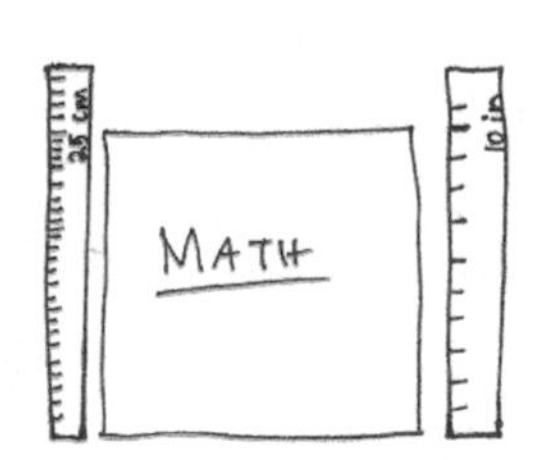

a. Inch is a larger unit.
b. Measuring the book with centimeters yields a greater number.
c. It takes more of a smaller unit, cm, to measure the same item with a larger unit, in.

Materials: (S) Ruler, paper or math book (optional)

Measure the length of your paper or math book using a ruler. Your teacher will tell you whether to measure in inches or centimeters.

Assign partners different units. After students complete the measurement individually and compare answers with a partner, facilitate a discussion using the following suggestions.

a. Which is a larger unit—an inch or a centimeter?
b. Which would yield a greater number when measuring the book—inches or centimeters?
c. Measure at least 2 different items with your partner, again using different units. What do you notice?
d. Change units with your partner. Measure different items again.

Concept Development (32 minutes)

Materials: (T) 1—clear plastic cup full of colored water, 2—other identical clear plastic cups (empty), 2—12" × 1" strips of construction paper (S) 2—12" × 1" strips of construction paper, 12-inch ruler

Note: Students should save the fraction strips they create during this lesson for use in future Module 5 lessons.

NOTES ON MULTIPLE MEANS OF ACTION AND EXPRESSION:

Some students may benefit from a review of how to use a ruler to measure. Suggest the following steps:

1. Identify the 0 mark on the ruler.
2. Line up the 0 mark with the left end of the paper strip.
3. Push down on the ruler as you make your mark.

Part 1: Partition fraction strips into equal parts.

T: Measure your paper strip using inches. How long is it?
S: 12 inches.
T: Make a small mark at 6 inches at both the top and bottom of the strip. Connect the two points with a straight line.
T: (After students do so.) How many equal parts have I split the paper into now?
S: 2.
T: The **fractional unit** for 2 equal parts is halves. What fraction of the whole strip is one of the parts?
S: 1 half.
T: Point to the halves and count them with me. (Point to each half of the strip as students count "one half, two halves.") Discuss with your partner how we know these parts are equal.
S: When I fold the strip along the line, the two sides match perfectly. → I measured and saw that each part was 6 inches long. → The whole strip is 12 inches long. 12 divided by 2 is 6. → 6 times 2 or 6 plus 6 is 12, so they are equal in length.

Continue with fourths on the same strip.

Fourths: Repeat the same questions asked when measuring halves. (Students who benefit from a challenge can think about how to find eighths as well.)

T: Make a small mark at 3 inches and 9 inches at the top and bottom of your strip. Connect the two points with a straight line. How many equal parts do you have now?

S: 4.

T: The fractional unit for 4 equal parts is fourths. Count the fourths.

S: 1 fourth, 2 fourths, 3 fourths, 4 fourths.

T: Discuss with your partner how you know that these parts are equal.

NOTES ON MULTIPLE MEANS OF REPRESENTATION:

Review and post frequently used vocabulary, such as 1 fourth, accompanied by a picture of 1 fourth, 1 out of 4 equal parts, and $\frac{1}{4}$.

Distribute a second fraction strip, and repeat the process with thirds and sixths.

Thirds: Have the students mark points at 4 inches and 8 inches at the top and bottom of a new strip. Ask them to identify the fractional unit. Ask them how they know the parts are equal, and then have them count the equal parts, "1 third, 2 thirds, 3 thirds."

Sixths: Have the students mark points at 2 inches, 6 inches, and 10 inches. Repeat the same process as with halves, fourths, and thirds. Ask students to think about the relationship of the halves to the fourths and the thirds to the sixths.

Part 2: Partition a whole amount of liquid into equal parts.

T: Just as we measured a whole strip of paper with a ruler to make halves, let's now measure precisely to make 2 equal parts of a whole amount of liquid.

Lead a demonstration using the following steps (pictured to the right).

1. Present two identical glasses. Make a mark about 1 fourth of the way up the cup to the right.
2. Fill the cup to that mark.
3. Pour that amount of liquid into the cup on the left, and mark off the top of that amount of liquid.
4. Repeat the process. Fill the cup on the right to the mark again, and pour it into the cup on the left.
5. Mark the top of the liquid in the cup on the left. The cup on the left now shows the markings for half the amount of water and the whole amount of water.
6. Have students discuss how they can make sure the middle mark shows half of the liquid. Compare the strip showing a whole partitioned into 2 equal parts and the liquid partitioned into 2 equal parts. Have students discuss how they are the same and different.

Problem Set (10 minutes)

Students should do their personal best to complete the Problem Set within the allotted 10 minutes. Some problems do not specify a method for solving. This is an intentional reduction of scaffolding that invokes MP.5, Use Appropriate Tools Strategically. Students should solve these problems using the RDW approach used for Application Problems.

For some classes, it may be appropriate to modify the assignment by specifying which problems students should work on first. With this option, let the careful sequencing of the Problem Set guide the selections so that problems continue to be scaffolded. Balance word problems with other problem types to ensure a range of practice. Assign incomplete problems for homework or at another time during the day.

Name Gina Date

1. A beaker is considered full when the liquid reaches the fill line shown near the top. Estimate the amount of water in the beaker by shading the drawing as indicated. The first one is done for you.

1 half 1 fourth 1 third

2. Juanita cut her string cheese into equal pieces as shown in the rectangles below. In the blanks below, name the fraction of the string cheese represented by the shaded part.

1 third

1 sixth

1 fourth

Student Debrief (8 minutes)

Lesson Objective: Specify and partition a whole into equal parts, identifying and counting unit fractions using concrete models.

The Student Debrief is intended to invite reflection and active processing of the total lesson experience.

Invite students to review their solutions for the Problem Set. They should check work by comparing answers with a partner before going over answers as a class. Look for misconceptions or misunderstandings that can be addressed in the Debrief. Guide students in a conversation to debrief the Problem Set and process the lesson.

Any combination of the questions below may be used to lead the discussion.

- Encourage students to use the words **fractional units**, *equal parts*, *fraction*, *whole*, *halves*, *fourths*, *thirds*, and *sixths*.
- The whole in Problem 2 never changes. What happened to the size of an equal part when the string cheese was divided into more parts?
- In Problem 1, which was the harder fraction for you to draw well?

3.

a. In the space below, draw a small rectangle. Estimate to split it into 2 equal parts. How many lines did you draw to make 2 equal parts? What is the name of each fractional unit?

I drew one line to make 2 equal parts. Each part is 1 half.

b. Draw another small rectangle. Estimate to split it into 3 equal parts. How many lines did you draw to make 3 equal parts? What is the name of each fractional unit?

I drew two lines to make 3 equal parts. Each part is 1 third.

c. Draw another small rectangle. Estimate to split it into 4 equal parts. How many lines did you draw to make 4 equal parts? What is the name of each fractional unit?

I drew 3 lines to make 4 equal parts. Each part is 1 fourth.

4. Each rectangle represents 1 sheet of paper.

a. Estimate to show how you would cut the paper into fractional units as indicated below.

sevenths ninths

b. What do you notice? How many lines do you think you would draw to make a rectangle with 20 equal parts?

I noticed that the number of lines is one less than the number of equal parts. For 20 equal parts, I would draw 19 lines.

5. Rochelle has a strip of wood 12 inches long. She cuts it into pieces that are each 6 inches in length. What fraction of the wood is one piece? Use your yellow strip from the lesson to help you. Draw a picture to show the piece of wood and how Rochelle cut it.

12 in

1 half | 1 half

6 in

One piece is 1 half of the wood.

- Using our method with the cups, how could we make a cup that showed thirds?
- In Problem 2, what do you notice about the thirds and sixths? When we marked our measurements on the strips, what did you remember about the measurement of 1 third of the strip and 1 sixth of the strip?
- In Problem 3, did you start drawing fourths by making a half? Can you do the same to draw eighths?
- Walk through the process of estimating to draw a half, then a half of a half to make fourths, etc.
- In Problem 5, let's look at two different solution strategies and compare them.

Exit Ticket (3 minutes)

After the Student Debrief, instruct students to complete the Exit Ticket. A review of their work will help with assessing students' understanding of the concepts that were presented in today's lesson and planning more effectively for future lessons. The questions may be read aloud to the students.

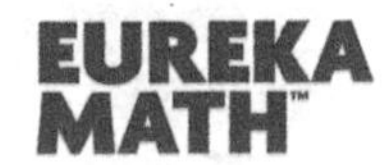

Name ______________________________ Date ______________

1. A beaker is considered full when the liquid reaches the fill line shown near the top. Estimate the amount of water in the beaker by shading the drawing as indicated. The first one is done for you.

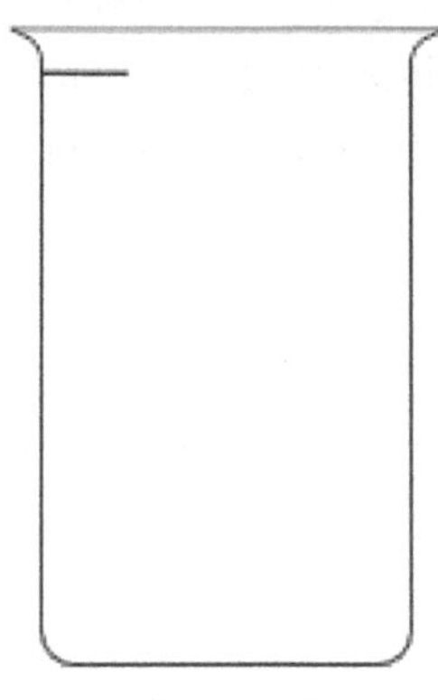

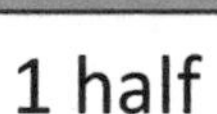
1 half

1 fourth

1 third

2. Juanita cut her string cheese into equal pieces as shown in the rectangles below. In the blanks below, name the fraction of the string cheese represented by the shaded part.

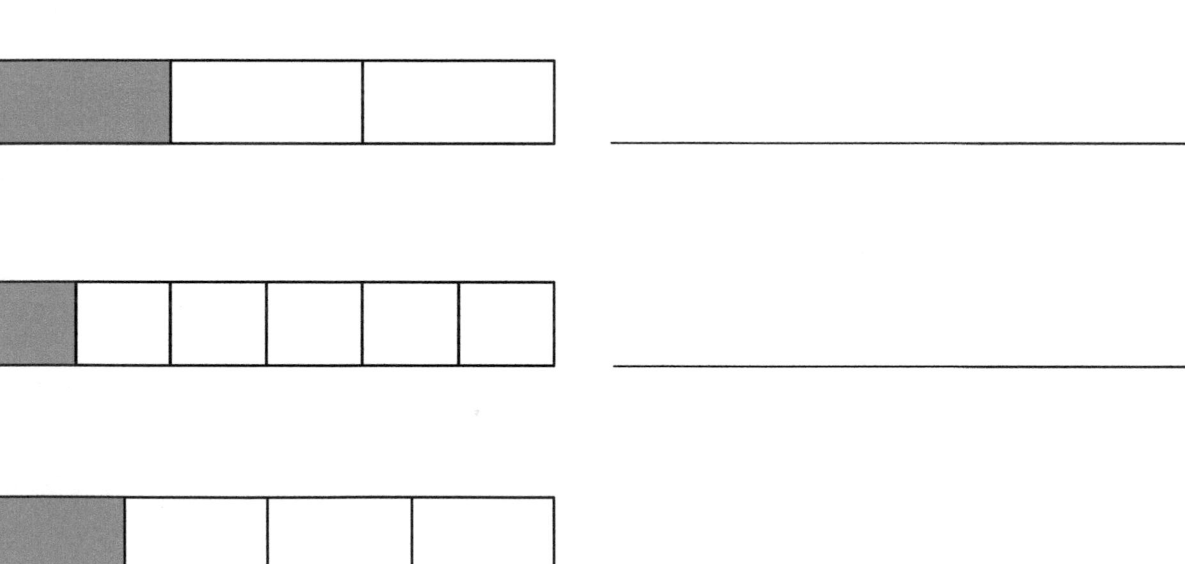

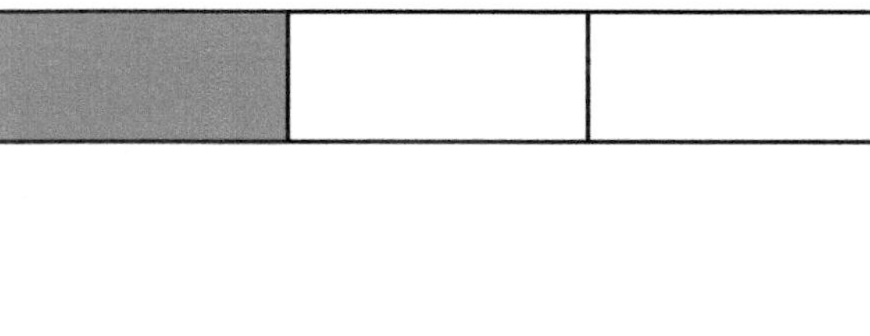 ______________________

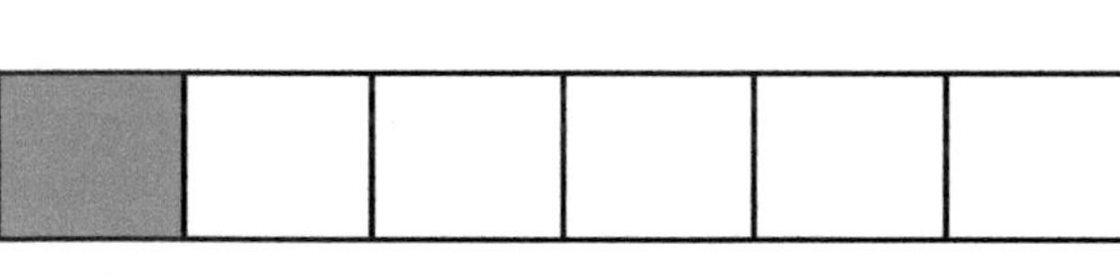 ______________________

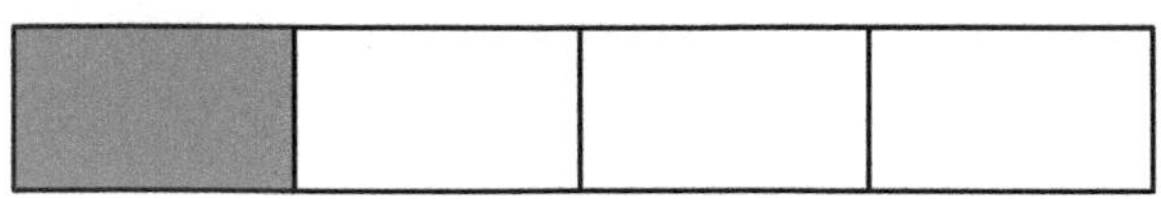 ______________________

3. a. In the space below, draw a small rectangle. Estimate to split it into 2 equal parts. How many lines did you draw to make 2 equal parts? What is the name of each fractional unit?

   b. Draw another small rectangle. Estimate to split it into 3 equal parts. How many lines did you draw to make 3 equal parts? What is the name of each fractional unit?

   c. Draw another small rectangle. Estimate to split it into 4 equal parts. How many lines did you draw to make 4 equal parts? What is the name of each fractional unit?

4. Each rectangle represents 1 sheet of paper.

   a. Estimate to show how you would cut the paper into fractional units as indicated below.

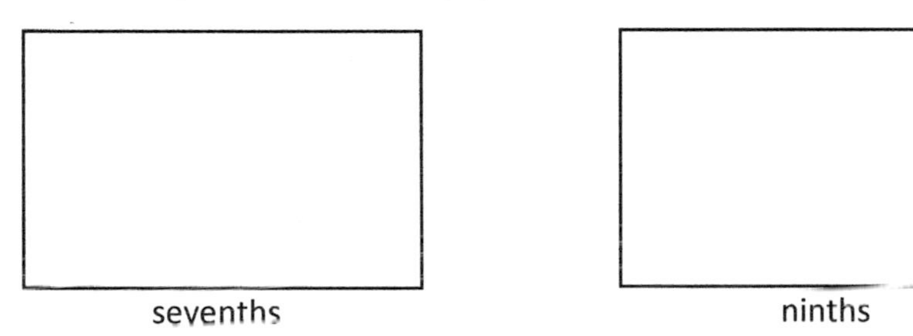

   b. What do you notice? How many lines do you think you would draw to make a rectangle with 20 equal parts?

5. Rochelle has a strip of wood 12 inches long. She cuts it into pieces that are each 6 inches in length. What fraction of the wood is one piece? Use your strip from the lesson to help you. Draw a picture to show the piece of wood and how Rochelle cut it.

Name ______________________________ Date ______________

1. Name the fraction that is shaded.

____

2. Estimate to partition the rectangle into thirds.

3. A plumber has 12 feet of pipe. He cuts it into pieces that are each 3 feet in length. What fraction of the pipe would one piece represent? (Use your strip from the lesson to help you.)

Name ______________________________ Date ________________

1. A beaker is considered full when the liquid reaches the fill line shown near the top. Estimate the amount of water in the beaker by shading the drawing as indicated. The first one is done for you.

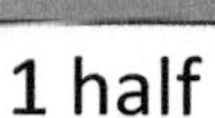

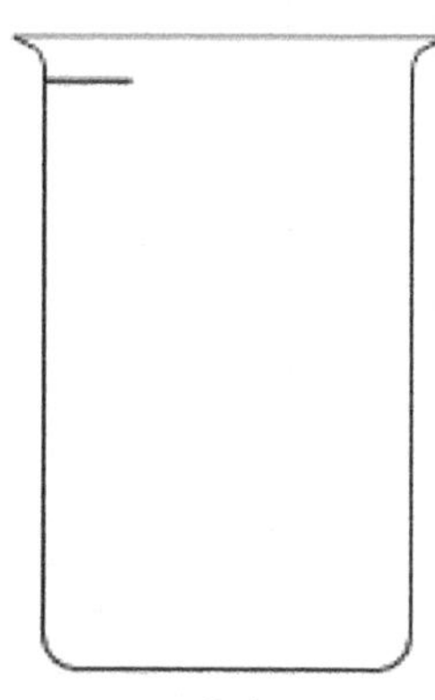

1 fifth

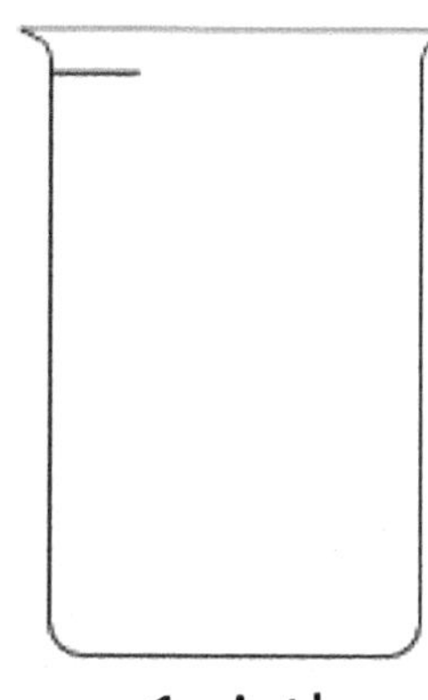

1 sixth

2. Danielle cut her candy bar into equal pieces as shown in the rectangles below. In the blanks below, name the fraction of candy bar represented by the shaded part.

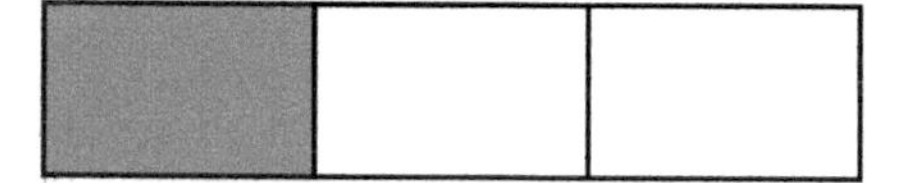
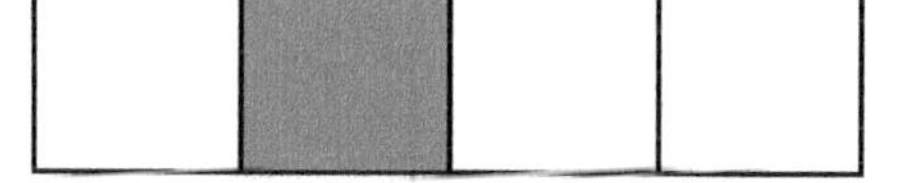
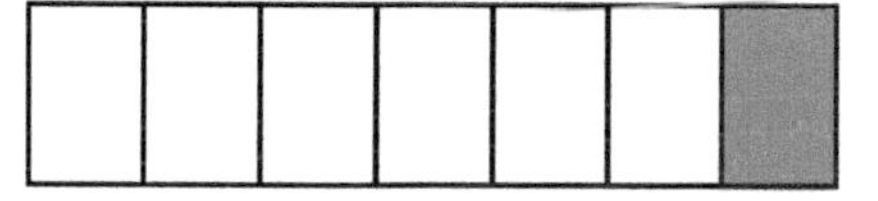

______________ ______________ ______________

3. Each circle represents 1 whole pie. Estimate to show how you would cut the pie into fractional units as indicated below.

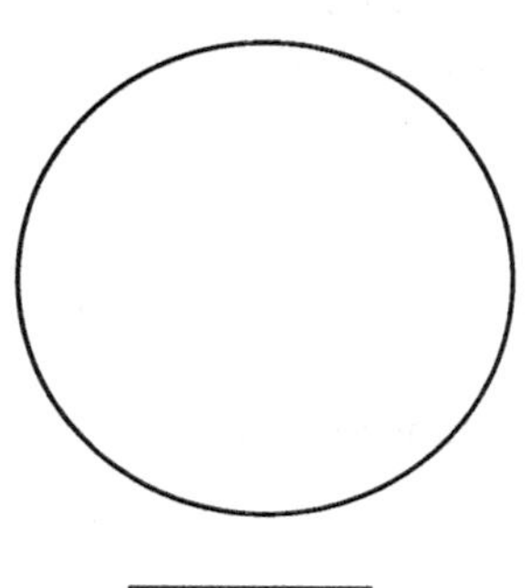

halves

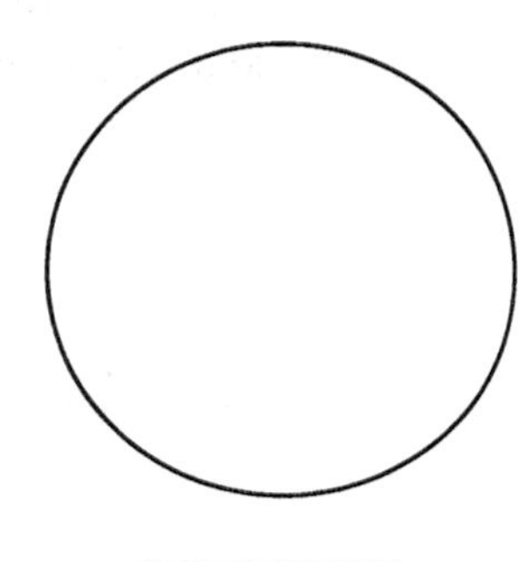

thirds

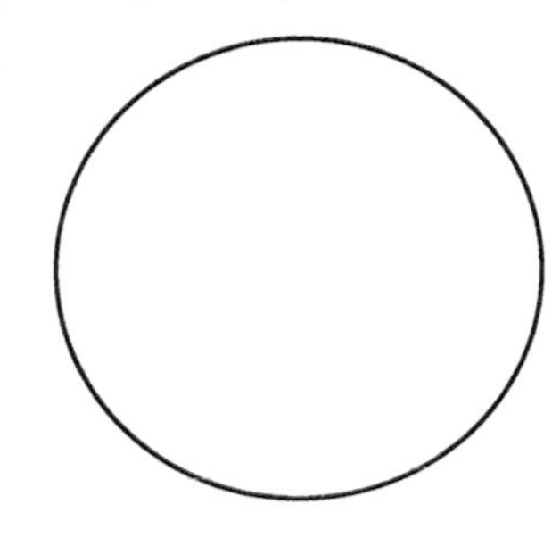

sixths

4. Each rectangle represents 1 sheet of paper. Estimate to draw lines to show how you would cut the paper into fractional units as indicated below.

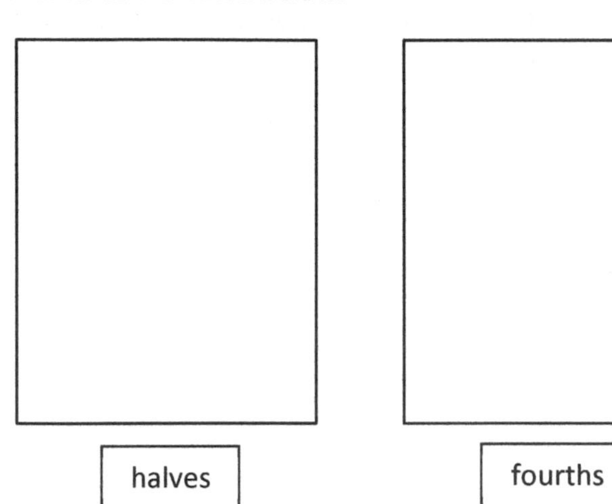

5. Each rectangle represents 1 sheet of paper. Estimate to draw lines to show how you would cut the paper into fractional units as indicated below.

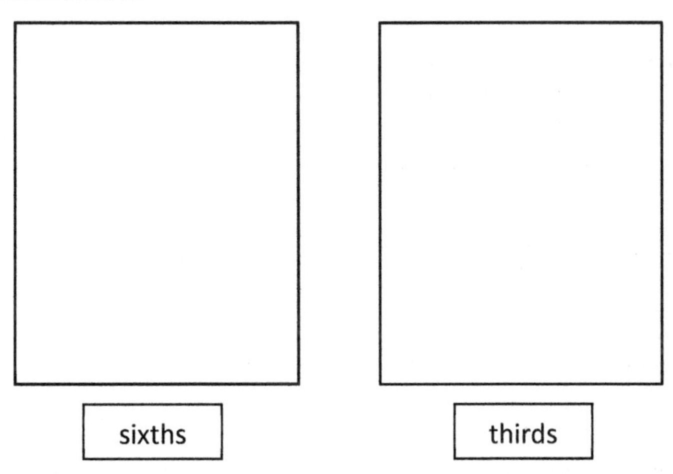

6. Yuri has a rope 12 meters long. He cuts it into pieces that are each 2 meters long. What fraction of the rope is one piece? Draw a picture. (You might fold a strip of paper to help you model the problem.)

7. Dawn bought 12 grams of chocolate. She ate half of the chocolate. How many grams of chocolate did she eat?

Lesson 2

Objective: Specify and partition a whole into equal parts, identifying and counting unit fractions by folding fraction strips.

Suggested Lesson Structure

Fluency Practice	(12 minutes)
Application Problem	(5 minutes)
Concept Development	(35 minutes)
Student Debrief	(8 minutes)
Total Time	**(60 minutes)**

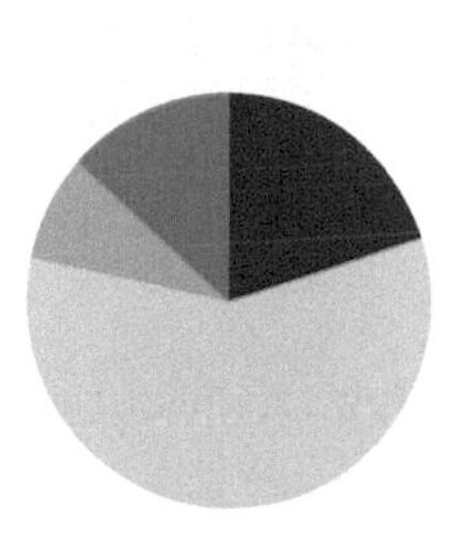

Fluency Practice (12 minutes)

- Group Counting **3.OA.1** (6 minutes)
- Multiplication by Three and Six **3.OA.4** (6 minutes)

Group Counting (6 minutes)

Materials: (S) Personal white board

Note: Group counting reviews interpreting multiplication as repeated addition.

Count forward and backward by threes twice. Pause between each counting sequence so that students see improvement on the second try. After doing the threes twice, have students underline the multiples of 6 (e. g., 0, 3, 6, 9, 12, 15, 18, 21, 24, 27, 30, 27, 24, 21, 18, 15, 12, 9, 6, 3, 0.) Then, count forward and backward by sixes twice, pausing between each counting sequence to analyze weak points.

Multiplication by Three and Six (6 minutes)

Materials: (S) Personal white board (optional)

Note: Choose a mode of delivery (e.g., oral work, personal white boards). This activity reviews multiplication using units of three and six.

Guide students to write and pair facts of 3 and 6 and uncover the doubling:

$2 \times 3 = 6$	$2 \times 6 = 12$
$3 \times 3 = 9$	$3 \times 6 = 18$
$4 \times 3 = 12$	$4 \times 6 = 24$

Application Problem (5 minutes)

Anu needs to cut a piece of paper into 6 equal parts. Draw at least 3 pictures to show how Anu can cut her paper so that all the parts are equal. (Early finishers can do the same thing with halves, fourths, or eighths.)

Note: This problem reviews the concept of equal parts from Lesson 1.

Concept Development (35 minutes)

Materials: (S) 8 paper strips sized $4\frac{1}{4}'' \times 1''$ (vertically cut an $8\frac{1}{2}'' \times 11''$ paper down the middle), pencil, crayon

Note: Students should save the fraction strips they create during this lesson for use in future Module 5 lessons.

Have students take one strip and fold it to make halves. (They might fold it one of two ways. This is correct, but for the purpose of this lesson, it is best to fold as pictured below.)

T: How many equal parts do you have in the whole?

S: Two.

T: What fraction of the whole is 1 part?

S: 1 half.

T: Draw a line to show where you folded your paper. Write the name of the fraction on each equal part.

Use the following sentence frames with the students chorally.

1. There are ____________ equal parts in all.
2. 1 equal part is called ____________________.

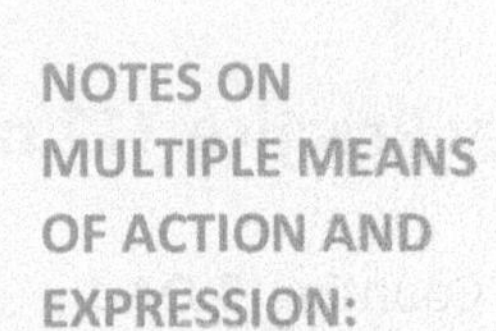

NOTES ON MULTIPLE MEANS OF ACTION AND EXPRESSION:

For English language learners and others, sentence frames support English language acquisition. Students are able to form complete sentences while providing details about the fraction they are analyzing.

Ask students working above grade level for a possible method to partition the whole into ninths (e.g., after partitioning thirds).

Students should fold and label strips showing fourths and eighths to start, followed by thirds and sixths and fifths and tenths. Some students may create more strips than others.

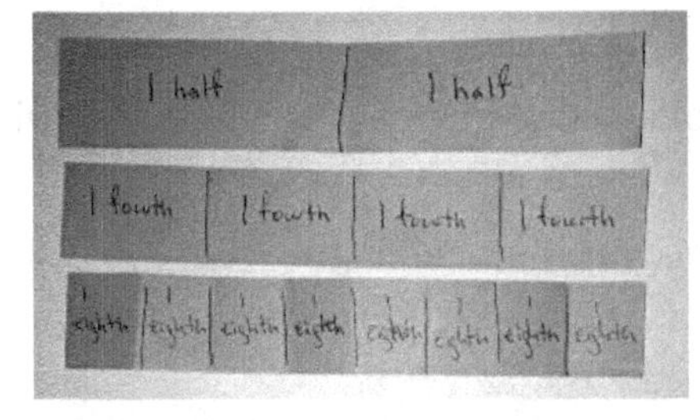

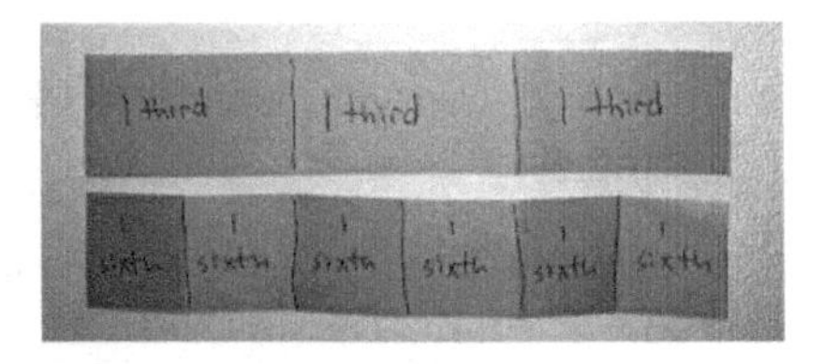

While circulating, watch for students who are not folding in equal parts. Encourage students to try specific strategies for folding equal parts. A word wall would be helpful to support the correct spelling of the fractional units, especially eighths.

When the students have created their fraction strips, ask a series of questions such as the following:

- Look at your set of fraction strips. Imagine they are 4 pieces of delicious pasta. Raise the strip in the air that best shows how to cut 1 piece of pasta into equal parts with your fork.
- Look at your fraction strips. Imagine they are lengths of ribbon. Raise the strip in the air that best shows how to divide the ribbon into 3 equal parts.
- Look at your fraction strips. Imagine they are candy bars. Which best shows how to share your candy bar fairly with 1 person? Which shows how to share your half fairly with 3 people?

Problem Set (10 minutes)

Students should do their personal best to complete the Problem Set within the allotted 10 minutes. For some classes, it may be appropriate to modify the assignment by specifying which problems they work on first. Some problems do not specify a method for solving. Students should solve these problems using the RDW approach used for Application Problems.

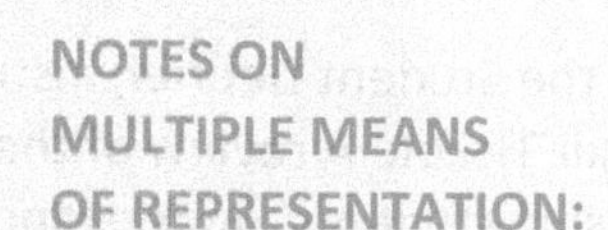

NOTES ON MULTIPLE MEANS OF REPRESENTATION:

Acting out word problems on the Problem Set using concrete materials may increase student understanding.

Student Debrief (8 minutes)

Lesson Objective: Specify and partition a whole into equal parts, identifying and counting unit fractions by folding fraction strips.

The Student Debrief is intended to invite reflection and active processing of the total lesson experience.

Invite students to review their solutions for the Problem Set. They should check work by comparing answers with a partner before going over answers as a class. Look for misconceptions or misunderstandings that can be addressed in the Debrief. Guide students in a conversation to debrief the Problem Set and process the lesson.

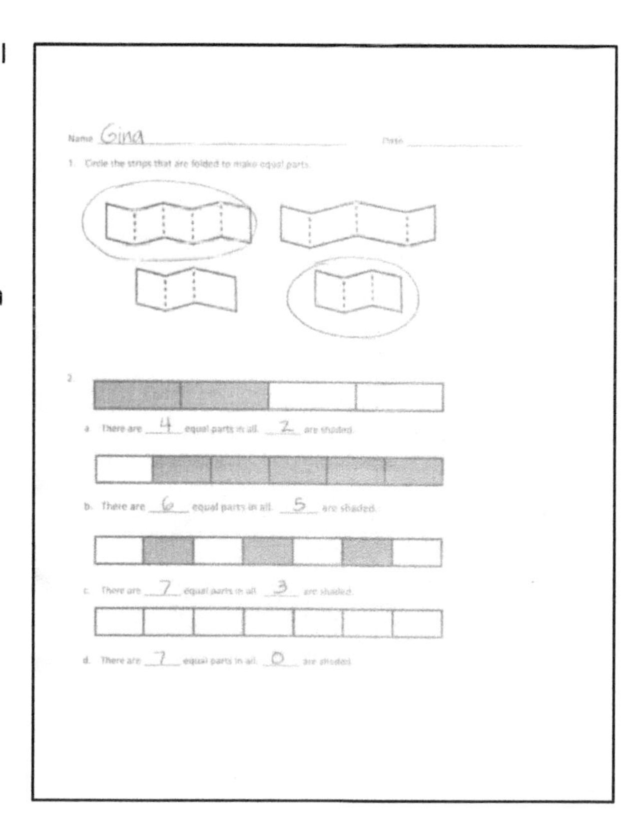
Name Gina Date

1. Circle the strips that are folded to make equal parts.

2.

a. There are 4 equal parts in all. 2 are shaded.

b. There are 6 equal parts in all. 5 are shaded.

c. There are 7 equal parts in all. 3 are shaded.

d. There are 7 equal parts in all. 0 are shaded.

Any combination of the questions below may be used to lead the discussion.

- If the size of the whole stays the same, what happens as you partition it into more and more parts?
- What is the relationship between the number of equal parts and the name of the fraction?
- What strategies did you use for folding different fractional parts?

- What is the relationship of halves to fourths? Of halves to eighths?
- What is the relationship of thirds to sixths?
- What is the relationship of halves, fourths, and eighths to thirds and sixths?
- How does today's Fluency Practice relate to the thirds and sixths we studied in the lesson?

Exit Ticket (3 minutes)

After the Student Debrief, instruct students to complete the Exit Ticket. A review of their work will help with assessing students' understanding of the concepts that were presented in today's lesson and planning more effectively for future lessons. The questions may be read aloud to the students.

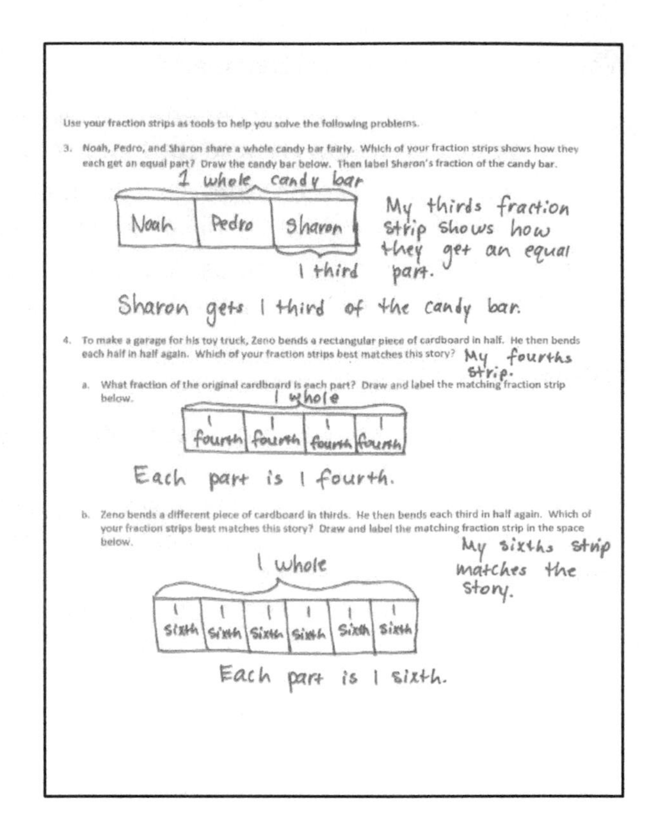

Use your fraction strips as tools to help you solve the following problems.

3. Noah, Pedro, and Sharon share a whole candy bar fairly. Which of your fraction strips shows how they each get an equal part? Draw the candy bar below. Then label Sharon's fraction of the candy bar.

4. To make a garage for his toy truck, Zeno bends a rectangular piece of cardboard in half. He then bends each half in half again. Which of your fraction strips best matches this story?

   a. What fraction of the original cardboard is each part? Draw and label the matching fraction strip below.

   b. Zeno bends a different piece of cardboard in thirds. He then bends each third in half again. Which of your fraction strips best matches this story? Draw and label the matching fraction strip in the space below.

Name ______________________________ Date ______________

1. Circle the strips that are folded to make equal parts.

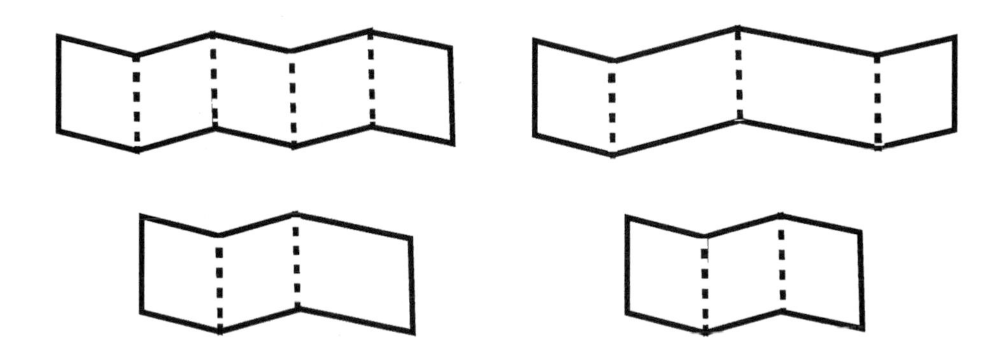

2. 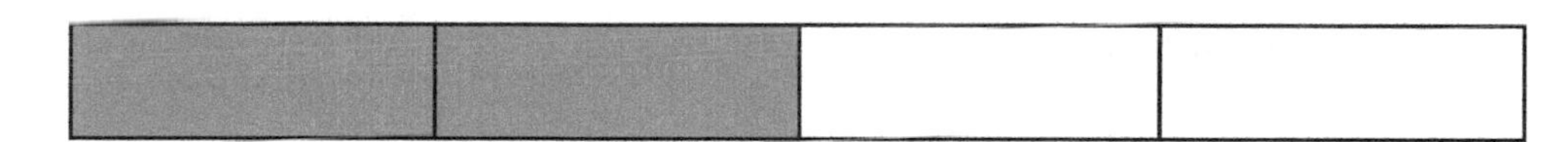

a. There are _______ equal parts in all. _______ are shaded.

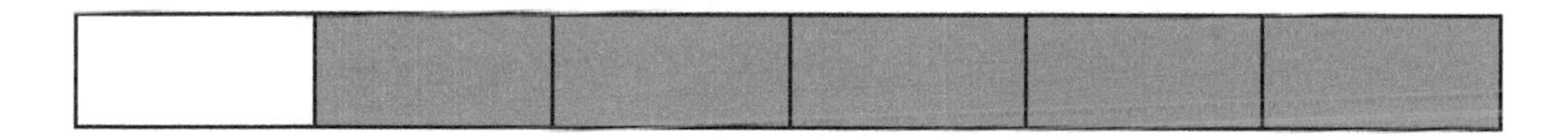

b. There are _______ equal parts in all. _______ are shaded.

c. There are _______ equal parts in all. _______ are shaded.

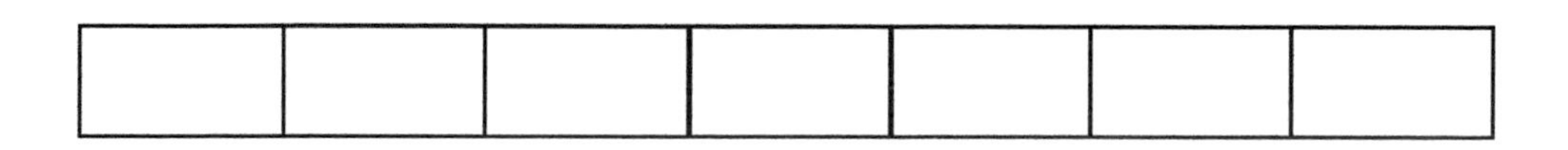

d. There are _______ equal parts in all. _______ are shaded.

Use your fraction strips as tools to help you solve the following problems.

3. Noah, Pedro, and Sharon share a whole candy bar fairly. Which of your fraction strips shows how they each get an equal part? Draw the candy bar below. Then, label Sharon's fraction of the candy bar.

4. To make a garage for his toy truck, Zeno bends a rectangular piece of cardboard in half. He then bends each half in half again. Which of your fraction strips best matches this story?

   a. What fraction of the original cardboard is each part? Draw and label the matching fraction strip below.

   b. Zeno bends a different piece of cardboard in thirds. He then bends each third in half again. Which of your fraction strips best matches this story? Draw and label the matching fraction strip in the space below.

Name ______________________________ Date ______________

1. Circle the model that correctly shows 1 third shaded.

2.

There are _______ equal parts in all. _______ are shaded.

3. Michael bakes a piece of garlic bread for dinner. He shares it equally with his 3 sisters. Show how Michael and his 3 sisters can each get an equal share of the garlic bread.

Name _______________________________________ Date ____________________

1. Circle the strips that are cut into equal parts.

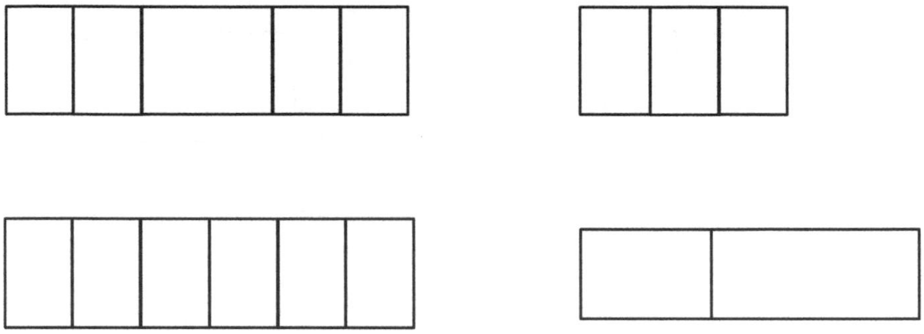

2.

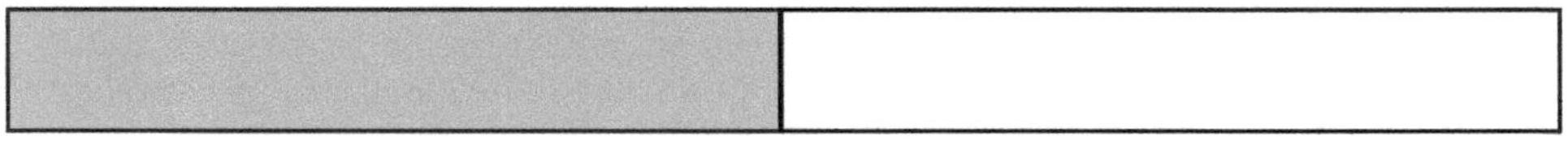

a. There are _______ equal parts in all. _______ is shaded.

b. There are _______ equal parts in all. _______ is shaded.

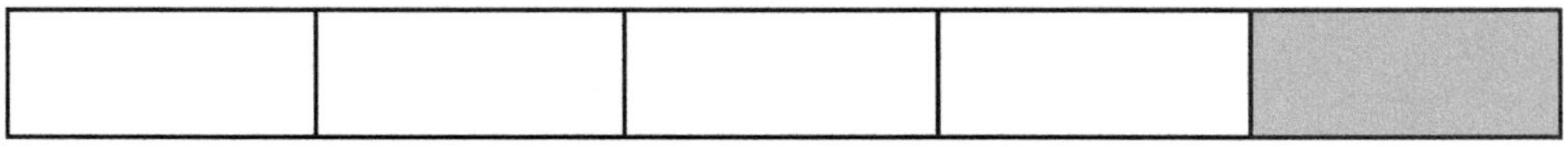

c. There are _______ equal parts in all. _______ is shaded.

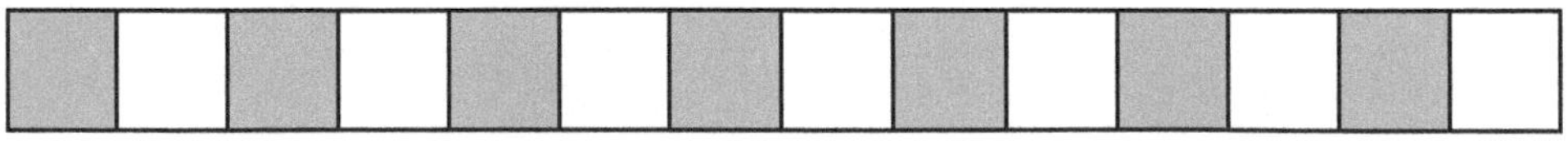

d. There are _______ equal parts in all. _______ are shaded.

3. Dylan plans to eat 1 fifth of his candy bar. His 4 friends want him to share the rest equally. Show how Dylan and his friends can each get an equal share of the candy bar.

4. Nasir baked a pie and cut it in fourths. He then cut each piece in half.

   a. What fraction of the original pie does each piece represent?

   b. Nasir ate 1 piece of pie on Tuesday and 2 pieces on Wednesday. What fraction of the original pie was not eaten?

Lesson 3

Objective: Specify and partition a whole into equal parts, identifying and counting unit fractions by drawing pictorial area models.

Suggested Lesson Structure

Fluency Practice	(12 minutes)
Application Problem	(10 minutes)
Concept Development	(28 minutes)
Student Debrief	(10 minutes)
Total Time	**(60 minutes)**

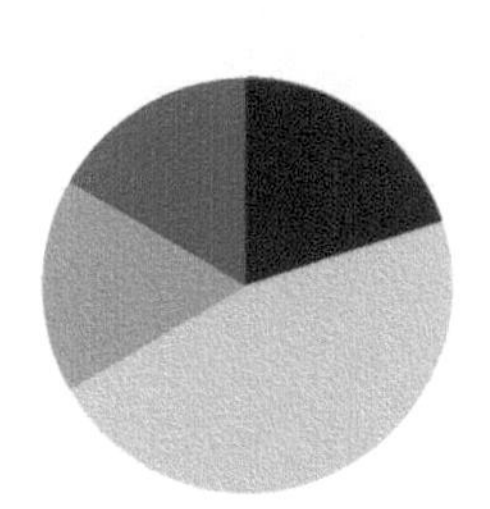

Fluency Practice (12 minutes)

- Sprint: Multiply with Six **3.OA.4** (10 minutes)
- Group Counting **3.OA.1** (2 minutes)

Sprint: Multiply with Six (10 minutes)

Materials: (S) Multiply with Six Sprint

Note: This Sprint supports fluency with multiplication using units of 6.

NOTE ON PACING FLUENCY:

Consider counting by sevens, eights, or nines between Sprints A and B so that both fluency activities can be completed within the 12 minutes allotted.

Group Counting (2 minutes)

Note: Group counting reviews interpreting multiplication as repeated addition.

Direct students to count forward and backward, occasionally changing the direction of the count.

- Sevens to 70
- Eights to 80
- Nines to 90

Application Problem (10 minutes)

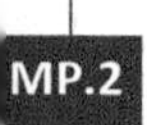

Marcos has a 1-liter jar of milk to share with his mother, father, and sister. Draw a picture to show how Marcos must share the milk so that everyone gets the same amount. What fraction of the milk does each person get?

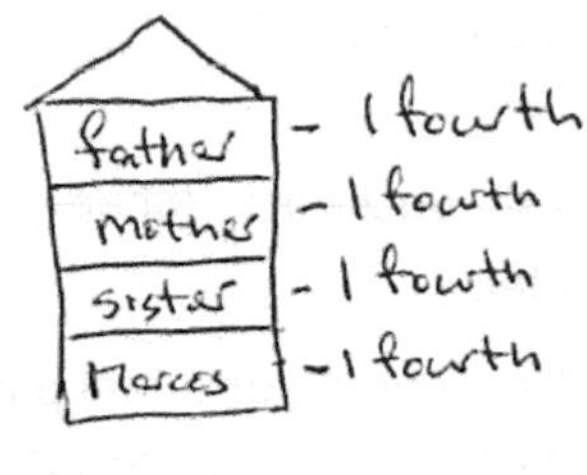

Each person gets 1 fourth of the milk.

Note: This problem reviews partitioning a whole into equal parts, as well as naming fractional parts of a whole.

Concept Development (28 minutes)

Materials: (T) Rectangular- and circular-shaped papers (S) Personal white board

T: I have a rectangle. I want to split it into 4 equal parts.

Fold the paper so the parts are not the same size. Then, open it up to draw the lines where it was folded and show the class. Invite the students to notice the inequality of the parts.

T: Let me try again. (Fold it into 4 equal parts.)
T: How many equal parts did I split the whole into?
S: 4.
T: What is the fractional unit for 4 equal parts?
S: Fourths.
T: What is each part called?
S: 1 fourth or 1 quarter.
T: I'm going to shade 3 **copies** of 1 fourth. (Shade 3 parts.) What fraction is shaded?
S: 3 fourths are shaded.
T: Let's count them.
S: 1 fourth, 2 fourths, 3 fourths.
T: I have a circle. I want to split it into 2 equal parts.

Fold the paper so the parts are not the same size. Then, open it up to draw the lines where it was folded and show the class. Again, invite the students to notice and analyze the inequality of the parts.

T: Let me try again. (Fold it into 2 equal parts.)
T: How many equal parts did I split the whole into?
S: 2.
T: What is the fractional unit for 2 equal parts?
S: Halves.

NOTES ON MULTIPLE MEANS OF ENGAGEMENT:

Increase the wait time for responses from English language learners and students with disabilities. Also, record student responses of the unit fraction and shaded amount on the board beside the model.

NOTES ON VOCABULARY:

Although the word *copies* may not be unfamiliar to students, its use in this context might be. The following excerpt from page 2 of the *3–5 Number and Operations—Fractions* progression describes the concept associated with *copies* in Module 5: "If a whole is partitioned into 4 equal parts, then each part is $\frac{1}{4}$ of the whole, and 4 copies of that part make the whole."

T: What's each part called?

S: 1 half.

T: I'm going to shade 1 part. (Shade 1 part.) What fraction is shaded?

S: 1 half is shaded.

Having established the meaning of equal parts, proceed to briskly analyze Shapes 1–4. Draw or project them, and then possibly use the brief sequence of questions elaborated for Shape 1:

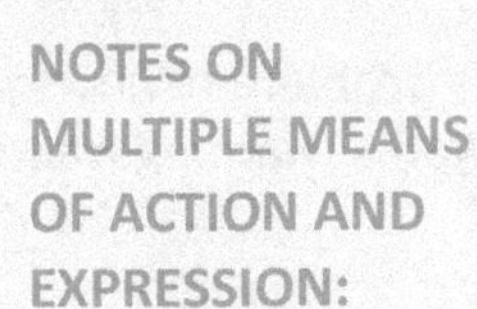

NOTES ON MULTIPLE MEANS OF ACTION AND EXPRESSION:

Some students may benefit from manipulating concrete models simultaneously as they work on the pictorial level.

T: How many equal parts are there in all?

S: 3.

T: What is the fractional unit for 3 equal parts?

S: Thirds.

T: What's each part called?

S: 1 third.

T: What fraction is shaded?

S: 2 thirds.

T: Count them.

S: 1 third, 2 thirds.

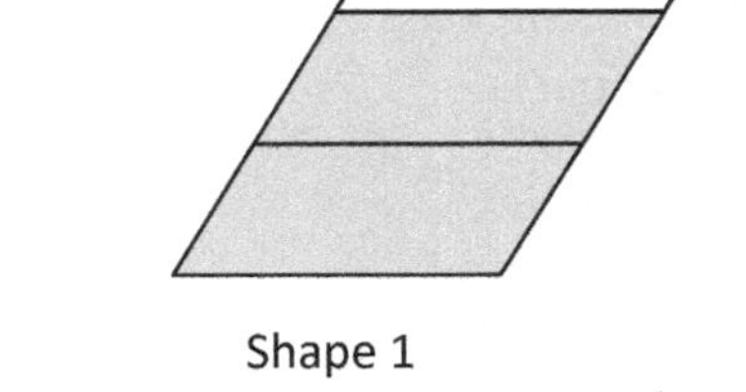

Shape 1

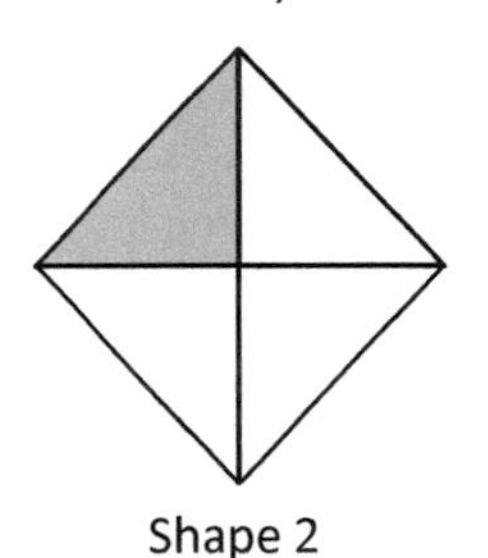

Shape 2

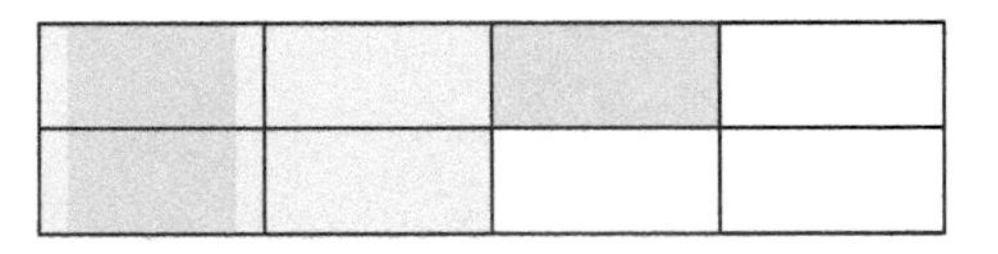

Shape 3

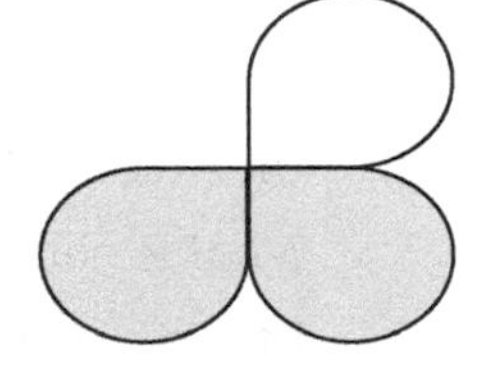

Shape 4

Repeat the steps and procedures with other shapes.

T: Take out your personal white board. We'll draw a few shapes and split them into smaller, equal parts.

T: Draw a rectangle and split it into thirds.

T: How many equal parts do we have altogether?

S: 3.

T: Shade 1 part. What fraction is shaded?

S: 1 third.

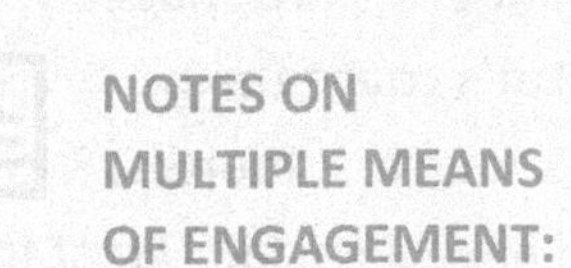

NOTES ON MULTIPLE MEANS OF ENGAGEMENT:

Open-ended activities, such as partitioning a whole into student-chosen fractional units, challenge students working above grade level.

Select a couple of student drawings to show the class.

Repeat the sequence to have students show 2 sixths of a square, 3 fourths of a line segment, and other examples as needed.

Problem Set (10 minutes)

Students should do their personal best to complete the Problem Set within the allotted 10 minutes. For some classes, it may be appropriate to modify the assignment by specifying which problems they work on first. Some problems do not specify a method for solving. Students should solve these problems using the RDW approach used for Application Problems.

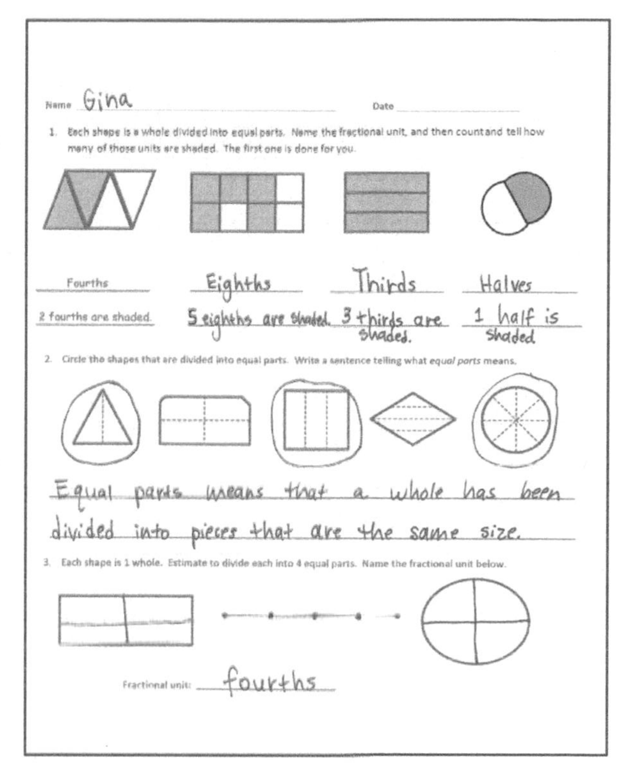
Name Gina Date

1. Each shape is a whole divided into equal parts. Name the fractional unit, and then count and tell how many of those units are shaded. The first one is done for you.

Fourths — 2 fourths are shaded.
Eighths — 5 eighths are shaded.
Thirds — 3 thirds are shaded.
Halves — 1 half is shaded

2. Circle the shapes that are divided into equal parts. Write a sentence telling what *equal parts* means.

Equal parts means that a whole has been divided into pieces that are the same size.

3. Each shape is 1 whole. Estimate to divide each into 4 equal parts. Name the fractional unit below.

Fractional unit: fourths

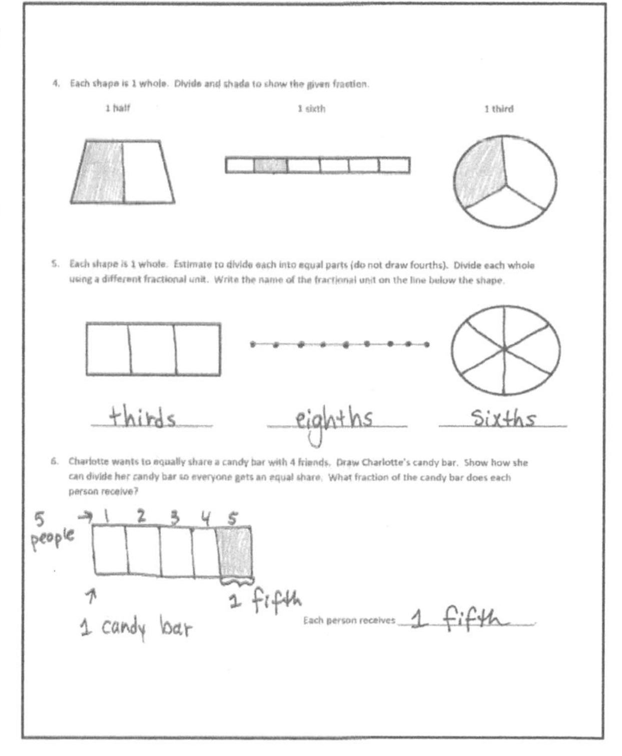
4. Each shape is 1 whole. Divide and shade to show the given fraction.

1 half 1 sixth 1 third

5. Each shape is 1 whole. Estimate to divide each into equal parts (do not draw fourths). Divide each whole using a different fractional unit. Write the name of the fractional unit on the line below the shape.

thirds eighths sixths

6. Charlotte wants to equally share a candy bar with 4 friends. Draw Charlotte's candy bar. Show how she can divide her candy bar so everyone gets an equal share. What fraction of the candy bar does each person receive?

Each person receives 1 fifth

Student Debrief (10 minutes)

Lesson Objective: Specify and partition a whole into equal parts, identifying and counting unit fractions by drawing pictorial area models.

The Student Debrief is intended to invite reflection and active processing of the total lesson experience.

Invite students to review their solutions for the Problem Set. They should check work by comparing answers with a partner before going over answers as a class. Look for misconceptions or misunderstandings that can be addressed in the Debrief. Guide students in a conversation to debrief the Problem Set and process the lesson.

Any combination of the questions below may be used to lead the discussion.

- What is the same about fair shares of a jug of milk and fair shares of a candy bar? What is different? (Though a fraction of a jug of milk and a fraction of a candy bar is clearly different, each might be represented by drawing a rectangle.)
- In Problem 6, how does drawing fourths help you draw fifths well?

Exit Ticket (3 minutes)

After the Student Debrief, instruct students to complete the Exit Ticket. A review of their work will help with assessing students' understanding of the concepts that were presented in today's lesson and planning more effectively for future lessons. The questions may be read aloud to the students.

A

Number Correct: _______

Multiply with Six

1.	1 × 6 =		23.	10 × 6 =	
2.	6 × 1 =		24.	9 × 6 =	
3.	2 × 6 =		25.	4 × 6 =	
4.	6 × 2 =		26.	8 × 6 =	
5.	3 × 6 =		27.	3 × 6 =	
6.	6 × 3 =		28.	7 × 6 =	
7.	4 × 6 =		29.	6 × 6 =	
8.	6 × 4 =		30.	6 × 10 =	
9.	5 × 6 =		31.	6 × 5 =	
10.	6 × 5 =		32.	6 × 4 =	
11.	6 × 6 =		33.	6 × 1 =	
12.	7 × 6 =		34.	6 × 9 =	
13.	6 × 7 =		35.	6 × 6 =	
14.	8 × 6 =		36.	6 × 3 =	
15.	6 × 8 =		37.	6 × 2 =	
16.	9 × 6 =		38.	6 × 7 =	
17.	6 × 9 =		39.	6 × 8 =	
18.	10 × 6 =		40.	11 × 6 =	
19.	6 × 10 =		41.	6 × 11 =	
20.	6 × 3 =		42.	12 × 6 =	
21.	1 × 6 =		43.	6 × 12 =	
22.	2 × 6 =		44.	13 × 6 =	

B

Number Correct: ______

Improvement: ______

Multiply with Six

1.	$6 \times 1 =$	
2.	$1 \times 6 =$	
3.	$6 \times 2 =$	
4.	$2 \times 6 =$	
5.	$6 \times 3 =$	
6.	$3 \times 6 =$	
7.	$6 \times 4 =$	
8.	$4 \times 6 =$	
9.	$6 \times 5 =$	
10.	$5 \times 6 =$	
11.	$6 \times 6 =$	
12.	$6 \times 7 =$	
13.	$7 \times 6 =$	
14.	$6 \times 8 =$	
15.	$8 \times 6 =$	
16.	$6 \times 9 =$	
17.	$9 \times 6 =$	
18.	$6 \times 10 =$	
19.	$10 \times 6 =$	
20.	$1 \times 6 =$	
21.	$10 \times 6 =$	
22.	$2 \times 6 =$	

23.	$9 \times 6 =$	
24.	$3 \times 6 =$	
25.	$8 \times 6 =$	
26.	$4 \times 6 =$	
27.	$7 \times 6 =$	
28.	$5 \times 6 =$	
29.	$6 \times 6 =$	
30.	$6 \times 5 =$	
31.	$6 \times 10 =$	
32.	$6 \times 1 =$	
33.	$6 \times 6 =$	
34.	$6 \times 4 =$	
35.	$6 \times 9 =$	
36.	$6 \times 2 =$	
37.	$6 \times 7 =$	
38.	$6 \times 3 =$	
39.	$6 \times 8 =$	
40.	$11 \times 6 =$	
41.	$6 \times 11 =$	
42.	$12 \times 6 =$	
43.	$6 \times 12 =$	
44.	$13 \times 6 =$	

Name ____________________________ Date ______________

1. Each shape is a whole divided into equal parts. Name the fractional unit, and then count and tell how many of those units are shaded. The first one is done for you.

Fourths
2 fourths are shaded.

2. Circle the shapes that are divided into equal parts. Write a sentence telling what *equal parts* means.

3. Each shape is 1 whole. Estimate to divide each into 4 equal parts. Name the fractional unit below.

Fractional unit: ______________

4. Each shape is 1 whole. Divide and shade to show the given fraction.

1 half 1 sixth 1 third

5. Each shape is 1 whole. Estimate to divide each into equal parts (do not draw fourths). Divide each whole using a different fractional unit. Write the name of the fractional unit on the line below the shape.

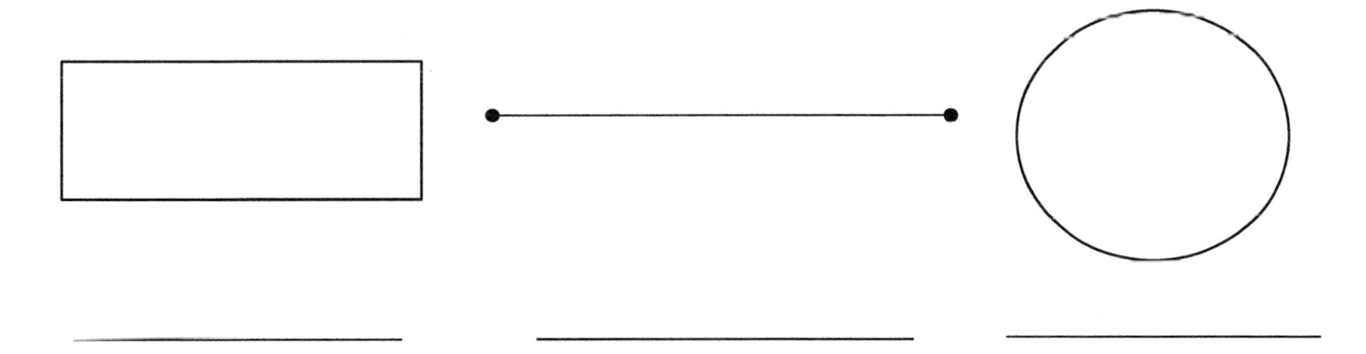

_______________ _______________ _______________

6. Charlotte wants to equally share a candy bar with 4 friends. Draw Charlotte's candy bar. Show how she can divide her candy bar so everyone gets an equal share. What fraction of the candy bar does each person receive?

Each person receives ____________________.

Name ______________________________ Date ________________

1. ______________ sevenths are shaded.

2. Circle the shapes that are divided into equal parts.

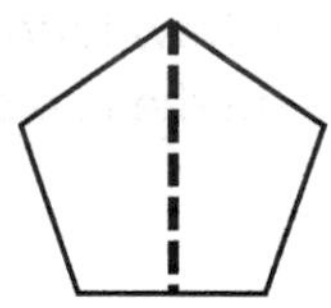

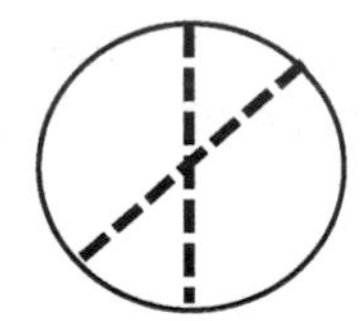

3. Steven wants to equally share his pizza with his 3 sisters. What fraction of the pizza does he and each sister receive?

He and each sister receive ____________________

Name ______________________________ Date ____________

1. Each shape is a whole divided into equal parts. Name the fractional unit, and then count and tell how many of those units are shaded. The first one is done for you.

Fourths ________ ________ ________

2 fourths are shaded. ________ ________ ________

2. Each shape is 1 whole. Estimate to divide each into equal parts. Divide each whole using a different fractional unit. Write the name of the fractional unit on the line below the shape.

________ ________ ________

3. Anita uses 1 sheet of paper to make a calendar showing each month of the year. Draw Anita's calendar. Show how she can divide her calendar so that each month is given the same space. What fraction of the calendar does each month receive?

Each month receives ____________.

Lesson 4

Objective: Represent and identify fractional parts of different wholes.

Suggested Lesson Structure

■ Fluency Practice	(11 minutes)
■ Application Problem	(4 minutes)
■ Concept Development	(35 minutes)
■ Student Debrief	(10 minutes)
Total Time	**(60 minutes)**

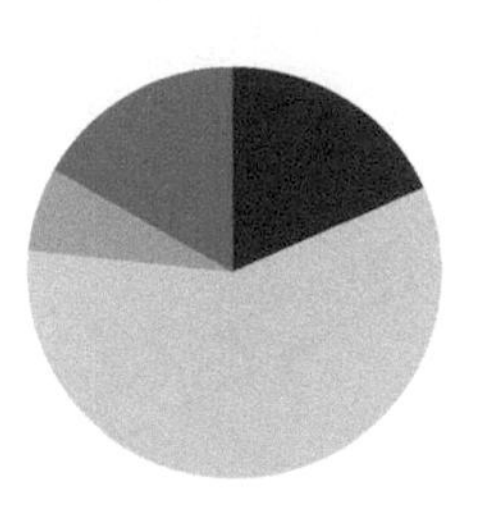

Fluency Practice (11 minutes)

- Sprint: Multiply and Divide by Six **3.OA.4** (9 minutes)
- Group Counting **3.OA.1** (2 minutes)

Sprint: Multiply and Divide by Six (9 minutes)

Materials: (S) Multiply and Divide by Six Sprint

Note: This Sprint supports fluency with multiplication and division using units of 6.

Group Counting (2 minutes)

Note: Group counting reviews interpreting multiplication as repeated addition.

Direct students to count forward and backward, occasionally changing the direction of the count.

- Sixes to 60
- Eights to 80
- Nines to 90

NOTES ON MULTIPLE MEANS OF ACTION AND EXPRESSION:

If students struggle with higher multiples, have them work with the first 3 or 4 multiples of each number (e.g., 8, 16, 24, 32, 24, 16, 8).

Stop before students become frustrated. End with success.

Application Problem (4 minutes)

Mr. Ramos sliced an orange into 8 equal pieces. He ate 1 slice. Draw a picture to represent the 8 slices of an orange. Shade in the slice Mr. Ramos ate. What fraction of the orange did Mr. Ramos eat? What fraction did he not eat?

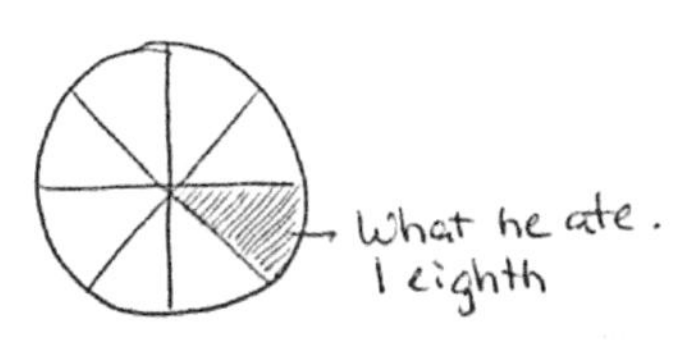

Mr. Ramos ate 1 eighth.
He did not eat 7 eighths.

Note: This problem reviews the skills learned in Topic A.

EUREKA MATH

Concept Development (35 minutes)

Materials: (S) Problem Set, see additional items for stations listed below

Exploration: Students work at stations to represent a given fractional unit using a variety of materials. Designate the following stations for groups of 3 students (more than 3 not suggested).

Station A: Halves	Station E: Sixths
Station B: Fourths	Station F: Ninths
Station C: Eighths	Station G: Fifths
Station D: Thirds	Station H: Tenths

> **NOTES ON MULTIPLE MEANS OF ENGAGEMENT:**
>
> Organize students working below grade level at the stations with easier fractional units and students working above grade level at stations with the most challenging fractional units. To create a greater challenge, make stations for sevenths and twelfths.

Equip each station with the following suggested materials:

- 1-meter length of yarn
- 1 rectangular piece of yellow construction paper (1" × 12")
- 1 piece of brown construction paper (candy bar) (2" × 6")
- 1 square piece of orange construction paper (4" × 4")
- A large cup containing a *whole* amount of water that corresponds to the denominator of the station's fractional unit (e.g., the *fourths* station gets a *whole* of 4 ounces of water)
- A number of small, clear plastic cups corresponding to the denominator of the station's fractional unit (e.g., the *fourths* station gets 4 cups)
- A 200-gram ball of clay or play dough (be sure to have precisely the same amount at each station)

To help students start, give as little direction as possible but enough depending on the particular class. It is suggested that students work without scissors or cutting. Paper and yarn can be folded. Pencil can be used on paper to designate equal parts rather than folding.

Below are some possible directions for students:

- You will partition each item and make a display at your station according to your fractional unit.
- Each item at your station represents 1 whole. You must use all of each whole. (For example, if showing thirds, all of the clay must be used.)
- Use your fractional unit to show each whole partitioned into equal parts.
- Partition the clay by dividing it into smaller equal pieces. (Possibly do this by forming the clay into equal-sized balls. If necessary, demonstrate.)
- Partition the whole amount of water by estimating to pour equal amounts from the large cup into each of the smaller cups. The water in each smaller cup represents an equal part of the whole.

Give students 15 minutes to create their display. Next, conduct a museum walk where they tour the work of the other stations.

Before the museum walk, chart and review the following points. If the analysis dwindles during the tour, circulate and refer students back to the chart. Students complete their Problem Sets as they move between stations; they may also use their Problem Sets as a guide.

- Identify the fractional unit.
- Think about how the units relate to each other at that station.
- Compare the yarn to the yellow strip.
- Compare the yellow strip to the brown paper or candy bar.
- Compare the water to the clay.
- Think about how that unit relates to your own and to other units.

NOTES ON MULTIPLE MEANS OF ACTION AND EXPRESSION:

As students move around the room during the museum walk, have them gently pick up the materials to encourage better analysis. This encourages more conversation, too.

Student Debrief (10 minutes)

Lesson Objective: Represent and identify fractional parts of different wholes.

The Student Debrief is intended to invite reflection and active processing of the total lesson experience.

Invite students to review their solutions for the Problem Set. They should check work by comparing answers with a partner before going over answers as a class. Look for misconceptions or misunderstandings that can be addressed in the Debrief. Guide students in a conversation to debrief the Problem Set and process the lesson.

Any combination of the questions below may be used to lead the discussion.

- What was the same at each station? What was different?
- What different fractional units did you see as you went from station to station?
- What did you notice about different fractional units at the stations?
- Which fractional units had the most equal parts?
- Which fractional units had the least equal parts?
- What surprised you when you were looking at the different fractional units?

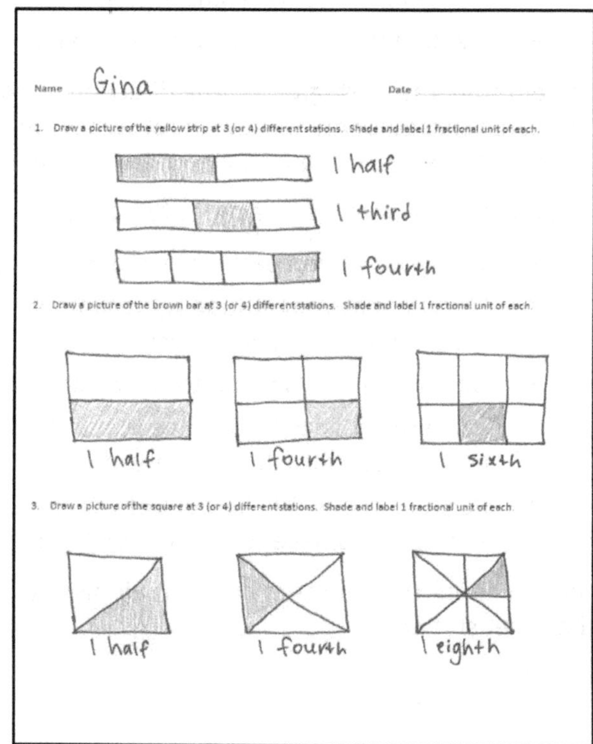
Name Gina Date

1. Draw a picture of the yellow strip at 3 (or 4) different stations. Shade and label 1 fractional unit of each.

1 half
1 third
1 fourth

2. Draw a picture of the brown bar at 3 (or 4) different stations. Shade and label 1 fractional unit of each.

1 half
1 fourth
1 sixth

3. Draw a picture of the square at 3 (or 4) different stations. Shade and label 1 fractional unit of each.

1 half
1 fourth
1 eighth

Exit Ticket (3 minutes)

After the Student Debrief, instruct students to complete the Exit Ticket. A review of their work will help with assessing students' understanding of the concepts that were presented in today's lesson and planning more effectively for future lessons. The questions may be read aloud to the students.

4. Draw a picture of the clay at 3 (or 4) different stations. Shade and label 1 fractional unit of each.

1 half
1 fourth
1 sixth

5. Draw a picture of the water at 3 (or 4) different stations. Shade and label 1 fractional unit of each.

1 half 1 third 1 fourth

6. Extension: Draw a picture of the yarn at 3 (or 4) different stations.

folded in half
folded in fourths
folded in thirds

A

Number Correct: _______

Multiply and Divide by Six

1.	2 × 6 =		23.	___ × 6 = 60	
2.	3 × 6 =		24.	___ × 6 = 12	
3.	4 × 6 =		25.	___ × 6 = 18	
4.	5 × 6 =		26.	60 ÷ 6 =	
5.	1 × 6 =		27.	30 ÷ 6 =	
6.	12 ÷ 6 =		28.	6 ÷ 6 =	
7.	18 ÷ 6 =		29.	12 ÷ 6 =	
8.	30 ÷ 6 =		30.	18 ÷ 6 =	
9.	6 ÷ 6 =		31.	___ × 6 = 36	
10.	24 ÷ 6 =		32.	___ × 6 = 42	
11.	6 × 6 =		33.	___ × 6 = 54	
12.	7 × 6 =		34.	___ × 6 = 48	
13.	8 × 6 =		35.	42 ÷ 6 =	
14.	9 × 6 =		36.	54 ÷ 6 =	
15.	10 × 6 =		37.	36 ÷ 6 =	
16.	48 ÷ 6 =		38.	48 ÷ 6 =	
17.	42 ÷ 6 =		39.	11 × 6 =	
18.	54 ÷ 6 =		40.	66 ÷ 6 =	
19.	36 ÷ 6 =		41.	12 × 6 =	
20.	60 ÷ 6 =		42.	72 ÷ 6 =	
21.	___ × 6 = 30		43.	14 × 6 =	
22.	___ × 6 = 6		44.	84 ÷ 6 =	

B

Number Correct: _______

Improvement: _______

Multiply and Divide by Six

1.	1 × 6 =			23.	___ × 6 = 12	
2.	2 × 6 =			24.	___ × 6 = 60	
3.	3 × 6 =			25.	___ × 6 = 18	
4.	4 × 6 =			26.	12 ÷ 6 =	
5.	5 × 6 =			27.	6 ÷ 6 =	
6.	18 ÷ 6 =			28.	60 ÷ 6 =	
7.	12 ÷ 6 =			29.	30 ÷ 6 =	
8.	24 ÷ 6 =			30.	18 ÷ 6 =	
9.	6 ÷ 6 =			31.	___ × 6 = 18	
10.	30 ÷ 6 =			32.	___ × 6 = 24	
11.	10 × 6 =			33.	___ × 6 = 54	
12.	6 × 6 =			34.	___ × 6 = 42	
13.	7 × 6 =			35.	48 ÷ 6 =	
14.	8 × 6 =			36.	54 ÷ 6 =	
15.	9 × 6 =			37.	36 ÷ 6 =	
16.	42 ÷ 6 =			38.	42 ÷ 6 =	
17.	36 ÷ 6 =			39.	11 × 6 =	
18.	48 ÷ 6 =			40.	66 ÷ 6 =	
19.	60 ÷ 6 =			41.	12 × 6 =	
20.	54 ÷ 6 =			42.	72 ÷ 6 =	
21.	___ × 6 = 6			43.	13 × 6 =	
22.	___ × 6 = 30			44.	78 ÷ 6 =	

Name ______________________________ Date ________________

1. Draw a picture of the yellow strip at 3 (or 4) different stations. Shade and label 1 fractional unit of each.

2. Draw a picture of the brown bar at 3 (or 4) different stations. Shade and label 1 fractional unit of each.

3. Draw a picture of the square at 3 (or 4) different stations. Shade and label 1 fractional unit of each.

4. Draw a picture of the clay at 3 (or 4) different stations. Shade and label 1 fractional unit of each.

5. Draw a picture of the water at 3 (or 4) different stations. Shade and label 1 fractional unit of each.

6. Extension: Draw a picture of the yarn at 3 (or 4) different stations.

Name ________________________________ Date ________________

Each shape is 1 whole. Estimate to equally partition the shape and shade to show the given fraction.

1. 1 fourth

2. 1 fifth ______________________________

3. The shape represents 1 whole. Write the fraction for the shaded part.

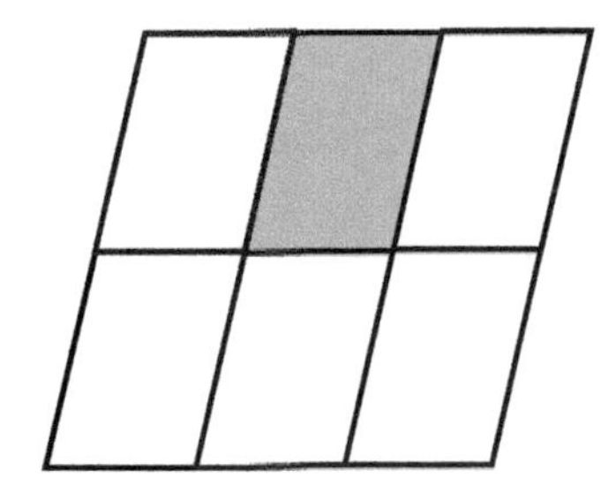

The shaded part is ____________________.

Name ______________________________ Date ________________

Each shape is 1 whole. Estimate to equally partition the shape and shade to show the given fraction.

1. 1 half

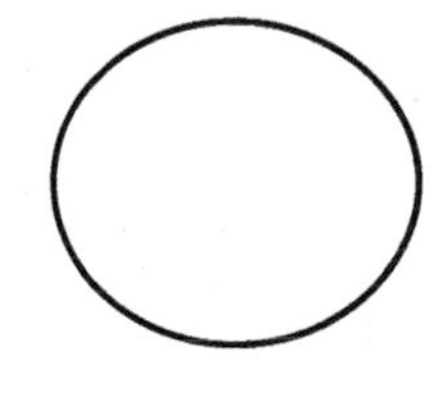

A

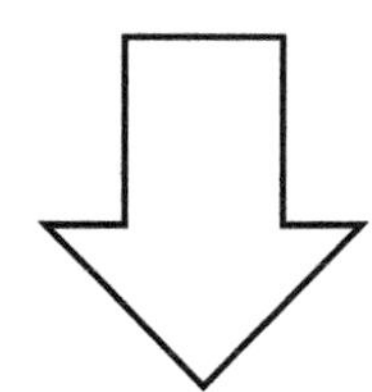

B

C

D

---

2. 1 fourth

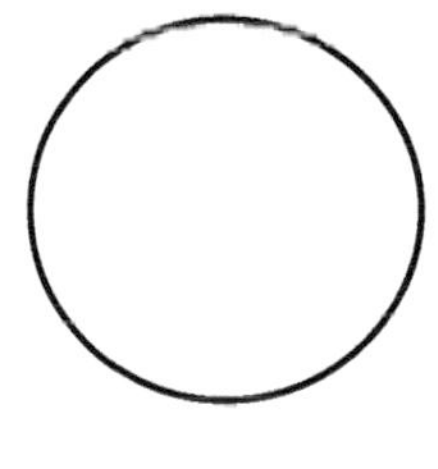

A

B

C

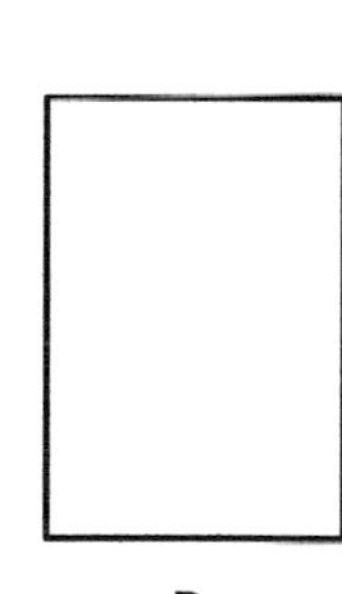

D

---

3. 1 third

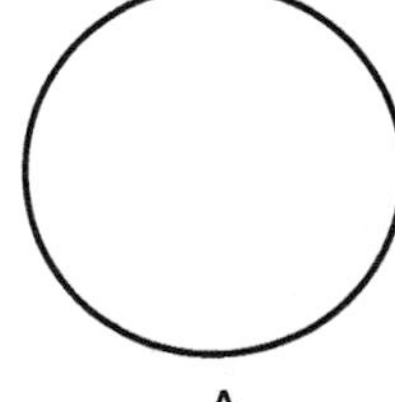

A

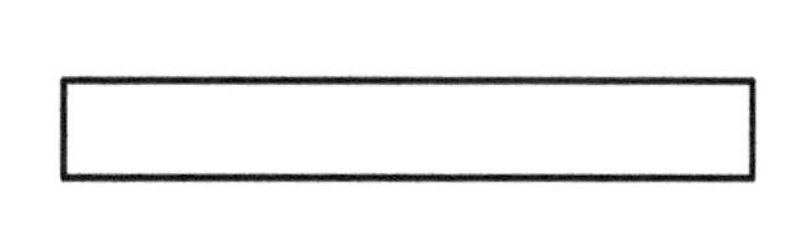

B

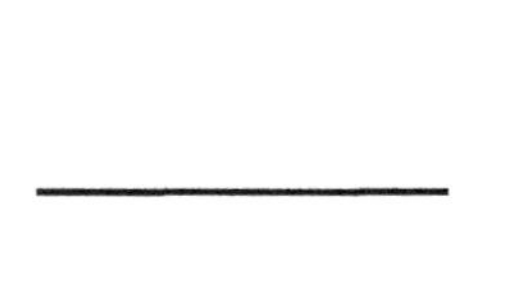

C

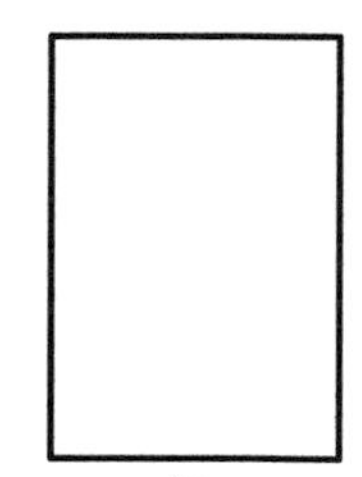

D

4. Each of the shapes represents 1 whole. Match each shape to its fraction.

1 fifth

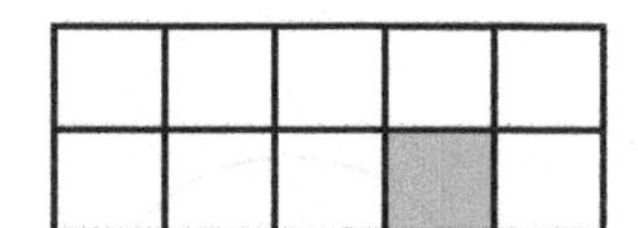

1 twelfth

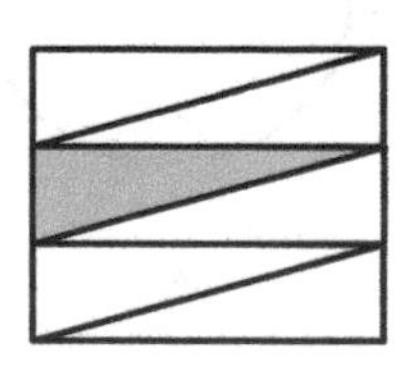

1 third

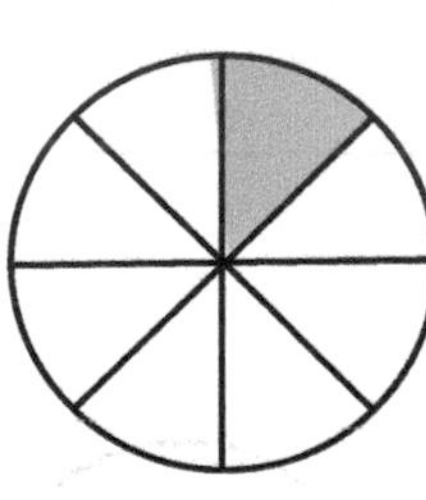

1 fourth

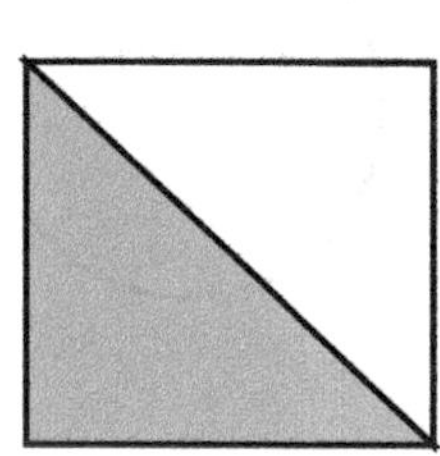

1 half

1 eighth

1 tenth

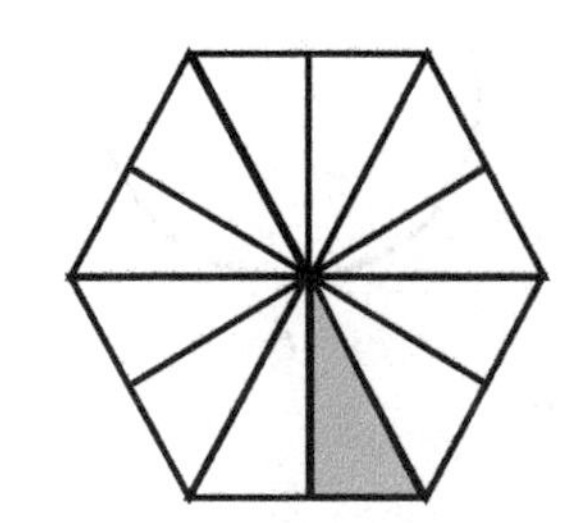

1 sixth

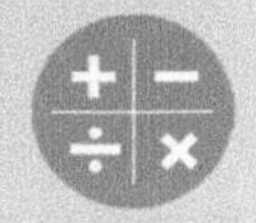

Topic B

Unit Fractions and Their Relation to the Whole

3.NF.1, 3.NF.3c, 3.G.2

Focus Standard:		3.NF.1	Understand a fraction $1/b$ as the quantity formed by 1 part when a whole is partitioned into b equal parts; understand a fraction a/b as the quantity formed by a parts of size $1/b$.
Instructional Days:		5	
Coherence	**-Links from:**	G2–M8	Time, Shapes, and Fractions as Equal Parts of Shapes
	-Links to:	G4–M5	Fraction Equivalence, Ordering, and Operations

In Topic A, students divided a given whole into equal parts to create fractional units (halves, thirds, fourths, etc.). Now, they associate one of the fractional units with a number called the unit fraction ($\frac{1}{2}, \frac{1}{3}, \frac{1}{4}$, etc.). This sets the foundation for students eventually understanding that a fraction is a number. Like any number, it corresponds to a point on the real number line and can be written in unit form or fraction form (e.g., 1 half or $\frac{1}{2}$).

An advantage of the term *fractional unit* is that it distinguishes the nature of the equal parts generated by partitioning a whole from the whole number division students studied in Modules 1 and 3. In Topic B, to avoid confusion, the term *fractional unit* is mostly replaced by the term *equal part*. The equal part is represented by the unit fraction. Students recognize that any non-unit fraction is composed of multiple copies of a unit fraction. They use number bonds to represent this. In particular, students construct fractions greater than 1 using multiple copies of a given unit fraction.

A Teaching Sequence Toward Mastery of Unit Fractions and Their Relation to the Whole

Objective 1: Partition a whole into equal parts and define the equal parts to identify the unit fraction numerically.
(Lesson 5)

Objective 2: Build non-unit fractions less than one whole from unit fractions.
(Lesson 6)

Objective 3: Identify and represent shaded and non-shaded parts of one whole as fractions.
(Lesson 7)

Objective 4: Represent parts of one whole as fractions with number bonds.
(Lesson 8)

Objective 5: Build and write fractions greater than one whole using unit fractions.
(Lesson 9)

Lesson 5

Objective: Partition a whole into equal parts and define the equal parts to identify the unit fraction numerically.

Suggested Lesson Structure

■ Fluency Practice	(15 minutes)
■ Application Problem	(10 minutes)
■ Concept Development	(25 minutes)
■ Student Debrief	(10 minutes)
Total Time	**(60 minutes)**

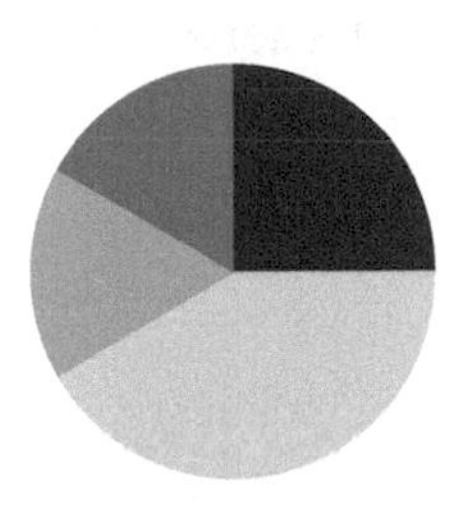

Fluency Practice (15 minutes)

- Count by Eight **3.OA.7** (5 minutes)
- Write the Fractional Unit **3.NF.1** (5 minutes)
- Partition Shapes **3.NF.1** (5 minutes)

Count by Eight (5 minutes)

Materials: (S) Personal white board

Note: This activity supports fluency with multiplication using units of 8.

1. Students count by eight as high as they can for 90 seconds. 0, 8, 16, 24, 32, 40, 48, 56, etc.
2. Correct by reading the multiples. Students practice for an additional minute after correction.
3. Students count by eight once again. Quickly celebrate improvement.

Write the Fractional Unit (5 minutes)

Materials: (S) Personal white board

Note: This activity reviews naming fractional units, as well as identifying shaded parts of a shape from Topic A.

T: (Draw a shape with 3 units, 2 shaded in.) Write the fractional unit on your personal white board.
S: (Write thirds.)
T: Blank thirds are shaded. Write the number that goes in the blank.
S: (Write 2.)

Continue with the following possible sequence: 3 fourths, 2 fifths, 5 sixths, 7 tenths, and 5 eighths.

Partition Shapes (5 minutes)

Materials: (S) Personal white board

Note: This activity reviews partitioning shapes into equal parts from Topic A.

T: Draw a square.

S: (Draw.)

T: (Write halves.) Estimate to partition the square into equal halves.

S: (Partition.)

Continue with the following possible sequence: line, fifths; circle, fourths; circle, eighths; bar, tenths; and bar, sixths.

Application Problem (10 minutes)

Ms. Browne cut a 6-meter rope into 3 equal-size pieces to make jump ropes. Mr. Ware cut a 5-meter rope into 3 equal size pieces to make jump ropes. Which class has longer jump ropes?

Extension: How long are the jump ropes in Ms. Browne's class?

Note: This problem reviews partitioning different wholes into equal parts from Lesson 4.

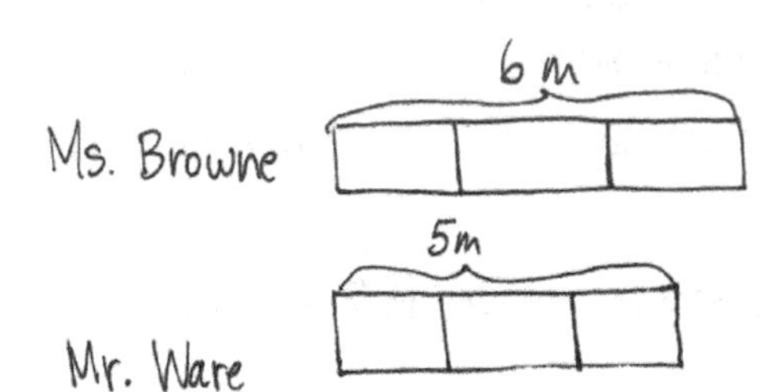

Ms. Browne's class gets longer jump ropes because the original rope was longer.

Extension: The jump ropes in Ms. Browne's class are 2 m long because I can count by 2 until I get to 6 m.

Concept Development (25 minutes)

Materials: (S) Personal white board

T: (Project or draw a circle, as shown below.) Whisper the name of this shape.

S: Circle.

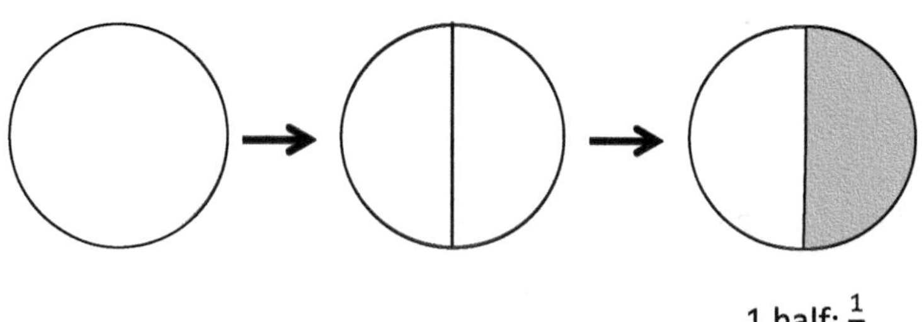

1 half; $\frac{1}{2}$

NOTES ON MULTIPLE MEANS OF REPRESENTATION:

While introducing the new terms—*unit form, fraction form,* and *unit fraction*—check for student understanding. English language learners may choose to discuss definitions of these terms in their first language with the teacher or their peers.

T: Watch as I partition the whole. (Draw a line to partition the circle into 2 equal parts, as shown.) How many equal parts are there?

S: 2 equal parts.

T: What's the name of each unit?

S: 1 half.

T: (Shade one unit.) What fraction is shaded?

S: 1 half.

T: Just like any number, we can write one half in many ways. This is the **unit form**. (Write *1 half* under the circle.) This is the **fraction form**. (Write $\frac{1}{2}$ under the circle.) Both of these refer to the same number, 1 out of 2 equal units. We call 1 half a **unit fraction** because it names one of the equal parts.

T: (Project or draw a square, as shown below.) What's the name of this shape?

S: It's a square.

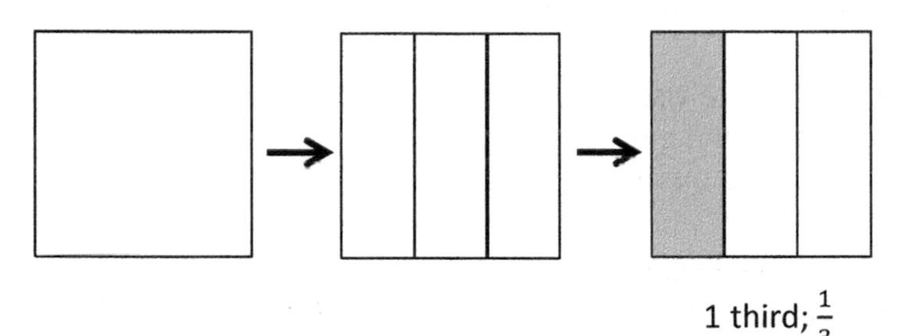

1 third; $\frac{1}{3}$

NOTES ON MULTIPLE MEANS OF ENGAGEMENT:

Students working above grade level may enjoy identifying fractions with an added challenge of each shape representing a *fraction* rather than the whole. For example, ask the following: "If the square is 1 third, name the shaded region" (e.g., $\frac{3}{12}$ or $\frac{1}{4}$).

T: Draw it on your personal white board. (After students draw the square.) Estimate to partition the square into 3 equal parts.

S: (Partition.)

T: What's the name of each unit?

S: 1 third.

T: Shade one unit. Then, write the fraction for the shaded amount in unit form and fraction form on your board.

S: (Shade and write 1 third and $\frac{1}{3}$.)

T: Talk to a partner: Is the number that you wrote to represent the shaded part a unit fraction? Why or why not?

S: (Discuss.)

Continue the process with more shapes as needed. The following suggested shapes include examples of both shaded and non-shaded unit fractions. Alter language accordingly.

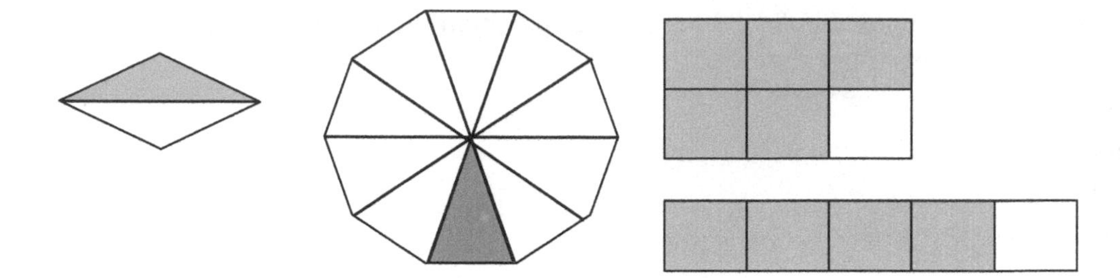

T: (Project or draw the following image.) Discuss with your partner: Does the shape have equal parts? How do you know?

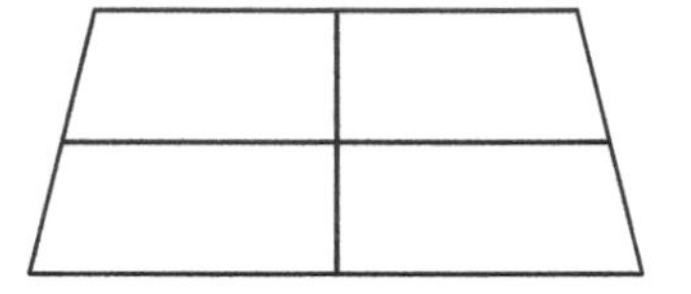

MP.6

S: No. The parts are not the same size. → They're also not exactly the same shape. → The parts are not equal because the bottom parts are larger. The lines on the sides lean in at the top.

T: Most agree that the parts are not equal. How could you partition the shape to make the parts equal?

S: I can cut it into 2 equal parts. You have to cut it right down the middle going up and down. The lines aren't all the same length like in a square.

T: Turn and talk: If the parts are not equal, can we call these fourths? Why or why not?

S: (Discuss.)

NOTES ON MULTIPLE MEANS OF ENGAGEMENT:

Review personal goals with students. For example, if students working below grade level chose to solve one word problem (per lesson) last week, encourage them to work toward completing two word problems by the end of this week.

Problem Set (10 minutes)

Students should do their personal best to complete the Problem Set within the allotted 10 minutes. For some classes, it may be appropriate to modify the assignment by specifying which problems they work on first. Some problems do not specify a method for solving. Students should solve these problems using the RDW approach used for Application Problems.

Student Debrief (10 minutes)

Lesson Objective: Partition a whole into equal parts and define the equal parts to identify the unit fraction numerically.

The Student Debrief is intended to invite reflection and active processing of the total lesson experience.

Invite students to review their solutions for the Problem Set. They should check work by comparing answers with a partner before going over answers as a class. Look for misconceptions or misunderstandings that can be addressed in the Debrief. Guide students in a conversation to debrief the Problem Set and process the lesson.

Any combination of the questions below may be used to lead the discussion.

- Are the numbers in Problem 1 unit fractions? How do you know?
- Use the following possible introduction to start a discussion about Problem 4: Let's imagine we're at Andre's birthday party. Who would rather have an eighth of the cake? Who would rather have a tenth? Why? The following are some suggested sentence frames:
  - "I would rather have a __________ because ______________."
  - "I agree/disagree because ____________."
- Guide students to begin understanding that a greater number of parts results in smaller pieces.

Exit Ticket (3 minutes)

After the Student Debrief, instruct students to complete the Exit Ticket. A review of their work will help with assessing students' understanding of the concepts that were presented in today's lesson and planning more effectively for future lessons. The questions may be read aloud to the students.

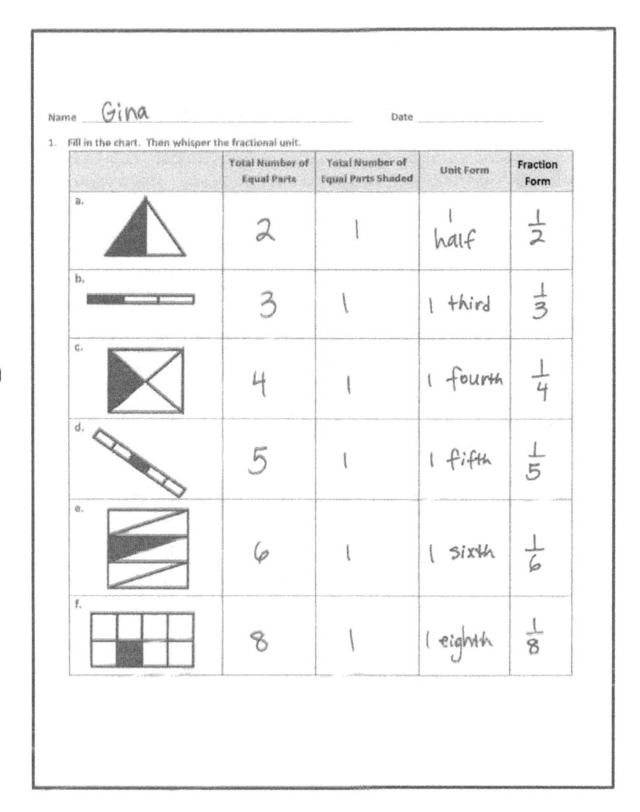
Name Gina Date

1. Fill in the chart. Then whisper the fractional unit.

	Total Number of Equal Parts	Total Number of Equal Parts Shaded	Unit Form	Fraction Form
a.	2	1	1 half	$\frac{1}{2}$
b.	3	1	1 third	$\frac{1}{3}$
c.	4	1	1 fourth	$\frac{1}{4}$
d.	5	1	1 fifth	$\frac{1}{5}$
e.	6	1	1 sixth	$\frac{1}{6}$
f.	8	1	1 eighth	$\frac{1}{8}$

2. Andre's mom baked his 2 favorite cakes for his birthday party. The cakes were the exact same size. Andre cut his first cake into 8 pieces for him and his 7 friends. The picture below shows how he cut it. Did Andre cut the cake into eighths? Explain your answer.

No, he didn't cut the cake into eighths. There are eight pieces, but they aren't all the same size. To be eighths, they have to be the same size.

3. Two of Andre's friends came late to his party. They decide they will all share the second cake. Show how Andre can slice the second cake so that he and his nine friends can each get an equal amount with none leftover. What fraction of the second cake will they each receive?

$\frac{1}{10}$	$\frac{1}{10}$	$\frac{1}{10}$	$\frac{1}{10}$	$\frac{1}{10}$
$\frac{1}{10}$	$\frac{1}{10}$	$\frac{1}{10}$	$\frac{1}{10}$	$\frac{1}{10}$

They will each receive $\frac{1}{10}$ of the cake.

4. Andre thinks it's strange that $\frac{1}{10}$ of the cake would be less than $\frac{1}{8}$ of the cake, since ten is bigger than eight. To explain to Andre, draw 2 identical rectangles to represent the cakes. Show 1 tenth shaded on one and 1 eighth shaded on the other. Label the unit fractions and explain to him which slice is bigger.

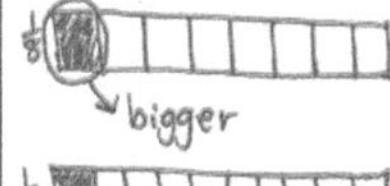

$\frac{1}{8}$ is bigger because when the cake is cut in 10 equal pieces, the pieces are smaller than when it's cut in 8 equal pieces. More equal pieces mean the pieces are smaller.

Name ______________________________ Date ______________

1. Fill in the chart. Each image is one whole.

	Total Number of Equal Parts	Total Number of Equal Parts Shaded	Unit Form	Fraction Form
a. 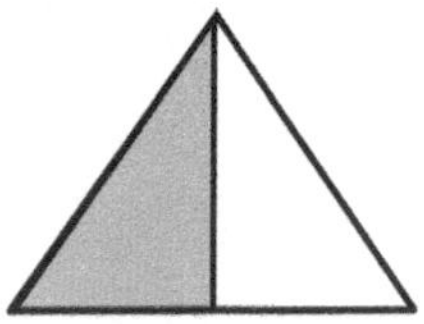				
b. 				
c. 				
d. 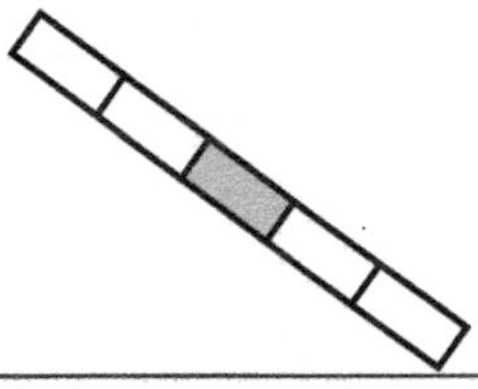				
e. 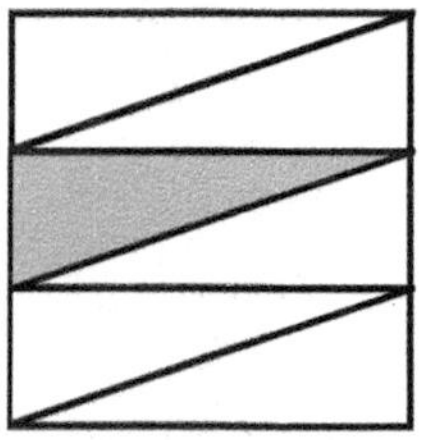				
f. 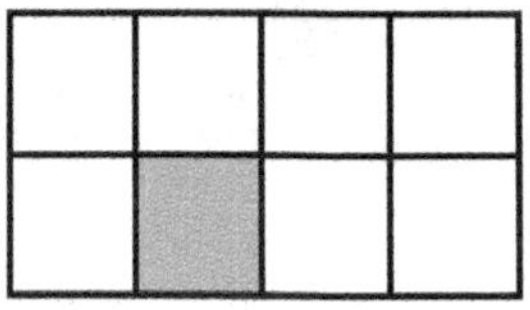				

2. Andre's mom baked his 2 favorite cakes for his birthday party. The cakes were the exact same size. Andre cut his first cake into 8 pieces for him and his 7 friends. The picture below shows how he cut it. Did Andre cut the cake into eighths? Explain your answer.

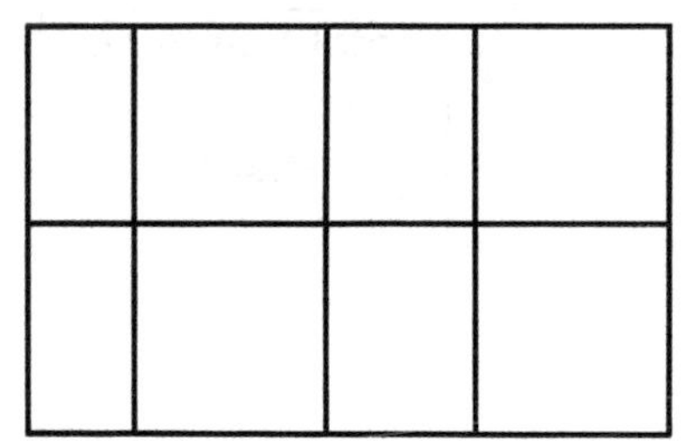

3. Two of Andre's friends came late to his party. They decide they will all share the second cake. Show how Andre can slice the second cake so that he and his nine friends can each get an equal amount with none leftover. What fraction of the second cake will they each receive?

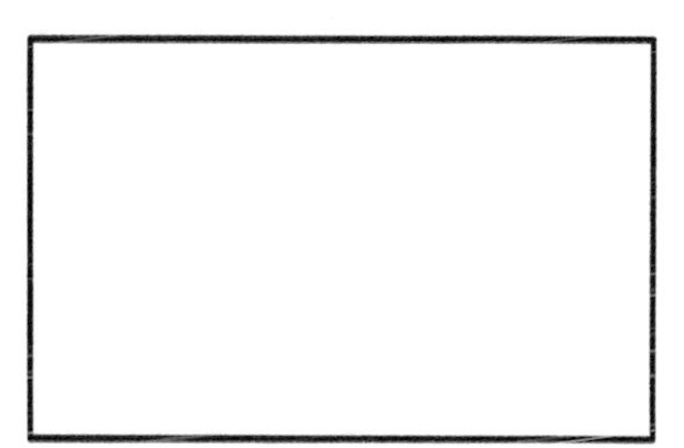

4. Andre thinks it's strange that $\frac{1}{10}$ of the cake would be less than $\frac{1}{8}$ of the cake since ten is bigger than eight. To explain to Andre, draw 2 identical rectangles to represent the cakes. Show 1 tenth shaded on one and 1 eighth shaded on the other. Label the unit fractions and explain to him which slice is bigger.

Name ________________________________ Date ________________

1. Fill in the chart.

	Total Number of Equal Parts	Total Number of Equal Parts Shaded	Unit Form	Fraction Form
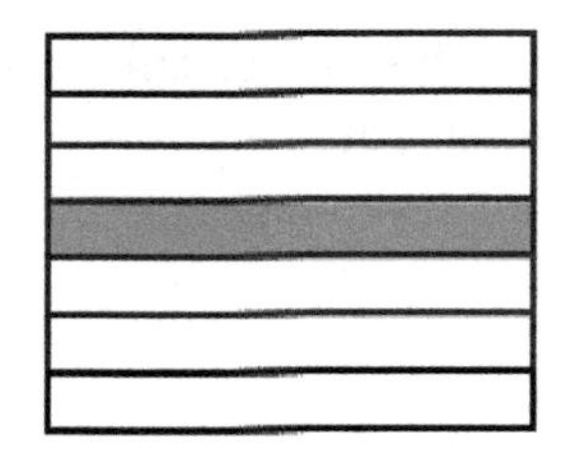				

2. Each image below is 1 whole. Write the fraction that is shaded.

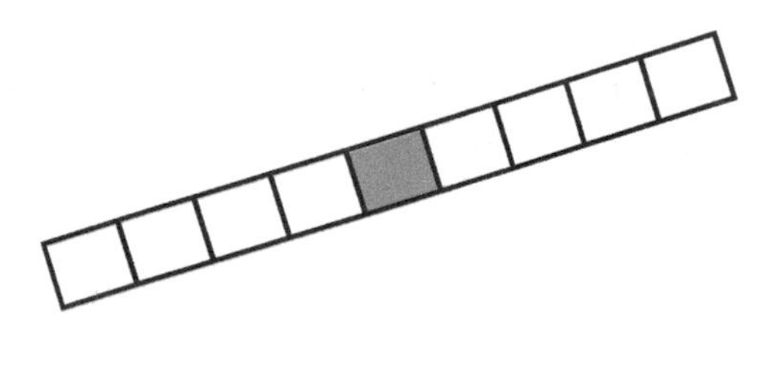

________________ ________________ ________________

3. Draw two identical rectangles. Partition one into 5 equal parts. Partition the other rectangle into 8 equal parts. Label the unit fractions and shade 1 equal part in each rectangle. Use your rectangles to explain why $\frac{1}{5}$ is bigger than $\frac{1}{8}$.

Name ______________________________ Date ______________

1. Fill in the chart. Each image is one whole.

	Total Number of Equal Parts	Total Number of Equal Parts Shaded	Unit Form	Fraction Form
a. 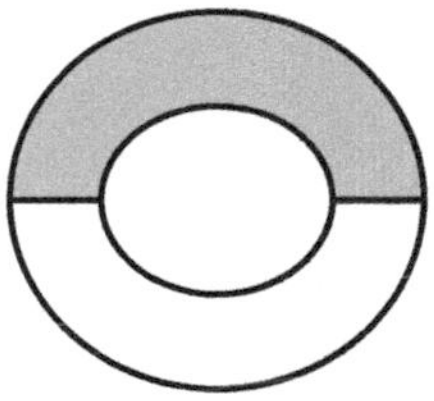				
b. 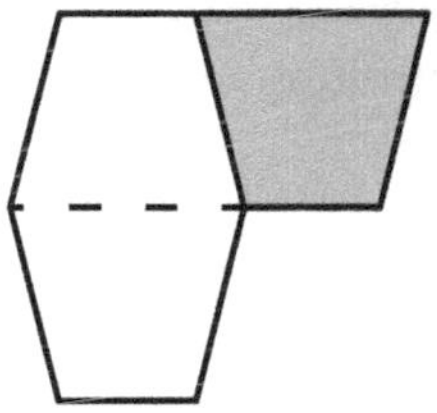				
c. 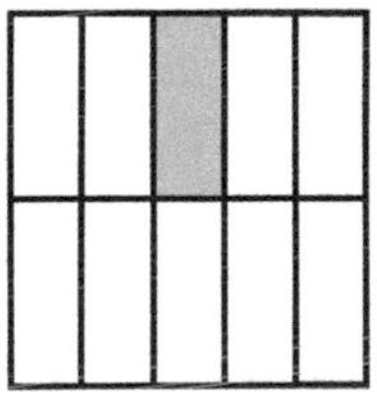				
d. 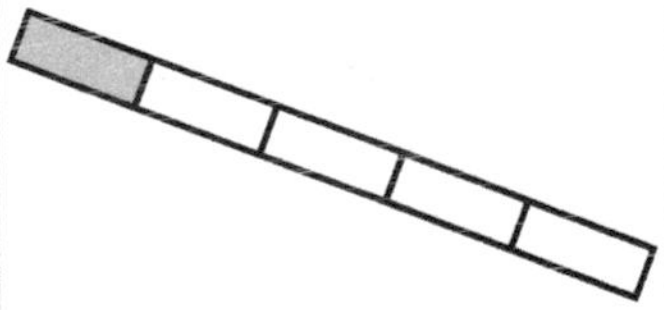				
e. 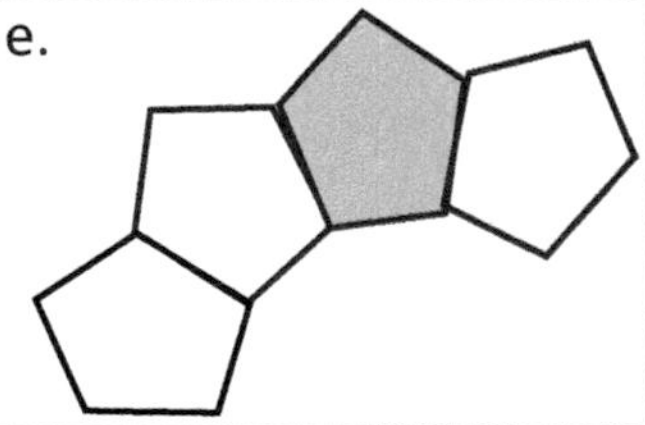				

2. This figure is divided into 6 parts. Are they sixths? Explain your answer.

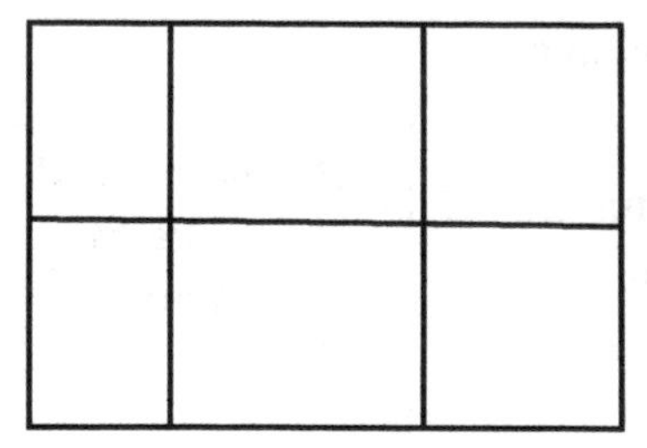

3. Terry and his 3 friends baked a pizza during his sleepover. They want to share the pizza equally. Show how Terry can slice the pizza so that he and his 3 friends can each get an equal amount with none left over.

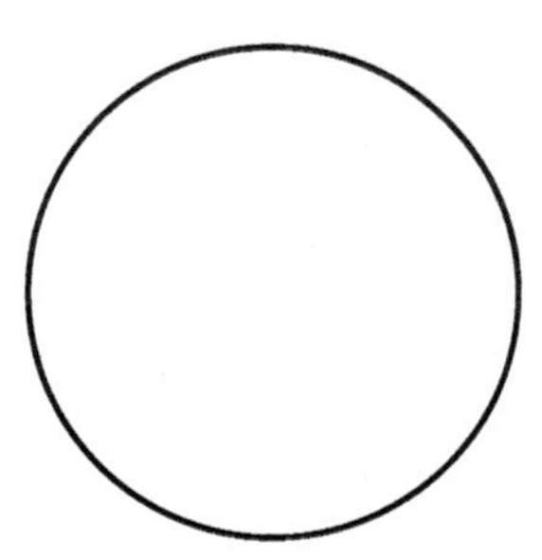

4. Draw two identical rectangles. Shade 1 seventh of one rectangle and 1 tenth of the other. Label the unit fractions. Use your rectangles to explain why $\frac{1}{7}$ is greater than $\frac{1}{10}$.

Lesson 6

Objective: Build non-unit fractions less than one whole from unit fractions.

Suggested Lesson Structure

■ Fluency Practice	(12 minutes)
■ Application Problem	(10 minutes)
■ Concept Development	(28 minutes)
■ Student Debrief	(10 minutes)
Total Time	**(60 minutes)**

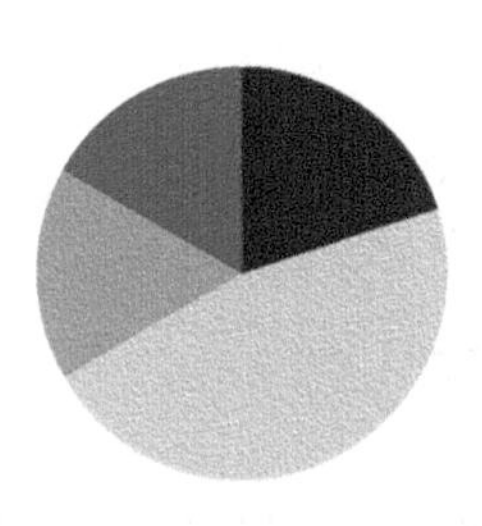

Fluency Practice (12 minutes)

- Sprint: Multiply with Seven **3.OA.4** (8 minutes)
- Write the Unit Fraction **3.G.2, 3.NF.1** (2 minutes)
- Find the Whole **3.NF.3d** (2 minutes)

Sprint: Multiply with Seven (8 minutes)

Materials: (S) Multiply with Seven Sprint

Note: This Sprint supports fluency with multiplication using units of 7.

Write the Unit Fraction (2 minutes)

Materials: (S) Personal white board

Note: This activity reviews naming unit fractions from Lesson 5.

T: (Draw a shape with $\frac{1}{2}$ shaded.) Write the unit fraction.

S: (Write $\frac{1}{2}$.)

Continue with the following possible sequence: $\frac{1}{4}, \frac{1}{8}, \frac{1}{6}, \frac{1}{10}$, and $\frac{1}{5}$.

Find the Whole (2 minutes)

Note: This activity prepares students for their work with non-unit fractions in this lesson.

T: (Project a number bond with parts $\frac{3}{5}$ and $\frac{2}{5}$.) Say the bigger part.

S: 3 fifths.

T: Say the smaller part.

S: 2 fifths.

T: How many fifths are in the whole?

S: 5 fifths.

T: (Write $\frac{5}{5}$ in the whole space.) Say the number sentence.

S: 3 fifths and 2 fifths equals 5 fifths.

Continue with the following possible sequence: $\frac{7}{10}$ and $\frac{3}{10}$, $\frac{5}{8}$ and $\frac{3}{8}$. Replace 8 eighths with 1 whole.

Application Problem (10 minutes)

Chloe's dad partitions his garden into 4 equal-sized sections to plant tomatoes, squash, peppers, and cucumbers. What fraction of the garden is available for growing tomatoes?

Extension: Chloe talked her dad into planting beans and lettuce, too. He used equal-sized sections for all the vegetables. What fraction do the tomatoes have now?

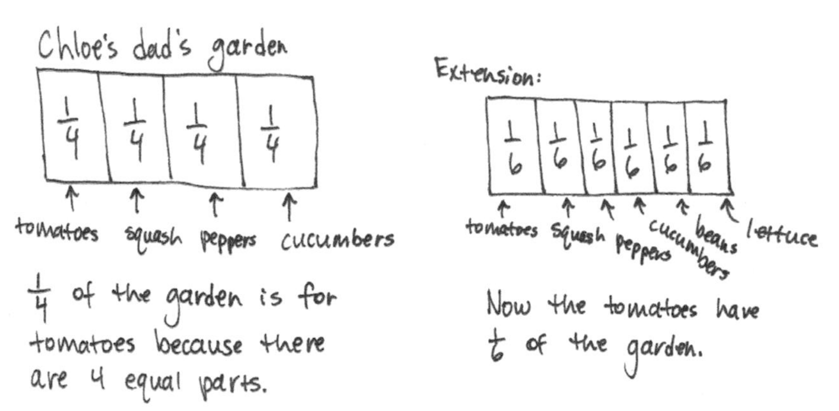

Note: This problem reviews partitioning shapes into equal parts and naming unit fractions.

Concept Development (28 minutes)

Materials: (S) Personal white board

T: Here is unit form. (Write 1 half.) Here it is written in fraction form. (Write $\frac{1}{2}$.) What does the 2 mean?

S: 2 is the number of equal parts that the whole is cut into.

T: What does the 1 mean?

S: We are talking about 1 of the equal parts.

NOTES ON MULTIPLE MEANS OF REPRESENTATION:

Recording choral responses on the board alongside the model supports English language acquisition.

Shape 1:

T: (Project or draw a circle partitioned into thirds.) This is 1 whole.

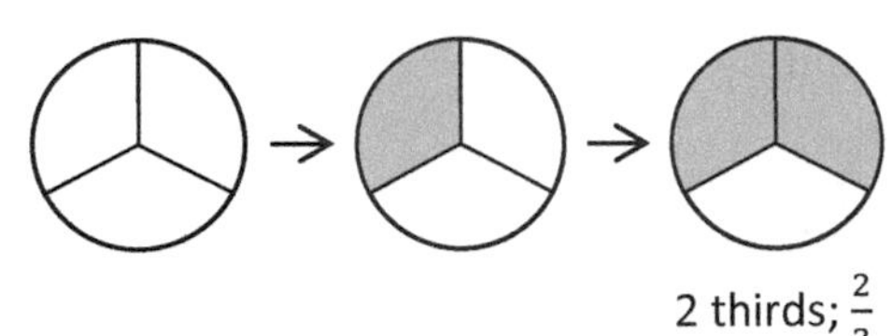

2 thirds; $\frac{2}{3}$

T: What unit is it partitioned into?

S: Thirds.

T: What is the unit fraction?

S: 1 third.

T: (Shade 1 third.) I'm going to make a **copy** of my shaded unit fraction. (Shade one more unit.) How many units are shaded now?

S: 2 thirds.

T: Let's count them.

S: 1 third, 2 thirds.

T: We can write the number 2 thirds in unit form (Write 2 thirds under the circle.) or fraction form. (Write $\frac{2}{3}$ under the circle.) What happened to our unit fraction when we made a copy? Turn and share.

S: We started with one unit shaded and then shaded in another unit to make a copy. 2 copies make 2 thirds. → True. That's why we changed 1 third to 2 thirds. Now we're talking about 2 copies.

T: Yes! Just like 2 copies of one make 2, we can make 2 copies of 1 third to make 2 thirds.

NOTES ON MULTIPLE MEANS OF REPRESENTATION:

To assist comprehension, develop multiple ways to ask the same question, for example, by changing the question, "What's happening to my parts?" to "How are my parts changing?" or "Do you notice an increase or decrease?" or "Is the amount growing or shrinking?"

Continue with the following suggested shapes. Students identify the unit fraction and then make copies to build the new fraction.

Shape 2:

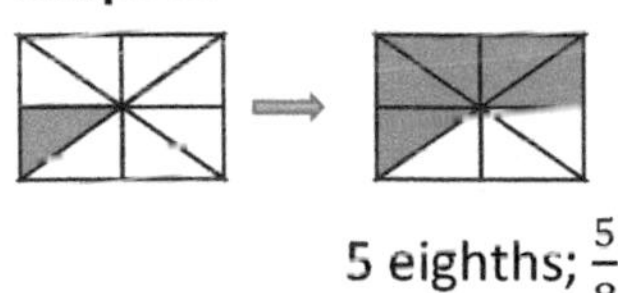

5 eighths; $\frac{5}{8}$

Shape 3:

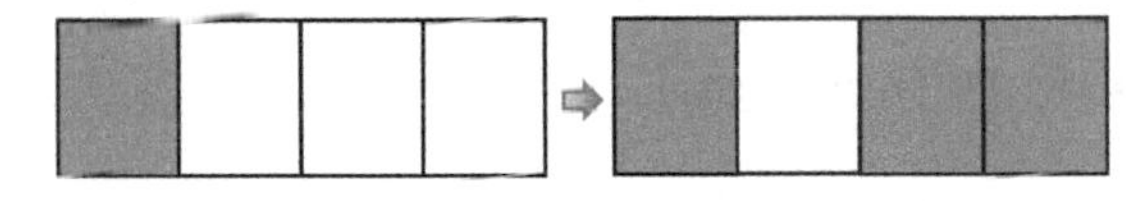

3 quarters or 3 fourths; $\frac{3}{4}$

Students transition into guided practice using personal white boards.

Give the following directions:

- Draw a unit fraction (select examples).
- Make copies of the unit fraction to build a new fraction.
- Count the unit fractions.
- Identify the new fraction both in unit form and fraction form.

NOTES ON MULTIPLE MEANS OF ENGAGEMENT:

Offer students working above grade level a Problem Set alternative of constructing written responses to open-ended questions, such as, "What do these wholes and fractions (pictured on the Problem Set) remind you of?"

Problem Set (10 minutes)

Students should do their personal best to complete the Problem Set within the allotted 10 minutes. For some classes, it may be appropriate to modify the assignment by specifying which problems they work on first. Some problems do not specify a method for solving. Students should solve these problems using the RDW approach used for Application Problems.

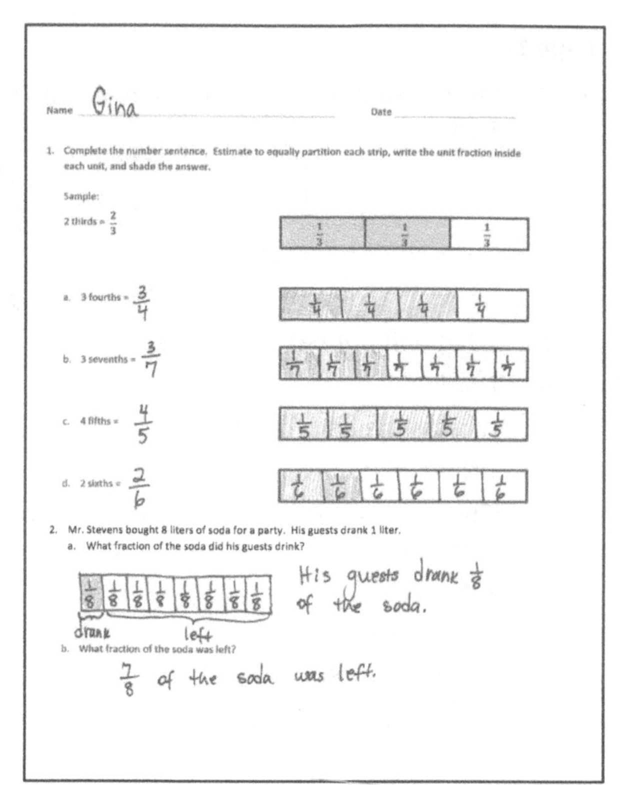

Name Gina Date

1. Complete the number sentence. Estimate to equally partition each strip, write the unit fraction inside each unit, and shade the answer.

Sample:

2 thirds = $\frac{2}{3}$

a. 3 fourths = $\frac{3}{4}$

b. 3 sevenths = $\frac{3}{7}$

c. 4 fifths = $\frac{4}{5}$

d. 2 sixths = $\frac{2}{6}$

2. Mr. Stevens bought 8 liters of soda for a party. His guests drank 1 liter.

a. What fraction of the soda did his guests drink?

His guests drank $\frac{1}{8}$ of the soda.

b. What fraction of the soda was left?

$\frac{7}{8}$ of the soda was left.

3. Fill in the chart.

	Total Number of Equal Parts	Total Number of Shaded Equal Parts	Unit Fraction	Fraction Shaded
Sample:	4	3	$\frac{1}{4}$	$\frac{3}{4}$
a.	9	5	$\frac{1}{9}$	$\frac{5}{9}$
b.	7	3	$\frac{1}{7}$	$\frac{3}{7}$
c.	5	4	$\frac{1}{5}$	$\frac{4}{5}$
d.	6	2	$\frac{1}{6}$	$\frac{2}{6}$
e.	8	8	$\frac{1}{8}$	$\frac{8}{8}$

Student Debrief (10 minutes)

Lesson Objective: Build non-unit fractions less than one whole from unit fractions.

The Student Debrief is intended to invite reflection and active processing of the total lesson experience.

Invite students to review their solutions for the Problem Set. They should check work by comparing answers with a partner before going over answers as a class. Look for misconceptions or misunderstandings that can be addressed in the Debrief. Guide students in a conversation to debrief the Problem Set and process the lesson.

- Through discussion, guide students to articulate the idea that to show non-unit fractions, they create **copies** of unit fractions. This resembles counting 3 ones to make 3, or counting by eighths to make copies of 8.

Exit Ticket (3 minutes)

After the Student Debrief, instruct students to complete the Exit Ticket. A review of their work will help with assessing students' understanding of the concepts that were presented in today's lesson and planning more effectively for future lessons. The questions may be read aloud to the students.

A

Number Correct: _______

Multiply with Seven

1.	1 × 7 =	
2.	7 × 1 =	
3.	2 × 7 =	
4.	7 × 2 =	
5.	3 × 7 =	
6.	7 × 3 =	
7.	4 × 7 =	
8.	7 × 4 =	
9.	5 × 7 =	
10.	7 × 5 =	
11.	6 × 7 =	
12.	7 × 6 =	
13.	7 × 7 =	
14.	8 × 7 =	
15.	7 × 8 =	
16.	9 × 7 =	
17.	7 × 9 =	
18.	10 × 7 =	
19.	7 × 10 =	
20.	7 × 3 =	
21.	1 × 7 =	
22.	2 × 7 =	

23.	10 × 7 =	
24.	9 × 7 =	
25.	4 × 7 =	
26.	8 × 7 =	
27.	7 × 3 =	
28.	7 × 7 =	
29.	6 × 7 =	
30.	7 × 10 =	
31.	7 × 5 =	
32.	7 × 6 =	
33.	7 × 1 =	
34.	7 × 9 =	
35.	7 × 4 =	
36.	7 × 3 =	
37.	7 × 2 =	
38.	7 × 7 =	
39.	7 × 8 =	
40.	11 × 7 =	
41.	7 × 11 =	
42.	12 × 7 =	
43.	7 × 12 =	
44.	13 × 7 =	

B

Number Correct: _______

Improvement: _______

Multiply with Seven

1.	7 × 1 =	
2.	1 × 7 =	
3.	7 × 2 =	
4.	2 × 7 =	
5.	7 × 3 =	
6.	3 × 7 =	
7.	7 × 4 =	
8.	4 × 7 =	
9.	7 × 5 =	
10.	5 × 7 =	
11.	7 × 6 =	
12.	6 × 7 =	
13.	7 × 7 =	
14.	7 × 8 =	
15.	8 × 7 =	
16.	7 × 9 =	
17.	9 × 7 =	
18.	7 × 10 =	
19.	10 × 7 =	
20.	1 × 7 =	
21.	10 × 7 =	
22.	2 × 7 =	

23.	9 × 7 =	
24.	3 × 7 =	
25.	8 × 7 =	
26.	4 × 7 =	
27.	7 × 7 =	
28.	5 × 7 =	
29.	6 × 7 =	
30.	7 × 5 =	
31.	7 × 10 =	
32.	7 × 1 =	
33.	7 × 6 =	
34.	7 × 4 =	
35.	7 × 9 =	
36.	7 × 2 =	
37.	7 × 7 =	
38.	7 × 3 =	
39.	7 × 8 =	
40.	11 × 7 =	
41.	7 × 11 =	
42.	12 × 7 =	
43.	7 × 12 =	
44.	13 × 7 =	

Name ______________________________ Date ______________

1. Complete the number sentence. Estimate to partition each strip equally, write the unit fraction inside each unit, and shade the answer.

Sample:

2 thirds = $\frac{2}{3}$

$\frac{1}{3}$	$\frac{1}{3}$	$\frac{1}{3}$

a. 3 fourths =

b. 3 sevenths =

c. 4 fifths =

d. 2 sixths =

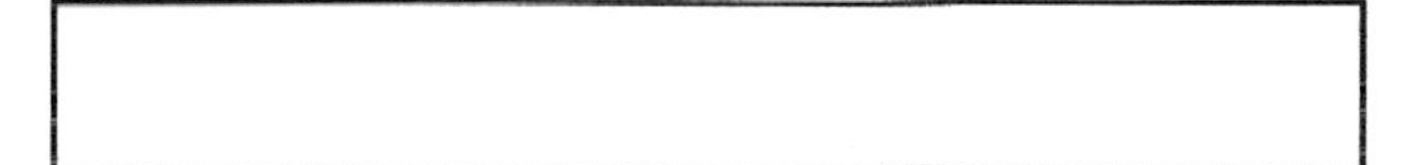

2. Mr. Stevens bought 8 liters of soda for a party. His guests drank 1 liter.

a. What fraction of the soda did his guests drink?

b. What fraction of the soda was left?

3. Fill in the chart.

	Total Number of Equal Parts	Total Number of Shaded Equal Parts	Unit Fraction	Fraction Shaded
Sample: 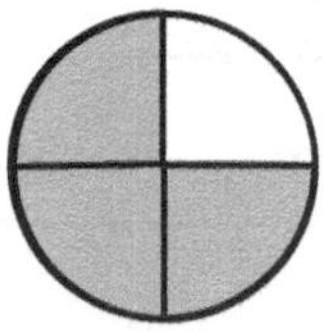	4	3	$\frac{1}{4}$	$\frac{3}{4}$
a. 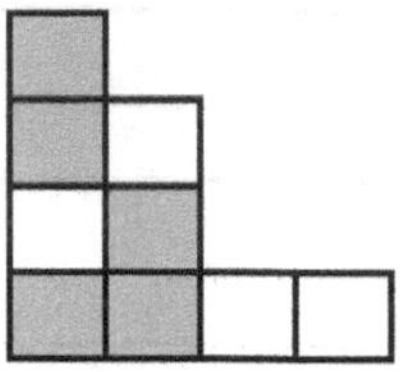				
b. 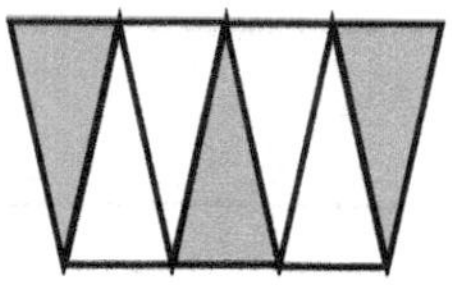				
c. 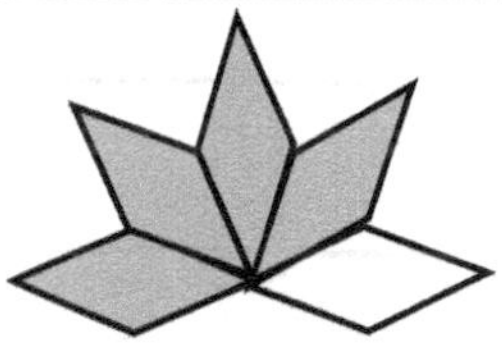				
d. 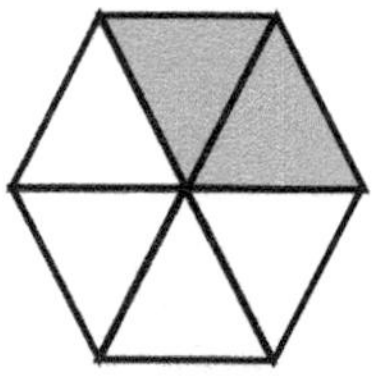				
e. 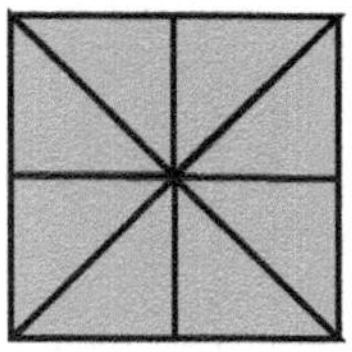				

Name ______________________________ Date ________________

1. Complete the number sentence. Estimate to partition the strip equally. Write the unit fraction inside each unit. Shade the answer.

   2 fifths =

2. 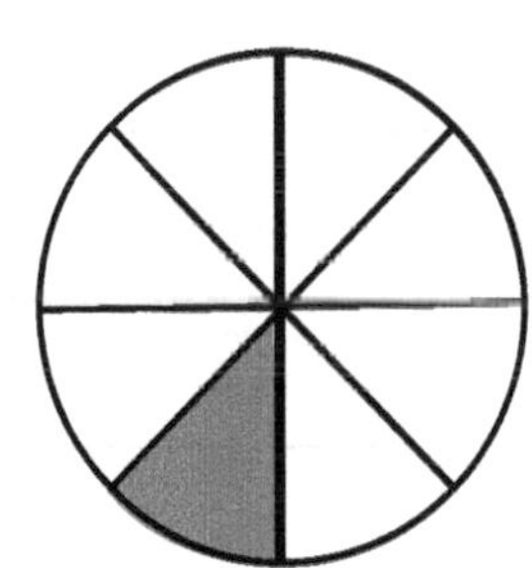

   a. What fraction of the circle is shaded?

   b. What fraction of the circle is not shaded?

3. Complete the chart.

	Total Number of Equal Parts	Total Number of Shaded Equal Parts	Unit Fraction	Fraction Shaded

Name ______________________________ Date ______________

1. Complete the number sentence. Estimate to partition each strip equally, write the unit fraction inside each unit, and shade the answer.

Sample:

3 fourths = $\frac{3}{4}$

$\frac{1}{4}$	$\frac{1}{4}$	$\frac{1}{4}$	$\frac{1}{4}$

a. 2 thirds =

b. 5 sevenths =

c. 3 fifths =

d. 2 eighths =

2. Mr. Abney bought 6 kilograms of rice. He cooked 1 kilogram of it for dinner.

a. What fraction of the rice did he cook for dinner?

b. What fraction of the rice was left?

3. Fill in the chart.

	Total Number of Equal Parts	Total Number of Shaded Equal Parts	Unit Fraction	Fraction Shaded
Sample: 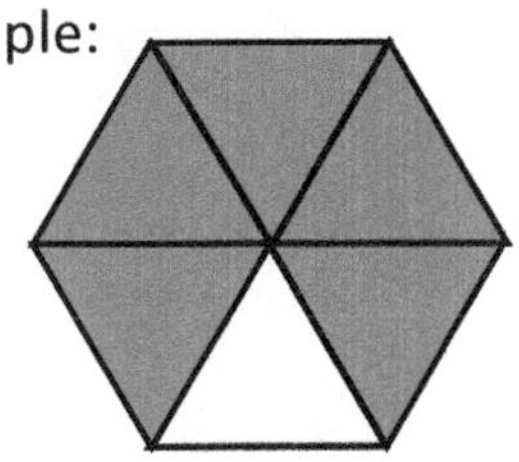	6	5	$\frac{1}{6}$	$\frac{5}{6}$
a. 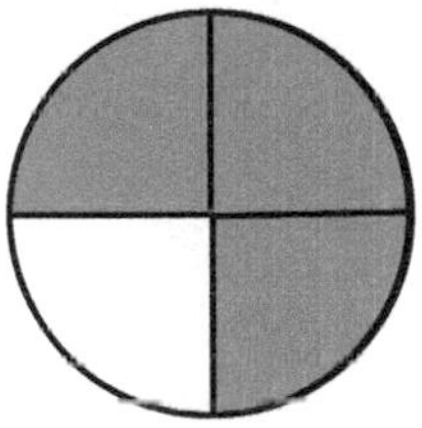				
b. 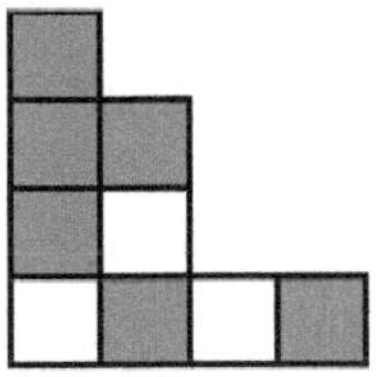				
c. 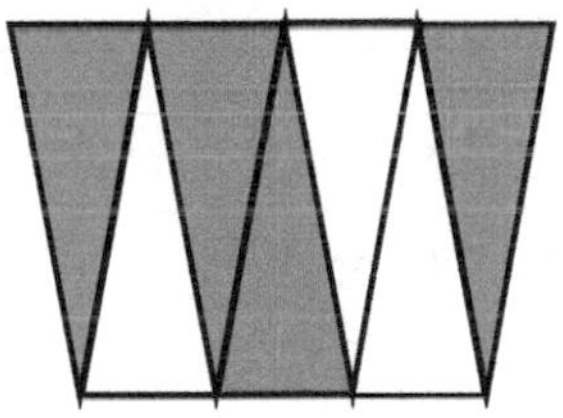				
d. 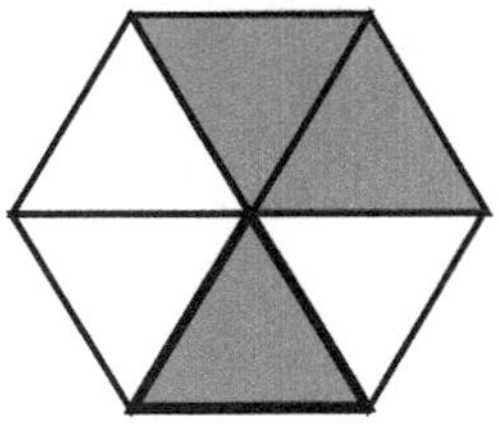				

Lesson 7

Objective: Identify and represent shaded and non-shaded parts of one whole as fractions.

Suggested Lesson Structure

■ Fluency Practice	(12 minutes)
■ Application Problem	(10 minutes)
■ Concept Development	(28 minutes)
■ Student Debrief	(10 minutes)
Total Time	**(60 minutes)**

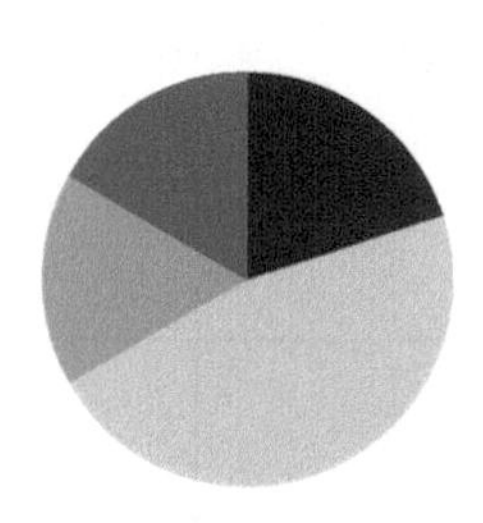

Fluency Practice (12 minutes)

- Group Counting **3.OA.1** (2 minutes)
- Sprint: Multiply and Divide by Seven **3.OA.4** (8 minutes)
- Skip-Count by Halves on the Clock **3.G.2, 3.NF.1** (2 minutes)

Group Counting (2 minutes)

Materials: (S) Personal white board

Note: Group counting reviews interpreting multiplication as repeated addition.

Direct students to count forward and backward by nine to 90 on their personal white boards.

T: Circle 27. How many nines did you count?
S: 3 nines.
T: What is 27 divided by 9?
S: 3.

Continue with the following possible sequence: 18, 81, 45, 36, 54, 72, 9, and 63.

Sprint: Multiply and Divide by Seven (8 minutes)

Materials: (S) Multiply and Divide by Seven Sprint

Note: This Sprint supports fluency with multiplication and division using units of 7.

Skip-Count by Halves on the Clock (2 minutes)

Materials: (T) Clock

Note: This activity reviews counting by halves on the clock from Module 2.

T: (Hold or project a clock.) Let's skip-count by halves on the clock starting with 1 o'clock.

S: 1, half past 1, 2, half past 2, 3, half past 3, 4, (switch direction), half past 3, 3, half past 2, 2, half past 1, 1.

Continue counting up and down.

Application Problem (10 minutes)

Robert ate half of the applesauce in a container. He split the remaining applesauce equally into 2 bowls for his mother and sister. Robert said, "I ate 1 half, and each of you gets 1 half." Is Robert right? Draw a picture to prove your answer.

Extension:

- What fraction of the applesauce did his mother get?
- What fraction of the applesauce did Robert's sister eat?

You can only have 2 halves in a whole. Robert is wrong!!! So 3 people cannot have $\frac{1}{2}$ each. His mom and sister got $\frac{1}{2}$ together!

Robert's mom and sister each ate $\frac{1}{4}$ of the applesauce.

Note: This problem reviews the concept that a whole is made of 2 halves. The extension challenges students to see the whole partitioned into halves and fourths.

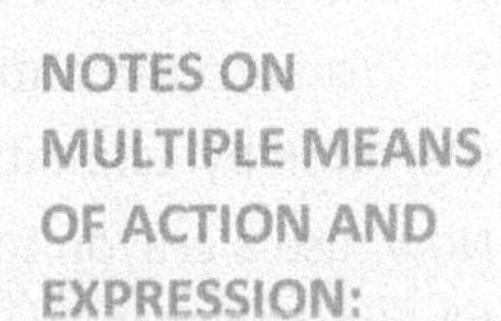

NOTES ON MULTIPLE MEANS OF ACTION AND EXPRESSION:

Give explicit steps for problem solving to students working below grade level. These steps can be organized as a checklist such as, "Underline important words, draw a model, label your model."

NOTES ON MULTIPLE MEANS OF REPRESENTATION:

These daily class discussions, as well as "Think-pair-share," support English language learners' English language acquisition. They offer students an opportunity to talk about their math ideas in English and actively use the language of mathematics.

Concept Development (28 minutes)

Materials: (T) 1-liter beaker, water (S) Paper, scissors, crayons, math journal

Show a beaker of liquid half full.

T: Whisper the fraction of liquid that you see to your partner.

S: 1 half.

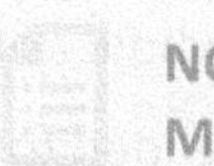

NOTES ON MATERIALS:

If a beaker is not available, use a clear container that has a consistent diameter from bottom to top, and measure the amount of liquid to precisely show the container half full.

T: What about the part that is not full? Talk to your partner: Could that be a fraction, too? Why or why not?

S: No, because there's nothing there. → I disagree. It's another part. It's just not full. → It's another half. Because half is full and half is empty. Two halves make one whole.

T: Even though parts might not be full or shaded, they are still part of the whole. Let's explore this idea some more. I'll give you 1 sheet of paper. Partition it into equal parts. Just be sure of these 3 things:

1. The parts must be equal.
2. There are no fewer than 5 and no more than 20 parts in all.
3. You use the entire sheet of paper.

S: (Partition by estimating to fold the paper into equal parts.)

T: Now, use a crayon to shade one unit.

S: (Shade one part.)

T: Next, you're going to cut your whole into parts by cutting along the lines you created when you folded the paper. You'll reassemble your parts into a unique piece of art for our fraction museum. As you make your art, make sure that all parts are touching but not on top of or under each other.

S: (Cut along the folds and reassemble pieces.)

T: As you tour our museum admiring the art, identify which unit fraction the artist chose and identify the fraction representing the unshaded equal parts of the art. Write both fractions in your journal next to each other.

S: (Walk around and collect data, which will be used in the Debrief portion of the lesson.)

Problem Set (10 minutes)

Students should do their personal best to complete the Problem Set within the allotted 10 minutes. For some classes, it may be appropriate to modify the assignment by specifying which problems they work on first. Some problems do not specify a method for solving. Students should solve these problems using the RDW approach used for Application Problems.

> **NOTES ON MULTIPLE MEANS OF ENGAGEMENT:**
>
> Offer students working above grade level a Problem Set alternative of constructing a word problem for one of the models (pictured in number 10 of the Problem Set). Constructively review errors with students who are accustomed to always scoring correctly or who may be perfectionists.

Student Debrief (10 minutes)

Lesson Objective: Identify and represent shaded and non-shaded parts of one whole as fractions.

The Student Debrief is intended to invite reflection and active processing of the total lesson experience.

Invite students to review their solutions for the Problem Set. They should check work by comparing answers with a partner before going over answers as a class. Look for misconceptions or misunderstandings that can be addressed in the Debrief. Guide students in a conversation to debrief the Problem Set and process the lesson.

Any combination of the questions below may be used to lead the discussion.

MP.3

- Show examples of student work for Problem 11. Avanti read 1 sixth of her book. What fraction of her book has she not read yet? Isn't Avanti's goal to read the whole book? (Guide students to notice that the whole book can be depicted as the part she has read and the part she has not read.)
- From the discussion above the teacher might briefly return to the *shaded* and *unshaded* figures in Problem 1 and help students notice that the whole can be expressed as two parts—the shaded and unshaded.
- Revisit students' art. Guide a discussion helping them recognize that while each student's art depicts a whole, each whole is composed of different fractional units (e.g., fourths, fifths, or sixths).

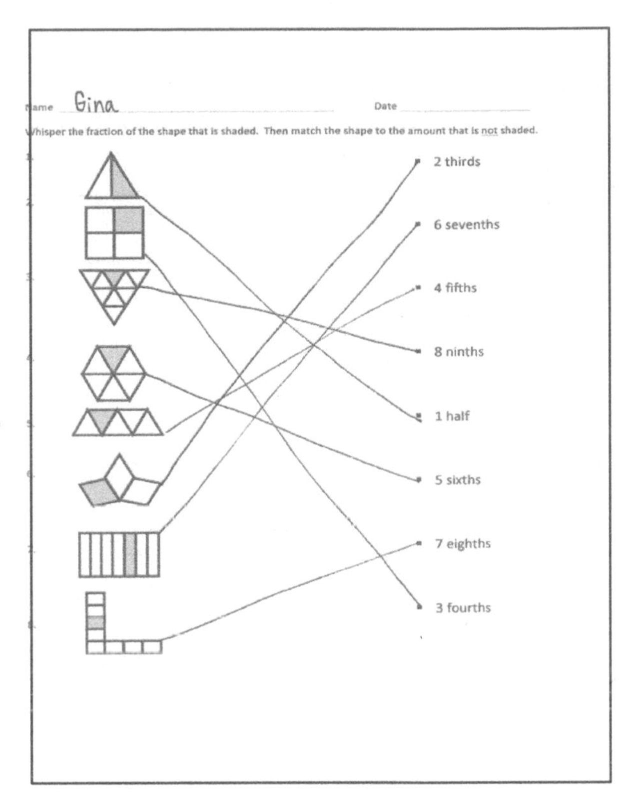

Exit Ticket (3 minutes)

After the Student Debrief, instruct students to complete the Exit Ticket. A review of their work will help with assessing students' understanding of the concepts that were presented in today's lesson and planning more effectively for future lessons. The questions may be read aloud to the students.

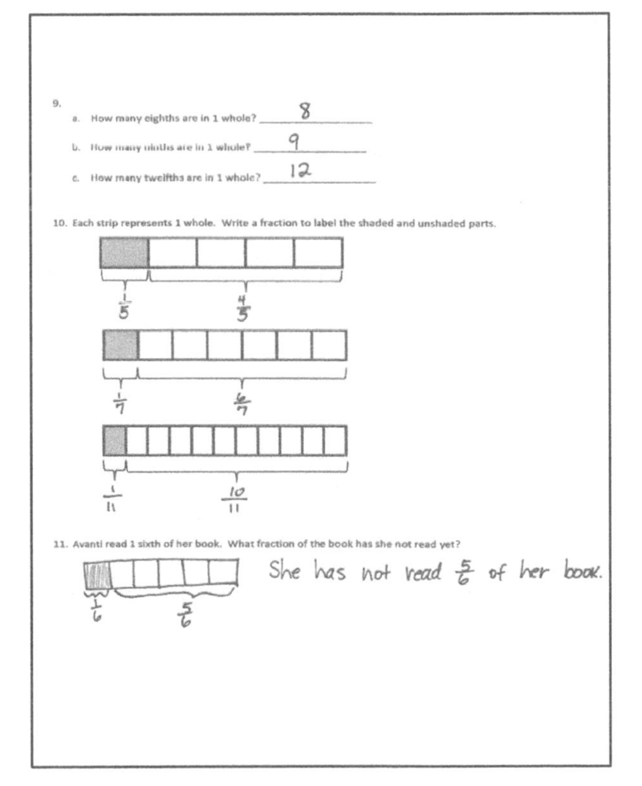

A

Number Correct: _______

Multiply and Divide by Seven

1.	$2 \times 7 =$		23.	$\_\_\_ \times 7 = 70$	
2.	$3 \times 7 =$		24.	$\_\_\_ \times 7 = 14$	
3.	$4 \times 7 =$		25.	$\_\_\_ \times 7 = 21$	
4.	$5 \times 7 =$		26.	$70 \div 7 =$	
5.	$1 \times 7 =$		27.	$35 \div 7 =$	
6.	$14 \div 7 =$		28.	$7 \div 7 =$	
7.	$21 \div 7 =$		29.	$14 \div 7 =$	
8.	$35 \div 7 =$		30.	$21 \div 7 =$	
9.	$7 \div 7 =$		31.	$\_\_\_ \times 7 = 42$	
10.	$28 \div 7 =$		32.	$\_\_\_ \times 7 = 49$	
11.	$6 \times 7 =$		33.	$\_\_\_ \times 7 = 63$	
12.	$7 \times 7 =$		34.	$\_\_\_ \times 7 = 56$	
13.	$8 \times 7 =$		35.	$49 \div 7 =$	
14.	$9 \times 7 =$		36.	$63 \div 7 =$	
15.	$10 \times 7 =$		37.	$42 \div 7 =$	
16.	$56 \div 7 =$		38.	$56 \div 7 =$	
17.	$49 \div 7 =$		39.	$11 \times 7 =$	
18.	$63 \div 7 =$		40.	$77 \div 7 =$	
19.	$42 \div 7 =$		41.	$12 \times 7 =$	
20.	$70 \div 7 =$		42.	$84 \div 7 =$	
21.	$\_\_\_ \times 7 = 35$		43.	$14 \times 7 =$	
22.	$\_\_\_ \times 7 = 7$		44.	$98 \div 7 =$	

B

Number Correct: _______

Improvement: _______

Multiply and Divide by Seven

1.	1 × 7 =	
2.	2 × 7 =	
3.	3 × 7 =	
4.	4 × 7 =	
5.	5 × 7 =	
6.	21 ÷ 7 =	
7.	14 ÷ 7 =	
8.	28 ÷ 7 =	
9.	7 ÷ 7 =	
10.	35 ÷ 7 =	
11.	10 × 7 =	
12.	6 × 7 =	
13.	7 × 7 =	
14.	8 × 7 =	
15.	9 × 7 =	
16.	49 ÷ 7 =	
17.	42 ÷ 7 =	
18.	56 ÷ 7 =	
19.	70 ÷ 7 =	
20.	63 ÷ 7 =	
21.	___ × 7 = 7	
22.	___ × 7 = 35	

23.	___ × 7 = 14	
24.	___ × 7 = 70	
25.	___ × 7 = 21	
26.	14 ÷ 7 =	
27.	7 ÷ 7 =	
28.	70 ÷ 7 =	
29.	35 ÷ 7 =	
30.	21 ÷ 7 =	
31.	___ × 7 = 21	
32.	___ × 7 = 28	
33.	___ × 7 = 63	
34.	___ × 7 = 49	
35.	56 ÷ 7 =	
36.	63 ÷ 7 =	
37.	42 ÷ 7 =	
38.	49 ÷ 7 =	
39.	11 × 7 =	
40.	77 ÷ 7 =	
41.	12 × 7 =	
42.	84 ÷ 7 =	
43.	13 × 7 =	
44.	91 ÷ 7 =	

Name ______________________________ Date ______________

Whisper the fraction of the shape that is shaded. Then, match the shape to the amount that is not shaded.

1. ▪ 2 thirds

2. ▪ 6 sevenths

3. ▪ 4 fifths

4. ▪ 8 ninths

5. ▪ 1 half

6.  ▪ 5 sixths

7. ▪ 7 eighths

8. 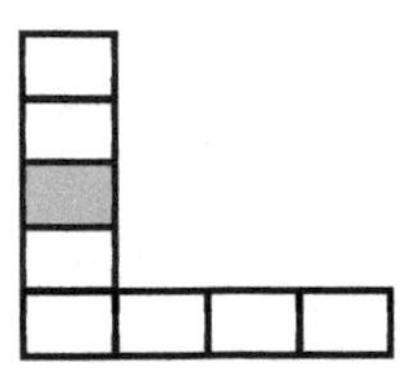 ▪ 3 fourths

9. a. How many eighths are in 1 whole? _______________

   b. How many ninths are in 1 whole? _________________

   c. How many twelfths are in 1 whole? _________________

10. Each strip represents 1 whole. Write a fraction to label the shaded and unshaded parts.

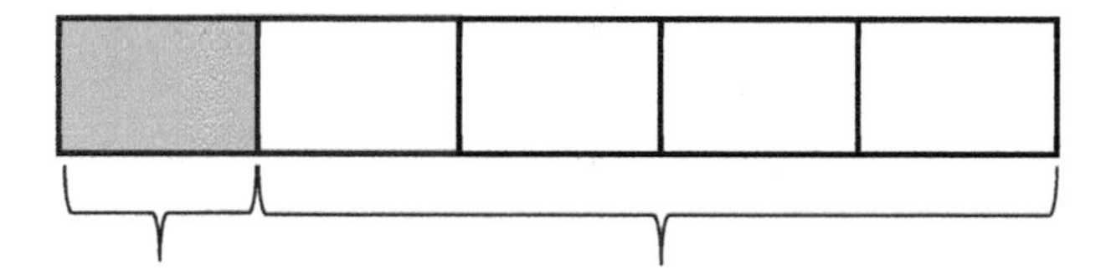

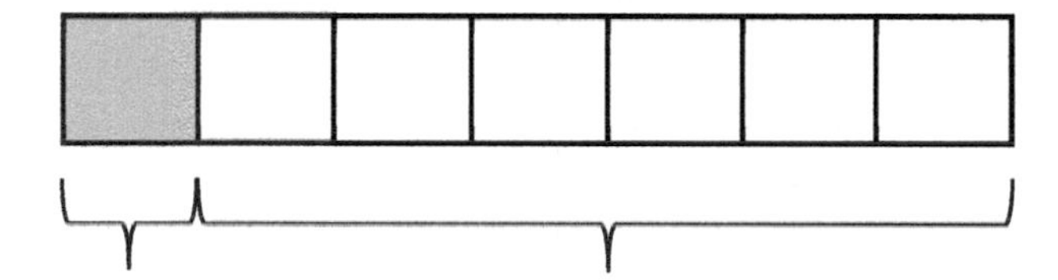

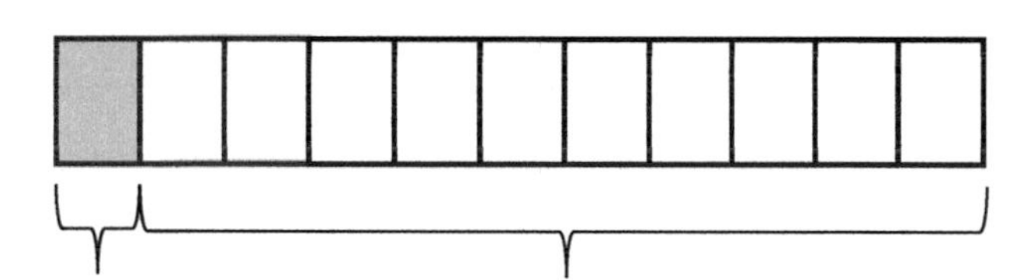

11. Avanti read 1 sixth of her book. What fraction of the book has she not read yet?

Name ______________________________ Date ______________

1. Write the fraction that is not shaded.

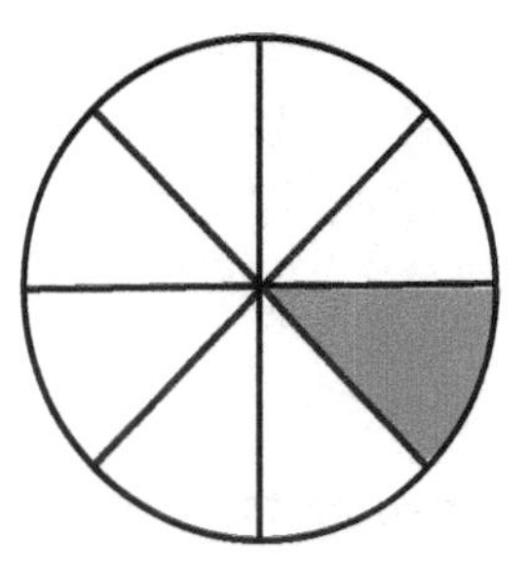

2. There are _________ sixths in 1 whole.

3. The fraction strip is 1 whole. Write fractions to label the shaded and unshaded parts.

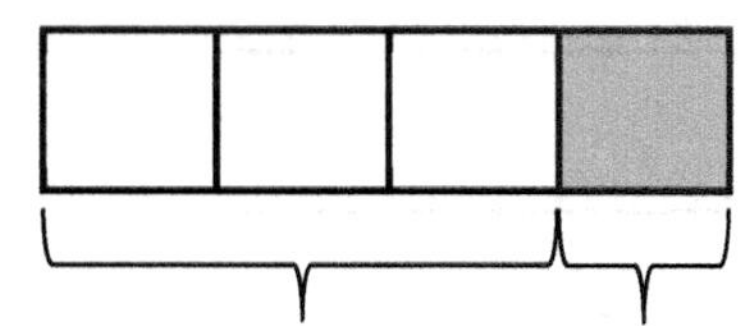

4. Justin mows part of his lawn. Then, his lawnmower runs out of gas. He has not mowed $\frac{9}{10}$ of the lawn. What part of his lawn is mowed?

Name ______________________________ Date ______________

Whisper the fraction of the shape that is shaded. Then, match the shape to the amount that is not shaded.

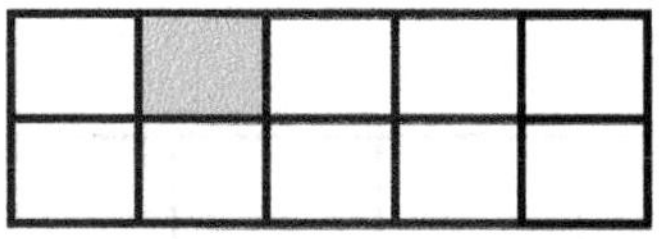

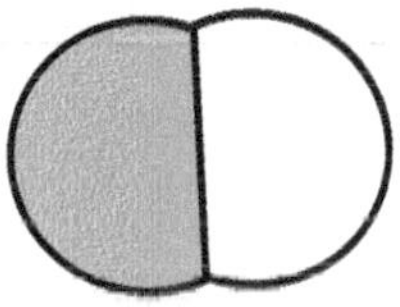

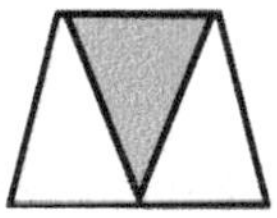

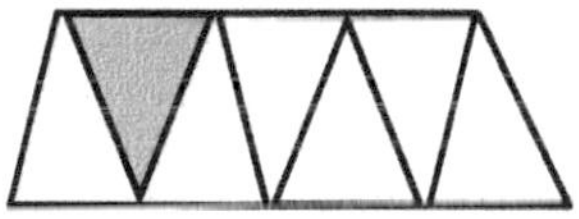

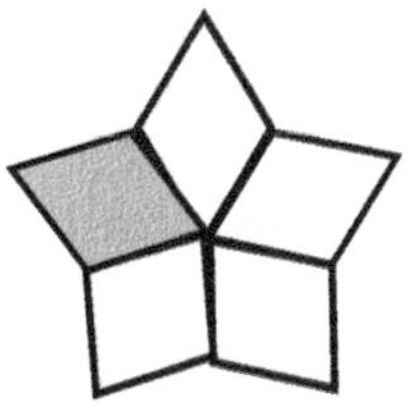

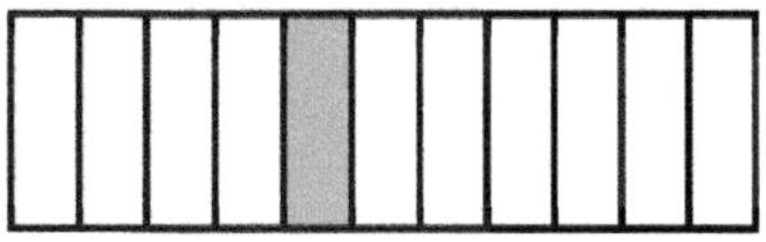

1.

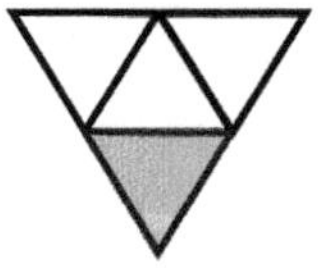

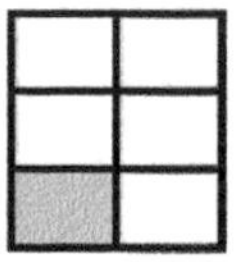

- 9 tenths

2.

- 4 fifths

3.

- 10 elevenths

4.

- 5 sixths

5.

- 1 half

6.

- 2 thirds

7.

- 3 fourths

8.

- 6 sevenths

9. Each strip represents 1 whole. Write a fraction to label the shaded and unshaded parts.

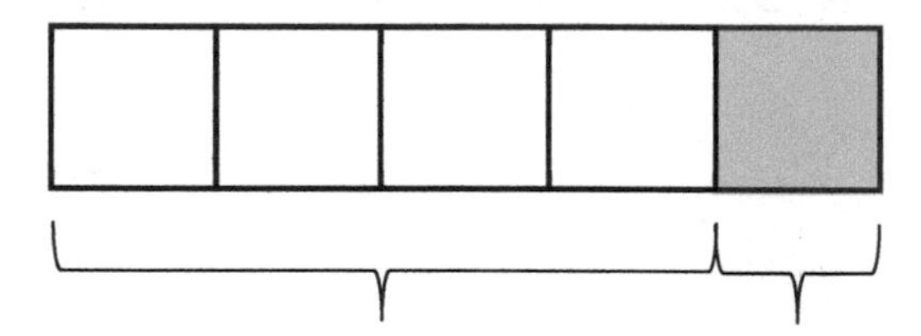

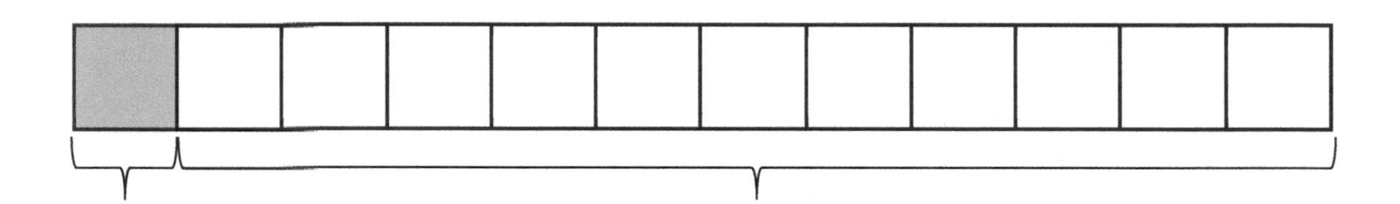

10. Carlia finished 1 fourth of her homework on Saturday. What fraction of her homework has she not finished? Draw and explain.

11. Jerome cooks 8 cups of oatmeal for his family. They eat 7 eighths of the oatmeal. What fraction of the oatmeal is uneaten? Draw and explain.

Lesson 8

Objective: Represent parts of one whole as fractions with number bonds.

Suggested Lesson Structure

■ Fluency Practice	(12 minutes)
■ Application Problem	(10 minutes)
■ Concept Development	(28 minutes)
■ Student Debrief	(10 minutes)
Total Time	**(60 minutes)**

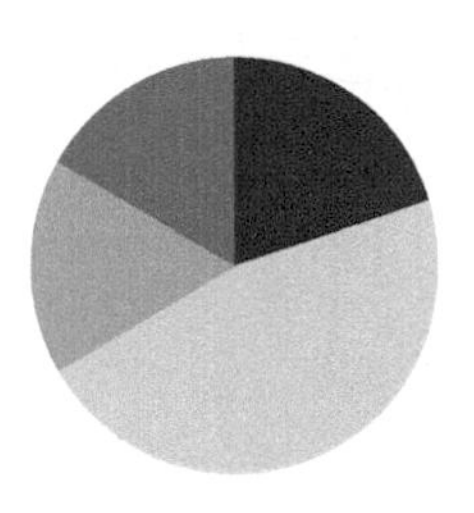

Fluency Practice (12 minutes)

- Unit and Non-Unit Fractions of 1 Whole **3.NF.1** (2 minutes)
- Sprint: Identify Fractions **3.G.2, 3.NF.2** (10 minutes)

Unit and Non-Unit Fractions of 1 Whole (2 minutes)

Materials: (S) Personal white board

Note: This activity reviews naming the shaded and unshaded equal parts of a whole.

T: (Draw a shape partitioned in halves with 1 half shaded.) Write the fraction that is shaded.

S: (Write $\frac{1}{2}$.)

T: Write the fraction that is not shaded.

S: (Write $\frac{1}{2}$.)

Continue with the following possible sequence of shaded and non-shaded parts: $\frac{2}{3}$ and $\frac{1}{3}$, $\frac{4}{5}$ and $\frac{1}{5}$, $\frac{9}{10}$ and $\frac{1}{10}$, and $\frac{7}{8}$ and $\frac{1}{8}$.

Sprint: Identify Fractions (10 minutes)

Materials: (S) Identify Fractions Sprint

Note: This Sprint supports fluency with identifying shaded parts of shapes. Have students keep Sprint B to use in the Concept Development.

Application Problem (10 minutes)

For breakfast, Mr. Schwartz spent 1 sixth of his money on a coffee and 1 sixth of his money on a bagel. What fraction of his money did Mr. Schwartz spend on breakfast?

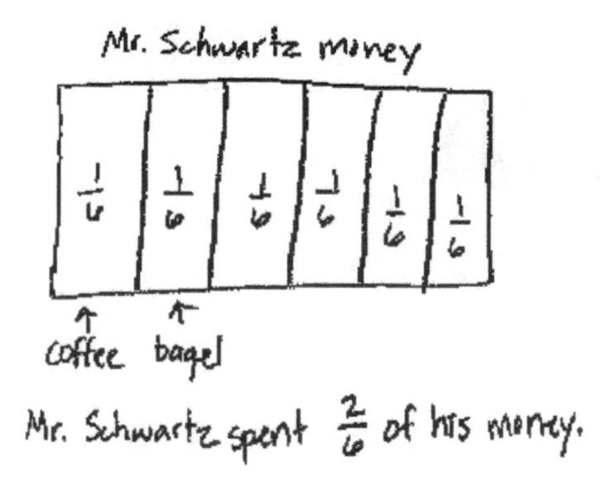

NOTES ON MULTIPLE MEANS OF ENGAGEMENT:

Challenge students working above grade level with extension questions, such as, "Did Mr. Schwartz spend more or less than 1 half of his money? How do you know?"

Note: This problem reviews building and naming non-unit fractions from Lesson 6.

Concept Development (28 minutes)

Materials: (S) Personal white board, Sprint B from the Fluency Practice

Problem 1: Decompose 4 into ones.

T: On your personal white board, draw a number bond decomposing 4 into 4 ones.

S: (Draw a number bond.)

T: Now, work with your partner to show a number bond decomposing 4 into 2 parts. One part should be composed of 3 ones.

S: (Work with a partner to draw the number bond.)

T: It took 3 copies of one to make 3. What are the two parts of your number bond? Please specify the unit.

S: 3 ones and 1 one.

T: Talk to your partner about the difference between these two number bonds.

S: The first bond has the ones all separated. → The second bond has 3 instead of 3 ones. → Both bonds are different ways of showing the same number—4. → You could also show 4 as one part 2 and one part 2. → The first bond has more parts than the second one.

NOTES ON MULTIPLE MEANS OF REPRESENTATION:

Emphasize key concepts and clarify unfamiliar words with gestures when speaking with English language learners. For example, illustrate the word *decompose* by showing hands held together. Then, indicate *breaking apart, separation, splitting,* or *partitioning* by using a downward motion to open the hands. Doing this can also help English speakers.

Problem 2: Decompose 1 into fourths.

MP.7

T: Draw a number bond decomposing 1 into 4 unit fractions.

S: (Draw a number bond.)

T: Now, work with your partner to show a number bond decomposing 1 into 2 parts. One part should be composed of 3 copies of the unit fraction.

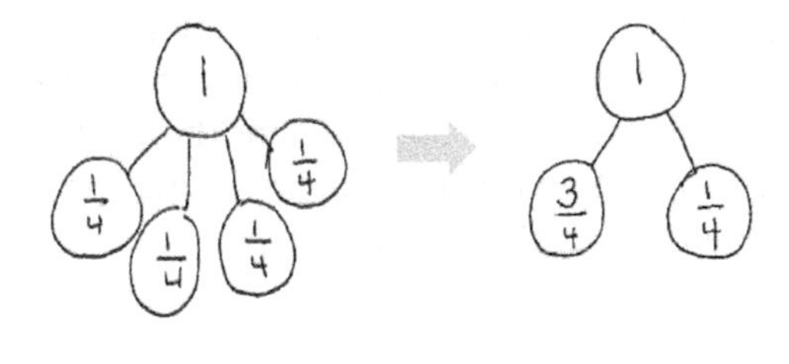

S: (Work with a partner to draw the number bond.)

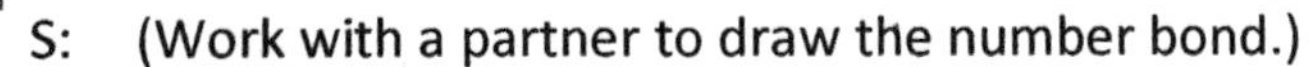

T: What unit fraction did we copy to make the number 3 fourths?

S: 1 fourth.

T: What are the two parts of your number bond? Please specify the unit.

S: 3 fourths and 1 fourth.

T: (Encourage students to compare the two number bonds just as they did with the number bond of 4.)

T: Look at your Sprint B. Discuss with your partner which of the figures on Sprint B match your number bond.

S: Numbers 3, 6, 9, 11, and 18–25 on Sprint B match my number bond.

Problem 3: Decompose 1 into fifths (2 non-unit fractions).

T: Draw a number bond decomposing 1 into 5 unit fractions.

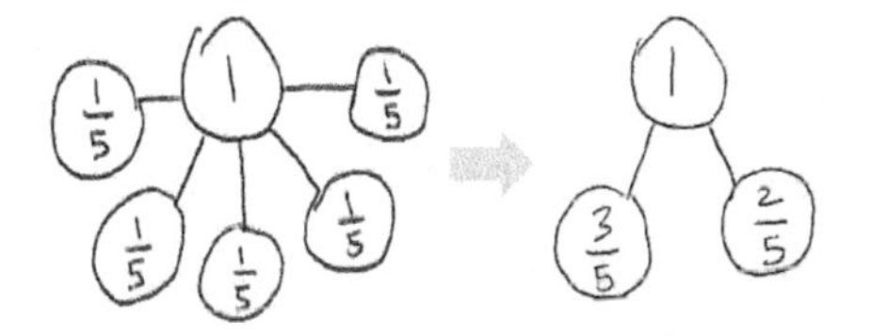

S: (Draw a number bond.)

T: Now, work with your partner to show a number bond decomposing 1 into 2 parts. One part should be composed of 2 copies of the unit fraction.

S: (Work with a partner to draw the number bond.)

T: What unit fraction did we copy to make the number 2 fifths?

S: 1 fifth.

T: What are the two parts of your number bond? Please specify the unit.

S: 2 fifths and 3 fifths.

T: Look at your Sprint B. Discuss with your partner which of these figures match your number bond.

S: Numbers 30–33 on Sprint B match my number bond.

T: Yes, 3 fifths can represent either the shaded or unshaded part.

After doing these three problems, having students use the same process to model Questions 1, 12, 28, 39, and 44 from Sprint B could be helpful. Ask them to find other models on the Sprint that are represented by the same bond.

NOTES ON MULTIPLE MEANS OF ACTION AND EXPRESSION:

When supporting a small group, go step-by-step. Avoid talking and doing at the same time. Draw a number bond in silence. Turn and face the group, and ask them to explain to a partner what they just saw. Then, perform the next action silently. Ask them to explain the action again. Doing this gives them the opportunity to analyze and reconstruct these actions so that they internalize a process they can use.

Problem Set (10 minutes)

Students should do their personal best to complete the Problem Set within the allotted 10 minutes. For some classes, it may be appropriate to modify the assignment by specifying which problems they work on first. Some problems do not specify a method for solving. Students should solve these problems using the RDW approach used for Application Problems.

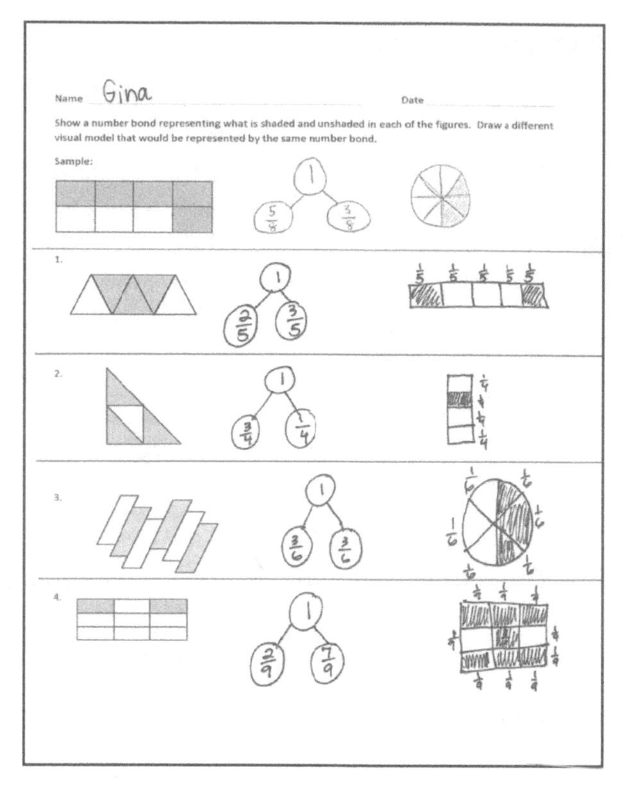

Name Gina Date

Show a number bond representing what is shaded and unshaded in each of the figures. Draw a different visual model that would be represented by the same number bond.

Sample:

1.

2.

3.

4.

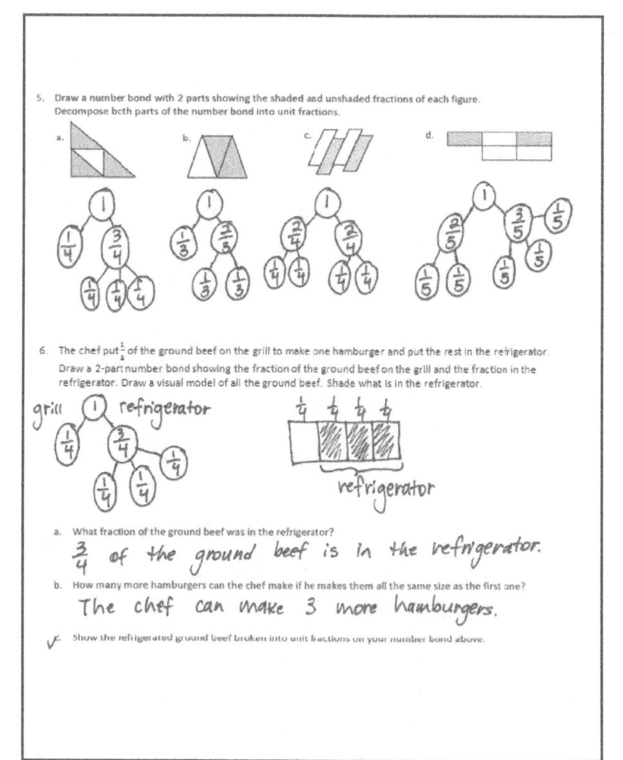

5. Draw a number bond with 2 parts showing the shaded and unshaded fractions of each figure. Decompose both parts of the number bond into unit fractions.

a. b. c. d.

6. The chef put $\frac{1}{4}$ of the ground beef on the grill to make one hamburger and put the rest in the refrigerator. Draw a 2-part number bond showing the fraction of the ground beef on the grill and the fraction in the refrigerator. Draw a visual model of all the ground beef. Shade what is in the refrigerator.

grill refrigerator

refrigerator

a. What fraction of the ground beef was in the refrigerator?

$\frac{3}{4}$ of the ground beef is in the refrigerator.

b. How many more hamburgers can the chef make if he makes them all the same size as the first one?

The chef can make 3 more hamburgers.

c. Show the refrigerated ground beef broken into unit fractions on your number bond above.

Student Debrief (10 minutes)

Lesson Objective: Represent parts of one whole as fractions with number bonds.

The Student Debrief is intended to invite reflection and active processing of the total lesson experience.

Invite students to review their solutions for the Problem Set. They should check work by comparing answers with a partner before going over answers as a class. Look for misconceptions or misunderstandings that can be addressed in the Debrief. Guide students in a conversation to debrief the Problem Set and process the lesson.

Any combination of the questions below may be used to lead the discussion.

- Share different representations for Problem 6 about the hamburger. Guide students to see that the chef's refrigerated meat can be made into 3 more burgers and that each of those burgers is $\frac{1}{4}$ of the meat.
- As in Lesson 7's Debrief, return to the shaded and unshaded figures so that students articulate that 1 whole can ultimately be decomposed into unit fractions. The number bond is a perfect tool for seeing the transition from 1 whole to 2 parts to unit fractions. It is analogous as well to the beginning problem, when the number 4 was decomposed into 4 ones.

Exit Ticket (3 minutes)

After the Student Debrief, instruct students to complete the Exit Ticket. A review of their work will help with assessing students' understanding of the concepts that were presented in today's lesson and planning more effectively for future lessons. The questions may be read aloud to the students.

A

Number Correct: _______

Identify Fractions.

1.		/
2.		/
3.		/
4.		/
5.		/
6.		/
7.		/
8.		/
9.		/
10.		/
11.		/
12.		/
13.		/
14.		/
15.		/
16.		/
17.		/
18.		/
19.		/
20.		/
21.		/
22.		/

23.		/
24.		/
25.		/
26.		/
27.		/
28.		/
29.		/
30.		/
31.		/
32.		/
33.		/
34.		/
35.		/
36.		/
37.		/
38.		/
39.		/
40.		/
41.		/
42.		/
43.		/
44.		/

B

Number Correct: _______

Improvement: _______

Identify Fractions.

1.		/
2.		/
3.		/
4.		/
5.		/
6.		/
7.		/
8.		/
9.		/
10.		/
11.		/
12.		/
13.		/
14.		/
15.		/
16.		/
17.		/
18.		/
19.		/
20.		/
21.		/
22.		/

23.		/
24.		/
25.		/
26.		/
27.		/
28.		/
29.		/
30.		/
31.		/
32.		/
33.		/
34.		/
35.		/
36.		/
37.		/
38.		/
39.		/
40.		/
41.		/
42.		/
43.		/
44.		/

Name ________________________________ Date ________________

Show a number bond representing what is shaded and unshaded in each of the figures. Draw a different visual model that would be represented by the same number bond.

Sample:

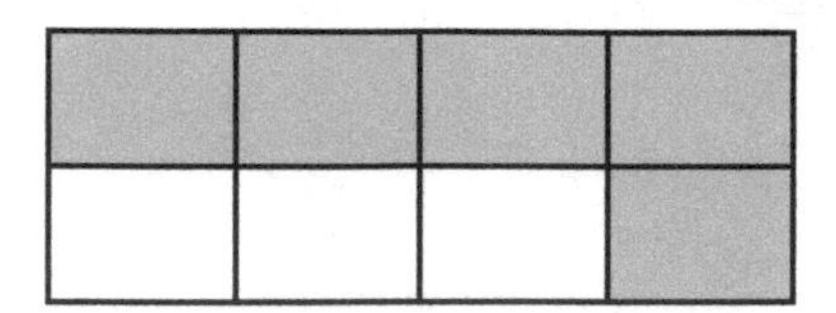

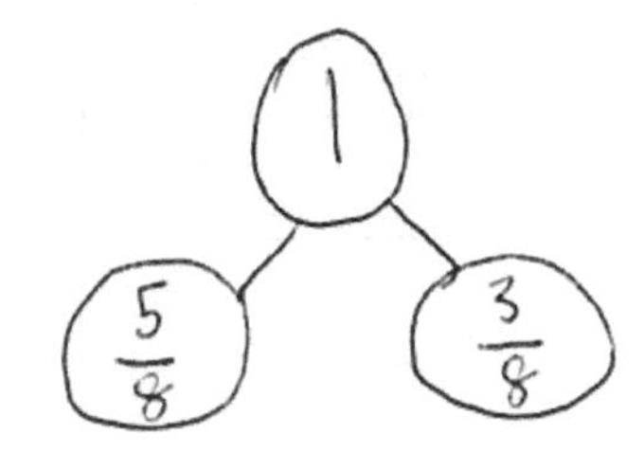

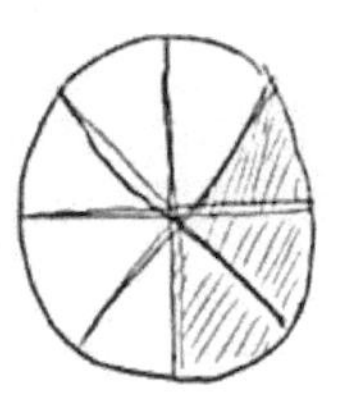

1.

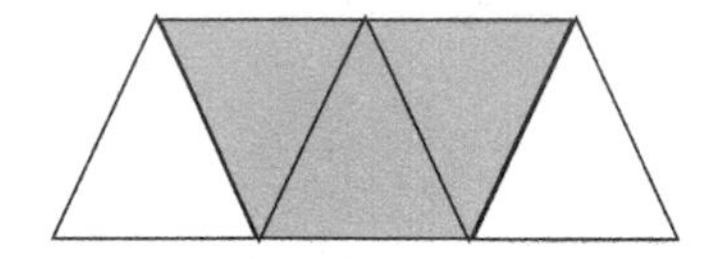

2.

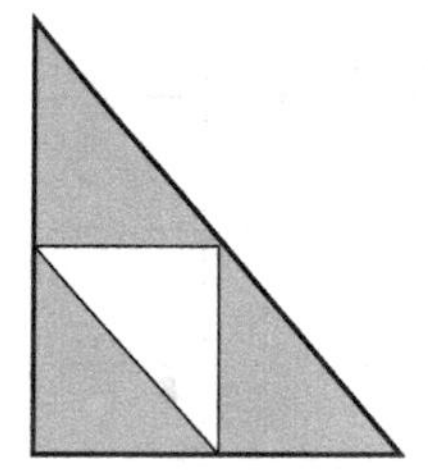

3.

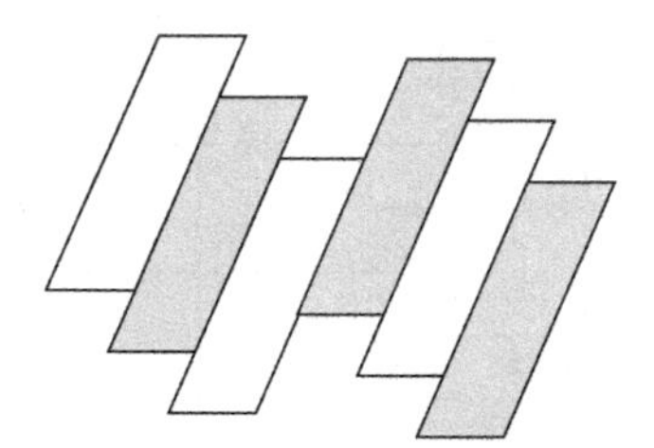

4.

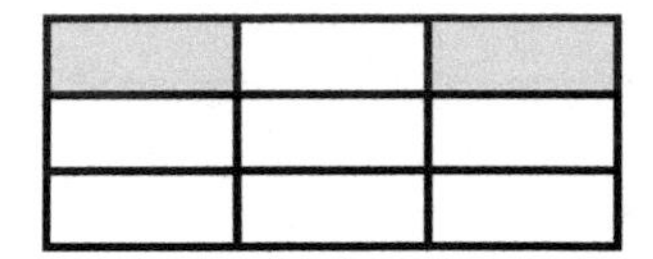

5. Draw a number bond with 2 parts showing the shaded and unshaded fractions of each figure. Decompose both parts of the number bond into unit fractions.

a. 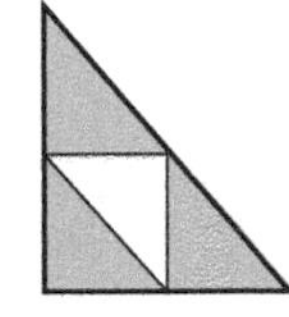

b. 

c. 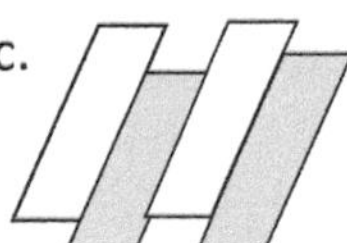

d. 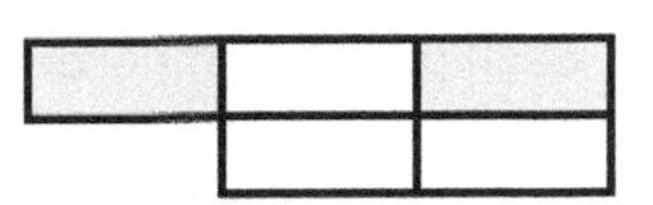

6. The chef put $\frac{1}{4}$ of the ground beef on the grill to make one hamburger and put the rest in the refrigerator. Draw a 2-part number bond showing the fraction of the ground beef on the grill and the fraction in the refrigerator. Draw a visual model of all the ground beef. Shade what is in the refrigerator.

a. What fraction of the ground beef was in the refrigerator?

b. How many more hamburgers can the chef make if he makes them all the same size as the first one?

c. Show the refrigerated ground beef broken into unit fractions on your number bond above.

Name ____ Date ____________________

1. Draw a number bond that shows the shaded and the unshaded parts of the shape below. Then, show each part decomposed into unit fractions.

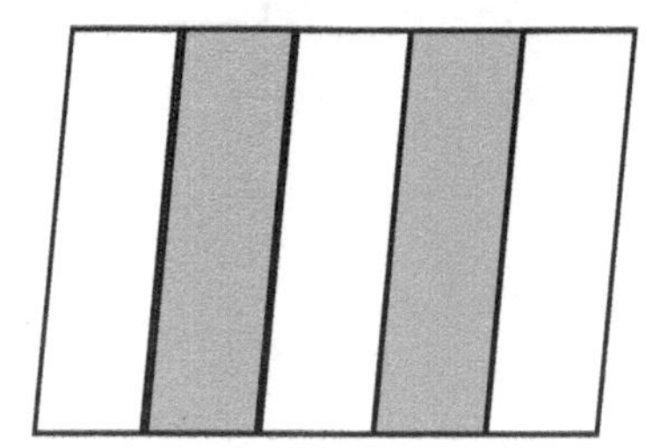

2. Complete the number bond. Draw a shape that has shaded and unshaded parts that match the completed number bond.

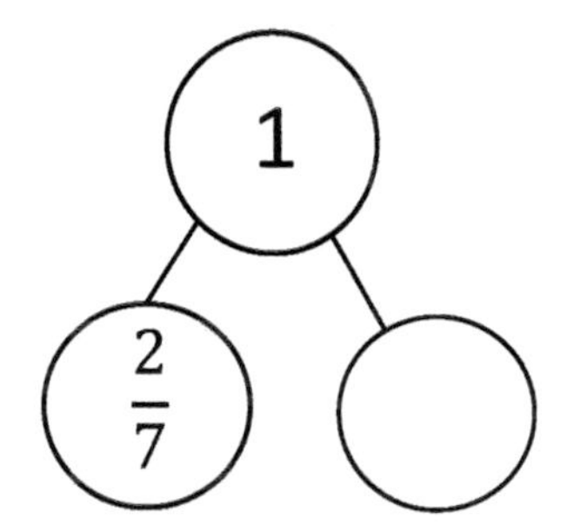

Name ______________________________ Date ______________

Show a number bond representing what is shaded and unshaded in each of the figures. Draw a different visual model that would be represented by the same number bond.

Sample:

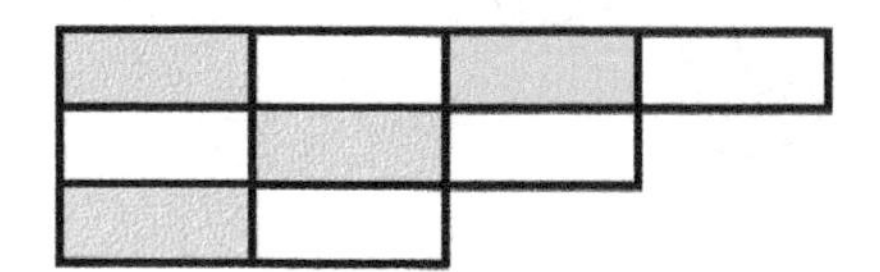

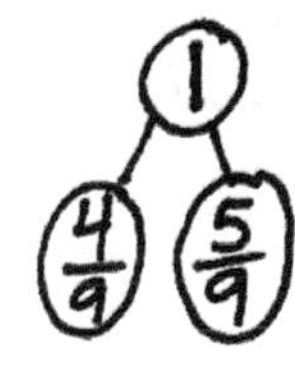

1.

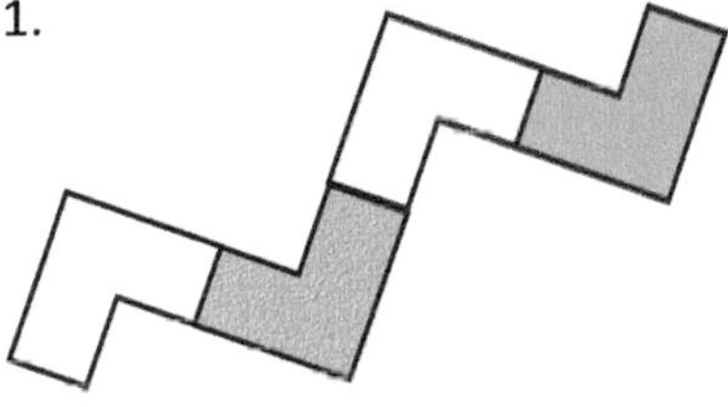

2.

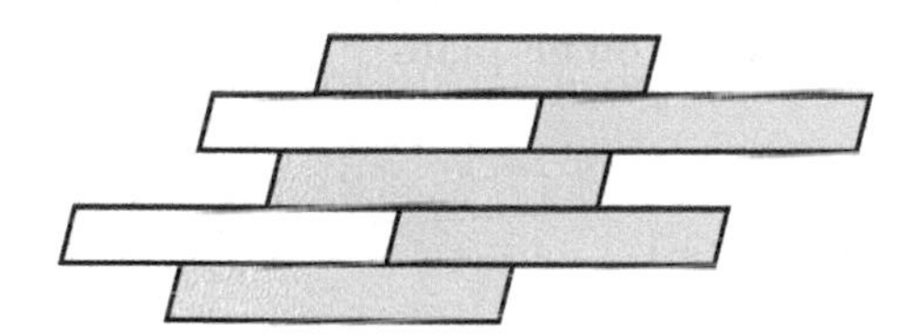

3.

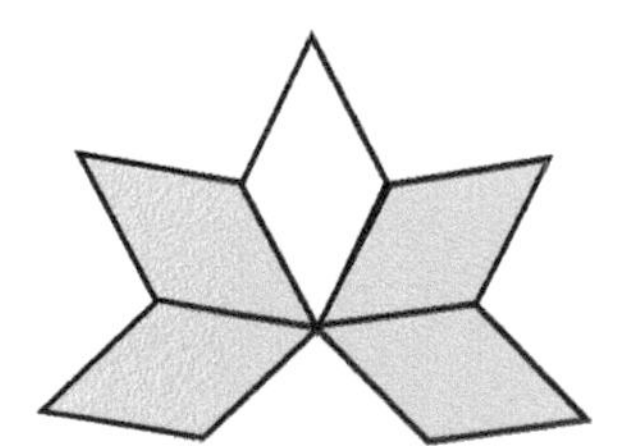

4.

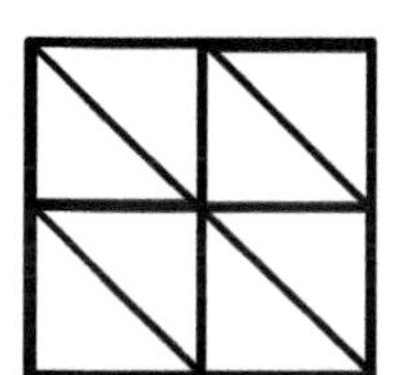

5. Draw a number bond with 2 parts showing the shaded and unshaded fractions of each figure. Decompose both parts of the number bond into unit fractions.

a.

b.

c. 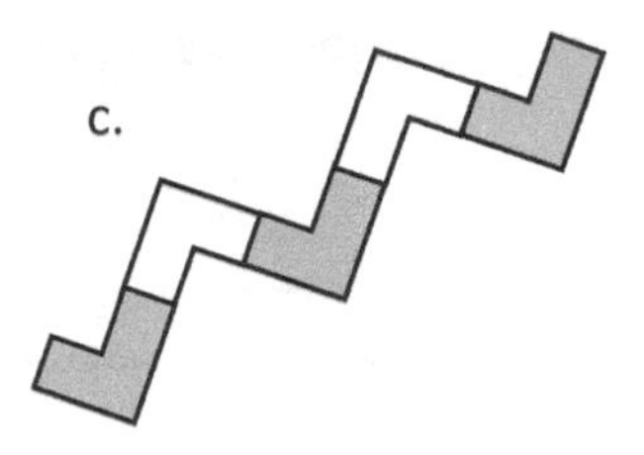

6. Johnny made a square peanut butter and jelly sandwich. He ate $\frac{1}{3}$ of it and left the rest on his plate. Draw a picture of Johnny's sandwich. Shade the part he left on his plate, and then draw a number bond that matches what you drew. What fraction of his sandwich did Johnny leave on his plate?

Lesson 9

Objective: Build and write fractions greater than one whole using unit fractions.

Suggested Lesson Structure

■ Fluency Practice	(12 minutes)
■ Application Problem	(10 minutes)
■ Concept Development	(28 minutes)
■ Student Debrief	(10 minutes)
Total Time	**(60 minutes)**

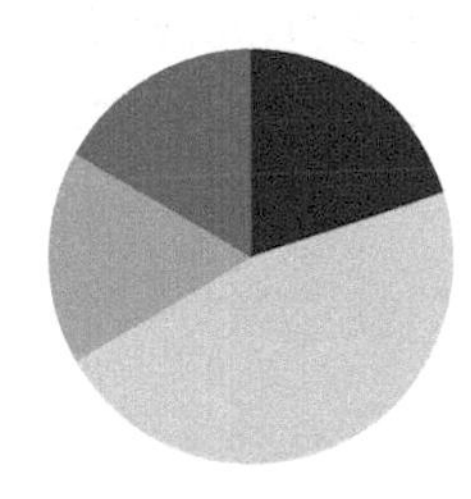

Fluency Practice (12 minutes)

- Sprint: Multiply with Eight **3.OA.2** (8 minutes)
- Find the Unknown Part **3.NF.3d** (2 minutes)
- Skip-Count by Halves on the Clock **3.G.2, 3.NF.1** (2 minutes)

Sprint: Multiply with Eight (8 minutes)

Materials: (S) Multiply with Eight Sprint

Note: This Sprint supports fluency with multiplication using units of 8.

Find the Unknown Part (2 minutes)

Note: This activity reviews representing parts of one whole as number bonds from Lesson 8.

T: (Project a number bond with $\frac{3}{3}$ as the whole and $\frac{2}{3}$ as a part.) Say the whole.

S: 3 thirds.

T: Say the known part.

S: 2 thirds.

T: Say the unknown part.

S: 1 third.

T: (Write $\frac{1}{3}$ in the unknown part.)

Continue with the following possible sequence: $\frac{6}{6}$ and $\frac{1}{6}$, $\frac{8}{8}$ and $\frac{3}{8}$, 1 whole and $\frac{3}{10}$, and 1 whole and $\frac{7}{12}$.

Skip-Count by Halves on the Clock (2 minutes)

Materials: (T) Clock

Note: This activity reviews counting by halves on the clock from Module 2.

T: (Hold or project a clock.) Let's skip-count by halves on the clock, starting with 5 o'clock.

S: 5, half past 5, 6, half past 6, 7.

T: Stop. Skip-count by halves backward, starting with 7.

S: Half past 6, 6, half past 5, 5, half past 4, 4, half past 3, 3.

Continue counting up and down.

Application Problem (10 minutes)

Julianne's friendship bracelet had 8 beads. When it broke, the beads fell off. She could only find 1 bead. To fix her bracelet, what fraction of the beads does she need to buy?

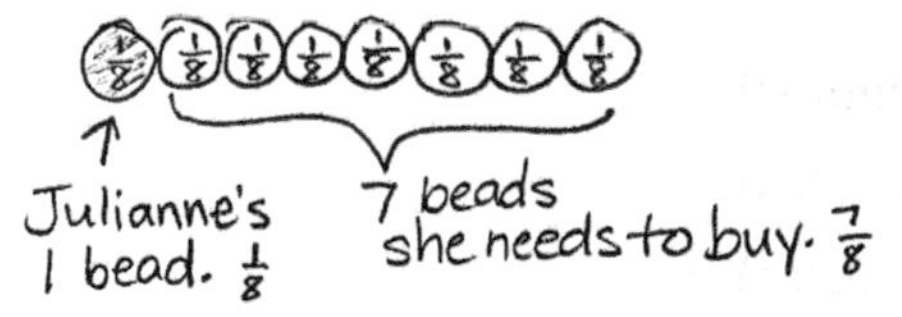

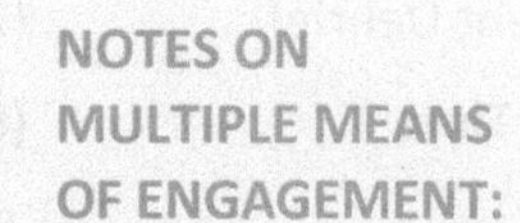

NOTES ON MULTIPLE MEANS OF ENGAGEMENT:

For students working above grade level, extend the Application Problem with an open-ended prompt such as, "If Julianne adds another bead of the same size and shape to her bracelet, what fraction would the new bead represent? Why do you think so?"

Note: Students may represent 1 eighth or 7 eighths as the shaded part of the whole. Invite students to share their models and discuss how both representations correctly model the problem.

Concept Development (28 minutes)

Materials: (S) Personal white board, fraction strips

T: I brought 2 oranges for lunch today. I cut each one into fourths so that I could eat them easily. Draw a picture on your personal white board to show how I cut my 2 oranges.

S: (Draw.)

T: If 1 orange represents 1 whole, how many copies of 1 fourth are in 1 whole?

S: 4 copies.

T: Then, what is our unit?

S: Fourths.

T: How many copies of 1 fourth are in two whole oranges?

S: 8 copies.

T: Let's count them.

S: 1 fourth, 2 fourths, 3 fourths, ..., 8 fourths.

T: Are you sure our unit is still fourths? Talk with your partner.

S: No, it's eighths because there are 8 pieces. → I disagree because the unit is fourths in each orange. → Remember, each orange is a whole, so the unit is fourths. 2 oranges aren't the whole!

T: I was so hungry I ate 1 whole orange and 1 piece of the second orange. Shade in the pieces I ate.

S: (Shade.)

T: How many pieces did I eat?

S: 5 pieces.

T: And what's our unit?

S: Fourths.

T: So we can say that I ate 5 fourths of an orange for lunch. Let's count them.

S: 1 fourth, 2 fourths, 3 fourths, 4 fourths, 5 fourths.

T: On your board, work together to show 5 fourths as a number bond of unit fractions.

S: (Work with a partner to draw a number bond.)

T: Compare the number of pieces I ate to 1 whole orange. What do you notice?

S: The number of pieces is larger! → You ate more pieces than the whole.

T: Yes. If the number of parts is greater than the number of equal parts in the whole, then you know that the fraction describes more than 1 whole.

T: Work with a partner to make a number bond with 2 parts. One part should show the pieces that make up the whole. The other part should show the pieces that are more than the whole.

S: (Work with a partner to draw a number bond.)

NOTES ON MULTIPLE MEANS OF ACTION AND EXPRESSION:

Turn and Talk is an excellent way for English language learners to use English to discuss their math thinking. Let English language learners choose the language they wish to use to discuss their math reasoning, particularly if their English language fluency is limited.

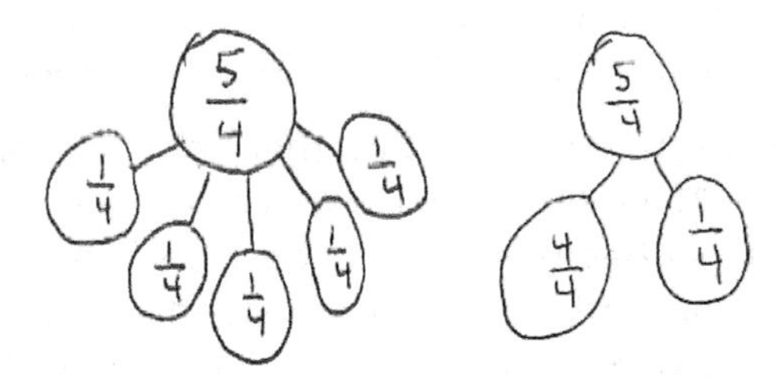

NOTES ON MULTIPLE MEANS OF ENGAGEMENT:

For students working below grade level, respectfully facilitate self-assessment of personal goals. Guide students to reflect upon questions such as, "Which fraction skills am I good at? What would I like to be better at? What is my plan to improve?" Celebrate improvement.

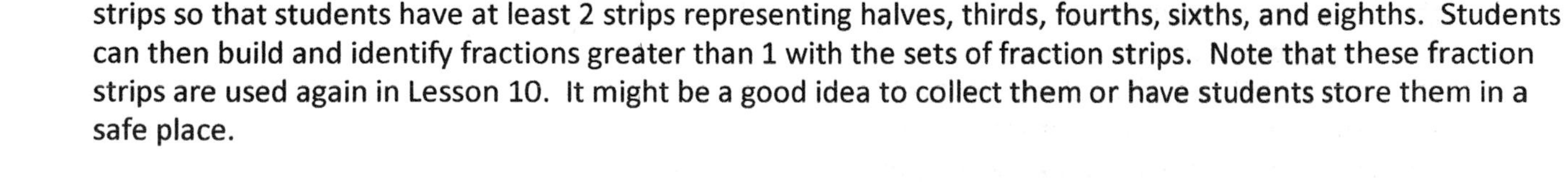

Demonstrate again using another concrete example. Follow by working with fraction strips. Fold fraction strips so that students have at least 2 strips representing halves, thirds, fourths, sixths, and eighths. Students can then build and identify fractions greater than 1 with the sets of fraction strips. Note that these fraction strips are used again in Lesson 10. It might be a good idea to collect them or have students store them in a safe place.

Problem Set (10 minutes)

Students should do their personal best to complete the Problem Set within the allotted 10 minutes. For some classes, it may be appropriate to modify the assignment by specifying which problems they work on first. Some problems do not specify a method for solving. Students should solve these problems using the RDW approach used for Application Problems.

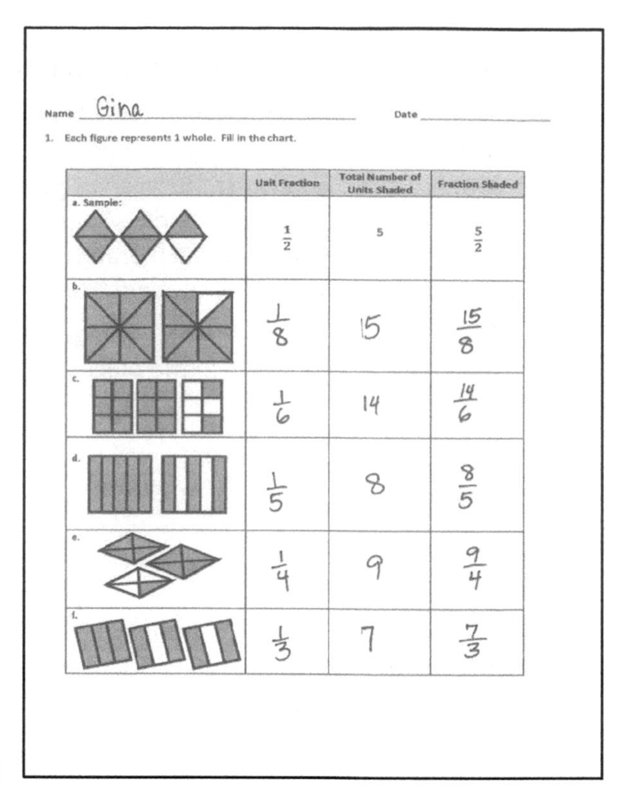

Name Gina Date

1. Each figure represents 1 whole. Fill in the chart.

	Unit Fraction	Total Number of Units Shaded	Fraction Shaded
a. Sample:	$\frac{1}{2}$	5	$\frac{5}{2}$
b.	$\frac{1}{8}$	15	$\frac{15}{8}$
c.	$\frac{1}{6}$	14	$\frac{14}{6}$
d.	$\frac{1}{5}$	8	$\frac{8}{5}$
e.	$\frac{1}{4}$	9	$\frac{9}{4}$
f.	$\frac{1}{3}$	7	$\frac{7}{3}$

Student Debrief (10 minutes)

Lesson Objective: Build and write fractions greater than one whole using unit fractions.

The Student Debrief is intended to invite reflection and active processing of the total lesson experience.

Invite students to review their solutions for the Problem Set. They should check work by comparing answers with a partner before going over answers as a class. Look for misconceptions or misunderstandings that can be addressed in the Debrief. Guide students in a conversation to debrief the Problem Set and process the lesson.

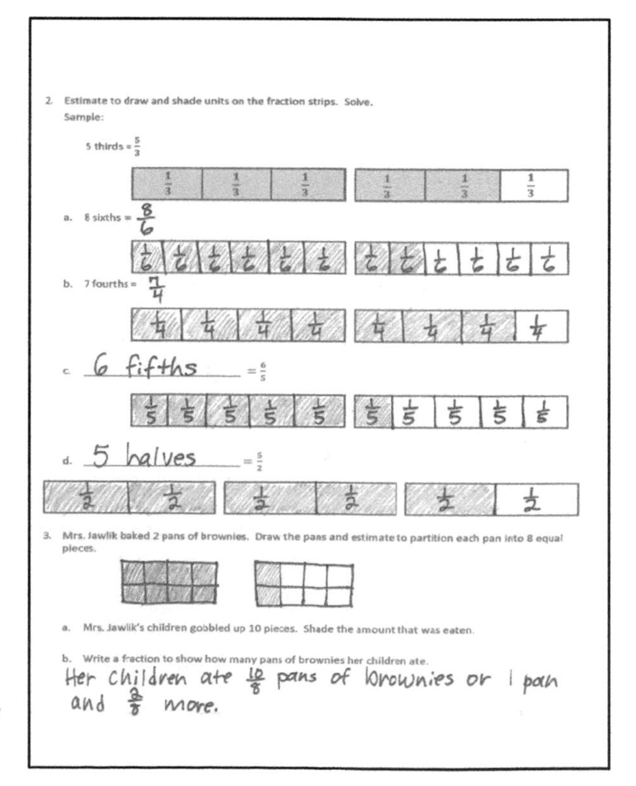

2. Estimate to draw and shade units on the fraction strips. Solve.

Sample:

5 thirds = $\frac{5}{3}$

a. 8 sixths = $\frac{8}{6}$

b. 7 fourths = $\frac{7}{4}$

c. 6 fifths = $\frac{6}{5}$

d. 5 halves = $\frac{5}{2}$

3. Mrs. Jawlik baked 2 pans of brownies. Draw the pans and estimate to partition each pan into 8 equal pieces.

a. Mrs. Jawlik's children gobbled up 10 pieces. Shade the amount that was eaten.

b. Write a fraction to show how many pans of brownies her children ate.

Her children ate $\frac{10}{8}$ pans of brownies or 1 pan and $\frac{2}{8}$ more.

Any combination of the questions below may be used to lead the discussion.

- Problem 3 is likely to be challenging and may result in confusion about whether the children ate $\frac{10}{8}$ or $\frac{10}{16}$. $\frac{10}{8}$ represents the number of pans of brownies they ate, and $\frac{10}{16}$ represents the number of brownies they ate. The question asks for the number of pans. Have students share their work to spark a discussion that helps clarify this. The student work sample shows 2 different ways to write the answer.
- Although students have not been introduced to mixed numbers, it may be an intuitive way for them to answer the question. If so, briefly examine and discuss the 2 *different* answers. Have students then clarify the lesson's objective. Have them discuss with a partner how to identify a fraction greater than one whole. If appropriate, advance to how they can identify a fraction greater than 2 wholes, etc.

Exit Ticket (3 minutes)

After the Student Debrief, instruct students to complete the Exit Ticket. A review of their work will help with assessing students' understanding of the concepts that were presented in today's lesson and planning more effectively for future lessons. The questions may be read aloud to the students.

A

Number Correct: ______

Multiply with Eight

1.	$8 \times 1 =$		23.	$9 \times 8 =$	
2.	$1 \times 8 =$		24.	$3 \times 8 =$	
3.	$8 \times 2 =$		25.	$8 \times 8 =$	
4.	$2 \times 8 =$		26.	$4 \times 8 =$	
5.	$8 \times 3 =$		27.	$7 \times 8 =$	
6.	$3 \times 8 =$		28.	$5 \times 8 =$	
7.	$8 \times 4 =$		29.	$6 \times 8 =$	
8.	$4 \times 8 =$		30.	$8 \times 5 =$	
9.	$8 \times 5 =$		31.	$8 \times 10 =$	
10.	$5 \times 8 =$		32.	$8 \times 1 =$	
11.	$8 \times 6 =$		33.	$8 \times 6 =$	
12.	$6 \times 8 =$		34.	$8 \times 4 =$	
13.	$8 \times 7 =$		35.	$8 \times 9 =$	
14.	$7 \times 8 =$		36.	$8 \times 2 =$	
15.	$8 \times 8 =$		37.	$8 \times 7 =$	
16.	$8 \times 9 =$		38.	$8 \times 3 =$	
17.	$9 \times 8 =$		39.	$8 \times 8 =$	
18.	$8 \times 10 =$		40.	$11 \times 8 =$	
19.	$10 \times 8 =$		41.	$8 \times 11 =$	
20.	$1 \times 8 =$		42.	$12 \times 8 =$	
21.	$10 \times 8 =$		43.	$8 \times 12 =$	
22.	$2 \times 8 =$		44.	$13 \times 8 =$	

B

Number Correct: _______

Improvement: _______

Multiply with Eight

1.	1 × 8 =		23.	10 × 8 =	
2.	8 × 1 =		24.	9 × 8 =	
3.	2 × 8 =		25.	4 × 8 =	
4.	8 × 2 =		26.	8 × 8 =	
5.	3 × 8 =		27.	8 × 3 =	
6.	8 × 3 =		28.	7 × 8 =	
7.	4 × 8 =		29.	6 × 8 =	
8.	8 × 4 =		30.	8 × 10 =	
9.	5 × 8 =		31.	8 × 5 =	
10.	8 × 5 =		32.	8 × 6 =	
11.	6 × 8 =		33.	8 × 1 =	
12.	8 × 6 =		34.	8 × 9 =	
13.	7 × 8 =		35.	8 × 4 =	
14.	8 × 7 =		36.	8 × 3 =	
15.	8 × 8 =		37.	8 × 2 =	
16.	9 × 8 =		38.	8 × 7 =	
17.	8 × 9 =		39.	8 × 8 =	
18.	10 × 8 =		40.	11 × 8 =	
19.	8 × 10 =		41.	8 × 11 =	
20.	8 × 3 =		42.	12 × 8 =	
21.	1 × 8 =		43.	8 × 12 =	
22.	2 × 8 =		44.	13 × 8 =	

Name ______________________________ Date ______________

1. Each figure represents 1 whole. Fill in the chart.

	Unit Fraction	Total Number of Units Shaded	Fraction Shaded
a. Sample: 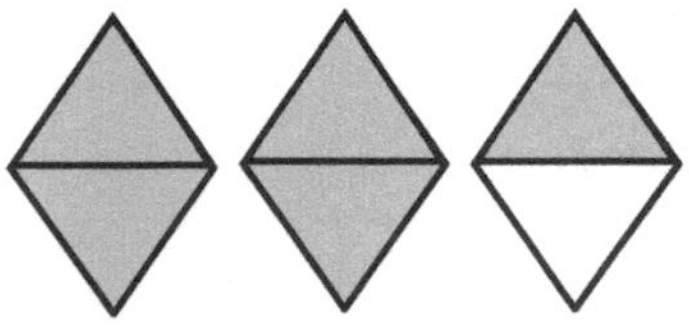	$\frac{1}{2}$	5	$\frac{5}{2}$
b. 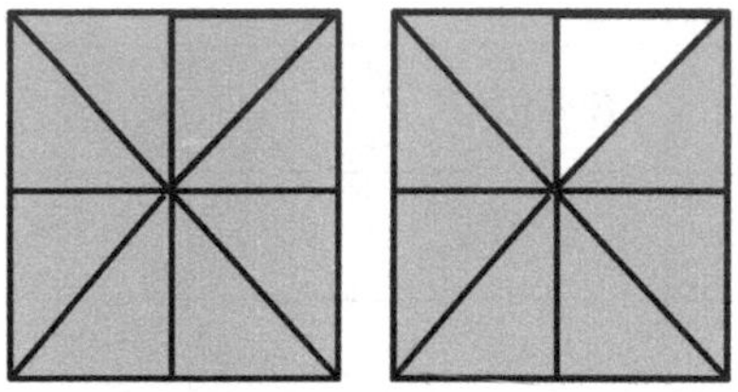			
c. 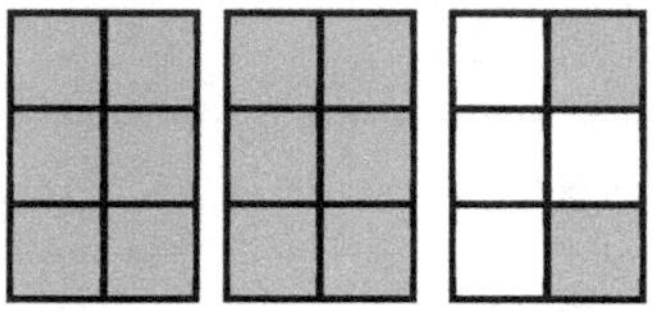			
d. 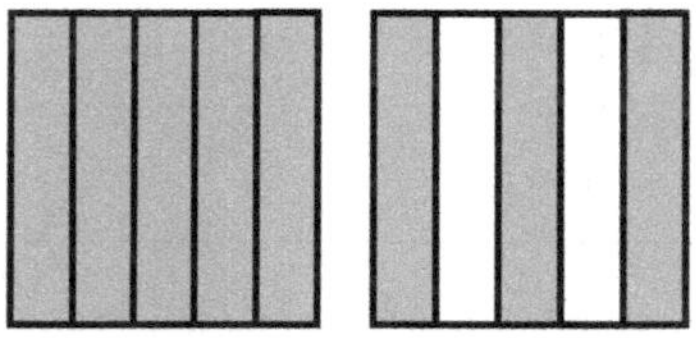			
e. 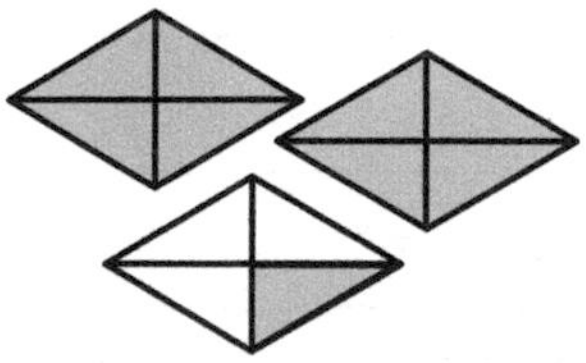			
f. 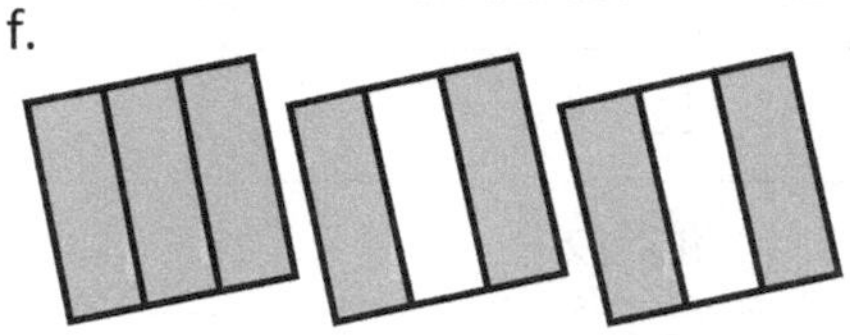			

2. Estimate to draw and shade units on the fraction strips. Solve.

   Sample:

   5 thirds = $\frac{5}{3}$

$\frac{1}{3}$	$\frac{1}{3}$	$\frac{1}{3}$	$\frac{1}{3}$	$\frac{1}{3}$	$\frac{1}{3}$

   a. 8 sixths =

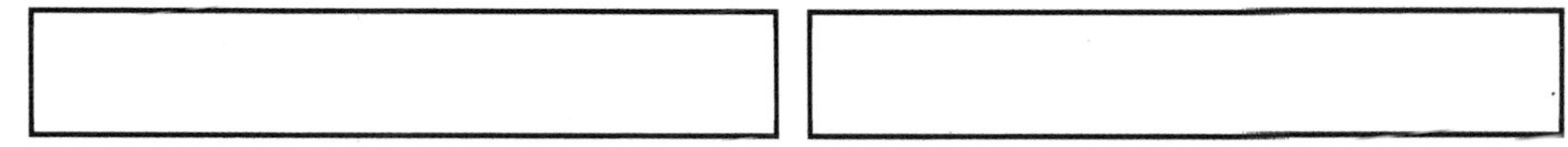

   b. 7 fourths =

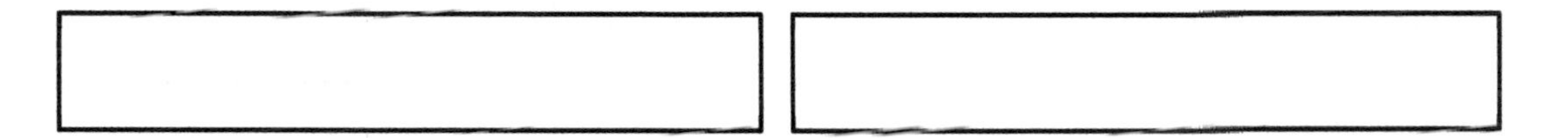

   c. ______________________ $= \frac{6}{5}$

   d. ______________________ $= \frac{5}{2}$

3. Mrs. Jawlik baked 2 pans of brownies. Draw the pans and estimate to partition each pan into 8 equal pieces.

   a. Mrs. Jawlik's children gobbled up 10 pieces. Shade the amount that was eaten.

   b. Write a fraction to show how many pans of brownies her children ate.

Name ____________________________________ Date ____________________

1. Each shape represents 1 whole. Fill in the chart.

	Unit Fraction	Total Number of Units Shaded	Fraction Shaded
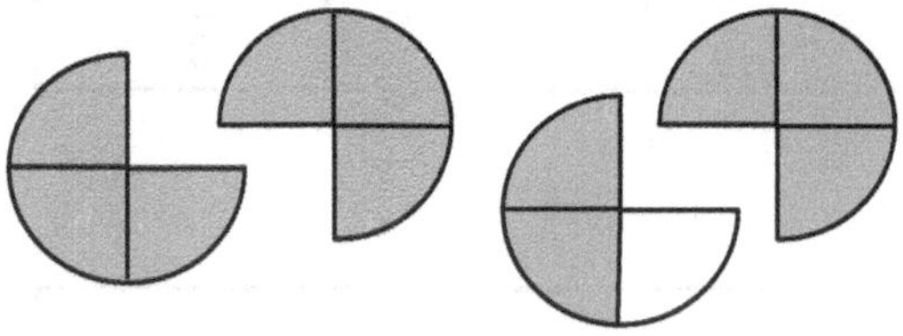			

2. Estimate to draw and shade units on the fraction strips. Solve.

a. 4 thirds =

b. ____________________ $= \frac{10}{4}$

Name ______________________________ Date ______________

1. Each shape represents 1 whole. Fill in the chart.

	Unit Fraction	Total Number of Units Shaded	Fraction Shaded
a. Sample:	$\frac{1}{2}$	3	$\frac{3}{2}$
b.			
c.			
d.			
e.			
f.			

2. Estimate to draw and shade units on the fraction strips. Solve.

Sample:

7 fourths = $\frac{7}{4}$

$\frac{1}{4}$	$\frac{1}{4}$	$\frac{1}{4}$	$\frac{1}{4}$	$\frac{1}{4}$	$\frac{1}{4}$	$\frac{1}{4}$	$\frac{1}{4}$

a. 5 thirds =

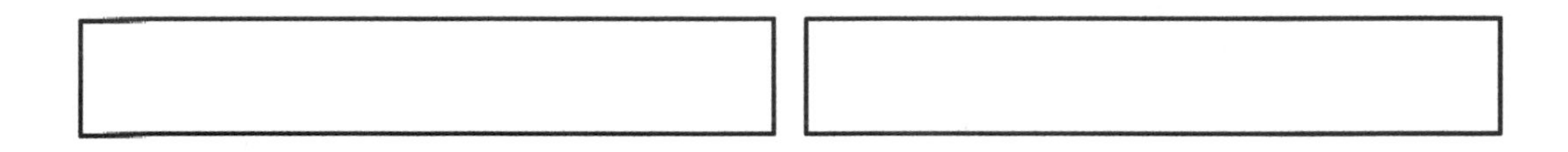

b. ______________________ $= \frac{9}{3}$

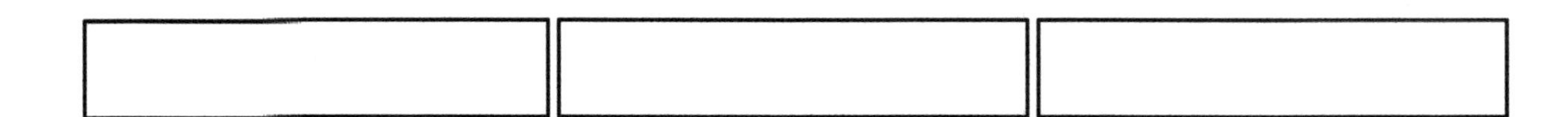

3. Reggie bought 2 candy bars. Draw the candy bars and estimate to partition each bar into 4 equal pieces.

a. Reggie ate 5 pieces. Shade the amount he ate.

b. Write a fraction to show how many candy bars Reggie ate.

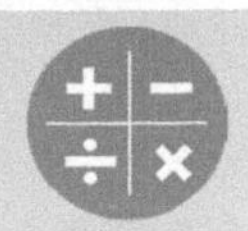

Topic C

Comparing Unit Fractions and Specifying the Whole

3.NF.3d, 3.NF.1, 3.NF.3a–c, 3.G.2

Focus Standard:		3.NF.3	Explain equivalence of fractions in special cases, and compare fractions by reasoning about their size. d. Compare two fractions with the same numerator or the same denominator by reasoning about their size. Recognize that comparisons are valid only when two fractions refer to the same whole. Record the results of comparisons with the symbols >, =, or <, and justify the conclusions, e.g., by using a visual fraction model.
Instructional Days:		4	
Coherence	**-Links from:**	G2–M8	Time, Shapes, and Fractions as Equal Parts of Shapes
	-Links to:	G4–M5	Fraction Equivalence, Ordering, and Operations

Students practiced identifying and labeling unit and non-unit fractions in Topic B. Now, in Topic C, they begin by comparing unit fractions. Using fraction strips, students recognize that, when the same whole is folded into more equal parts, each part is smaller. Next, using real-life examples and area models, students understand that, when comparing fractions, the whole must be the same size. Next, students create corresponding wholes based on a given unit fraction using similar materials to those in Lesson 4's exploration: clay, yarn, two rectangles, and a square. They conduct a *museum walk* to study the wholes, identifying the unit fractions and observing part–whole relationships. Finally, students learn that redefining the whole can change the unit fraction that describes the shaded part.

A Teaching Sequence Toward Mastery of Comparing Unit Fractions and Specifying the Whole

Objective 1: Compare unit fractions by reasoning about their size using fraction strips.
(Lesson 10)

Objective 2: Compare unit fractions with different-sized models representing the whole.
(Lesson 11)

Objective 3: Specify the corresponding whole when presented with one equal part.
(Lesson 12)

Objective 4: Identify a shaded fractional part in different ways depending on the designation of the whole.
(Lesson 13)

Lesson 10

Objective: Compare unit fractions by reasoning about their size using fraction strips.

Suggested Lesson Structure

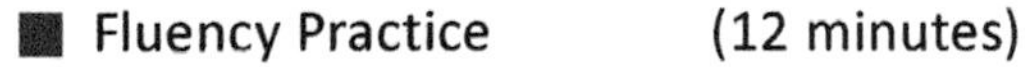

Fluency Practice	(12 minutes)
Application Problem	(6 minutes)
Concept Development	(32 minutes)
Student Debrief	(10 minutes)
Total Time	**(60 minutes)**

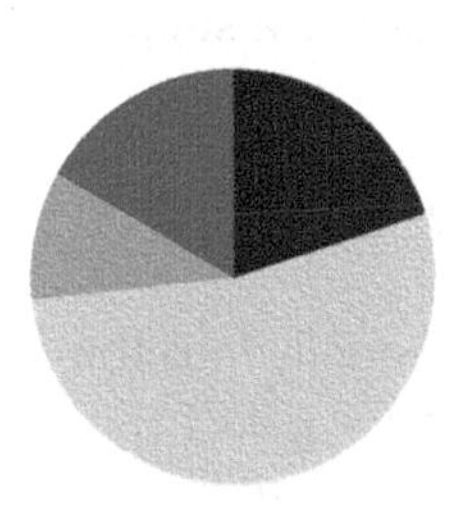

Fluency Practice (12 minutes)

- Sprint: Divide by Eight **3.OA.4** (9 minutes)
- Skip-Count by Fourths on the Clock **3.G.2, 3.NF.1** (2 minutes)
- Greater or Less Than 1 Whole **3.G.2, 3.NF.2** (1 minute)

Sprint: Multiply and Divide by Eight (9 minutes)

Materials: (S) Multiply and Divide by Eight Sprint

Note: This Sprint supports fluency with multiplication and division using units of 8.

Skip-Count by Fourths on the Clock (2 minutes)

Materials: (T) Clock

Note: This activity reviews counting by fourths on the clock from Module 2.

T: (Hold or project a clock.) Let's skip-count by fourths on the clock starting with 1 o'clock.

S: 1, 1:15, 1:30, 1:45, 2, 2:15, 2:30, 2:45, 3.

Continue with the following possible sequences:

- 1, 1:15, half past 1, 1:45, 2, 2:15, half past 2, 2:45, 3.
- 1, quarter past 1, half past 1, quarter 'til 2, 2, quarter past 2, half past 2, quarter 'til 3, 3.

Greater or Less Than 1 Whole (1 minute)

Note: This activity reviews identifying fractions greater and less than 1 whole.

T: (Write $\frac{1}{2}$.) Greater or less than 1 whole?

S: Less!

Continue with the following possible sequence: $\frac{3}{2}, \frac{5}{4}, \frac{3}{4}, \frac{3}{7}, \frac{5}{3}$, and $\frac{5}{2}$. It may be appropriate for some classes to draw responses on personal white boards for extra support.

Application Problem (6 minutes)

Sarah makes soup. She divides each batch equally into thirds to give away. Each family that she makes soup for gets 1 third of a batch. Sarah needs to make enough soup for 5 families. How much soup does Sarah give away? Write your answer in terms of batches.

Batch 1: Fam 1 $\frac{1}{3}$, Fam 2 $\frac{1}{3}$, Fam 3 $\frac{1}{3}$
Batch 2: Fam 4 $\frac{1}{3}$, Fam 5 $\frac{1}{3}$, $\frac{1}{3}$ ← left over

Sarah will give away $\frac{5}{3}$ batches of soup.
Extension: $\frac{1}{3}$ batches will be left over.

Extension: What fraction will be left over for Sarah?

Note: This problem reviews writing fractions greater than 1 whole from Lesson 9.

NOTES ON MULTIPLE MEANS OF ENGAGEMENT:

Scaffold solving the Application Problem for students working below grade level with step-by-step questioning. For example, ask the following:

- "How much soup does 1 family receive?" (1 third of the batch of soup.)
- "2 families?" (2 thirds.)
- "3 families?" (3 thirds or 1 whole batch of soup.)
- "Does Sarah have to make more than 1 batch?" (Yes.)
- "How much of the second batch will she give away?" (2 thirds.)
- "How much will remain?" (1 third.)

Concept Development (32 minutes)

Materials: (S) Folded fraction strips (halves, thirds, fourths, sixths, and eighths) from Lesson 9, personal white board, 1 set of <, >, = cards per pair

MP.2

T: Take out the fraction strips you folded yesterday.

S: (Take out strips folded into halves, thirds, fourths, sixths, and eighths.)

T: Look at the different units. Take a minute to arrange the strips in order from the largest to the smallest unit.

S: (Place the fraction strips in order: halves, thirds, fourths, sixths, and eighths.)

T: Turn and talk to your partner about what you notice.

S: Eighths are the smallest even though the number 8 is the biggest. → When the whole is folded into more units, each unit is smaller. I only folded once to get halves, and they're the biggest.

T: Look at 1 half and 1 third. Which unit fraction is larger?

S: 1 half.

T: Explain to your partner how you know.

S: I can just see 1 half is larger on the strip. → When you split it between 2 people, the pieces are larger than if you split it between 3 people. → There are fewer pieces, so the pieces are larger.

Continue with other examples using the fraction strips as necessary.

T: What happens when we aren't using fraction strips? What if we're talking about something round, like a pizza? Is 1 half still larger than 1 third? Turn and talk to your partner about why or why not.

S: I'm not sure. → Sharing a pizza among 3 people is not as good as sharing it between 2 people. I think pieces that are halves are still larger. → I agree because the number of parts doesn't change even if the shape of the whole changes.

T: Let's make a model and see what happens. Draw 5 circles that are the same size to represent pizzas on your personal white board.

S: (Draw.)

T: Estimate to partition the first circle into halves. Label the unit fraction.

S: (Draw and label.)

T: Estimate to partition the second circle into thirds. (Model if necessary.) Label the unit fraction.

S: (Draw and label.)

T: The more we cut, what's happening to our pieces?

S: They're getting smaller!

T: So, is 1 third still smaller than 1 half?

S: Yes!

T: Partition your remaining circles into fourths, sixths, and eighths. Label the unit fraction in each one.

S: (Draw and label.)

T: Compare your drawings to your fraction strips. Talk to a partner: Do you notice the same pattern as with your fraction strips?

S: (Discuss.)

Continue with other real world examples if necessary.

T: Let's compare unit fractions. For each turn, you and your partner will each choose any single fraction strip. Choose now.

S: (Choose a strip to play.)

T: Now, compare unit fractions by folding to show only the unit fraction. Then, place the appropriate symbol card (<, >, or =) on the table between your strips.

S: (Fold, compare, and place symbol cards.)

T: (Hold symbol cards face down.) I will flip one of my symbol cards to see if the unit fraction that is *greater than* or *less than* wins this round. If I flip *equals,* it's a tie. (Flip a card.)

Continue at a rapid pace for a few rounds.

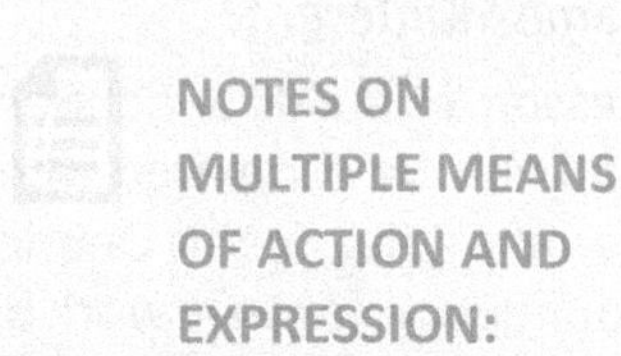

NOTES ON MULTIPLE MEANS OF ACTION AND EXPRESSION:

This partner activity benefits English language learners as it includes repeated use of math language in a reliable structure (e.g., "__ is greater than __"). It also offers the English language learner an opportunity to discuss the math with a peer, which may be more comfortable than speaking in front of the class or to the teacher.

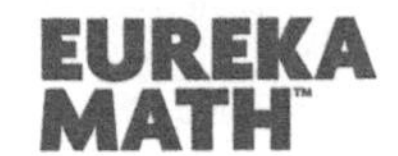

Problem Set (10 minutes)

Students should do their personal best to complete the Problem Set within the allotted 10 minutes. For some classes, it may be appropriate to modify the assignment by specifying which problems they work on first. Some problems do not specify a method for solving. Students should solve these problems using the RDW approach used for Application Problems.

Name Gina Date

1. Each fraction strip is 1 whole. All the fraction strips are equal in length. Color 1 fractional unit in each strip. Then, answer the questions below.

$\frac{1}{2}$

$\frac{1}{4}$

$\frac{1}{8}$

$\frac{1}{3}$

$\frac{1}{6}$

2. Circle *less than* or *greater than*. Whisper the complete sentence.

a. $\frac{1}{2}$ is less than / greater than $\frac{1}{4}$

b. $\frac{1}{6}$ is less than / greater than $\frac{1}{2}$

c. $\frac{1}{3}$ is less than / greater than $\frac{1}{2}$

d. $\frac{1}{3}$ is less than / greater than $\frac{1}{6}$

e. $\frac{1}{8}$ is less than / greater than $\frac{1}{6}$

f. $\frac{1}{8}$ is less than / greater than $\frac{1}{4}$

g. $\frac{1}{2}$ is less than / greater than $\frac{1}{8}$

h. 9 eighths is less than / greater than 2 halves

3. Lily needs $\frac{1}{3}$ cup of oil and $\frac{1}{4}$ cup of water to make muffins. Will Lily use more oil or more water? Explain your answer using pictures, numbers, and words.

$\frac{1}{3}$ oil $\frac{1}{4}$ water

She will use more oil. You can see from the drawing that the amount of water ($\frac{1}{4}$) is less than the oil ($\frac{1}{3}$). That's because when the same whole is partitioned into fourths, there are more "pieces" and the "pieces" are smaller.

4. Use >, <, or = to compare.

a. 1 third > 1 fifth

b. 1 seventh < 1 fourth

c. 1 sixth = $\frac{1}{6}$

d. 1 tenth > $\frac{1}{12}$

e. $\frac{1}{16}$ < 1 eleventh

f. 1 whole = 2 halves

Extension:

g. $\frac{1}{8}$ = 1 eighth < $\frac{1}{6}$ < $\frac{1}{3}$ < 2 halves = 1 whole

5. Your friend Eric says that $\frac{1}{6}$ is greater than $\frac{1}{5}$ because 6 is greater than 5. Is Eric correct? Use words and pictures to explain what happens to the size of a unit fraction when the number of parts gets larger.

He is wrong because if you have 1 whole and you make 6 pieces then each piece is smaller than if you only have 5 pieces. Fifths are bigger because when the number of parts is smaller, the pieces are bigger.

$\frac{1}{6}$ $\frac{1}{5}$

Student Debrief (10 minutes)

Lesson Objective: Compare unit fractions by reasoning about their size using fraction strips.

The Student Debrief is intended to invite reflection and active processing of the total lesson experience.

Invite students to review their solutions for the Problem Set. They should check work by comparing answers with a partner before going over answers as a class. Look for misconceptions or misunderstandings that can be addressed in the Debrief. Guide students in a conversation to debrief the Problem Set and process the lesson.

Any combination of the questions below may be used to lead the discussion.

- How did Problem 3 help you answer Problem 5?
- Compare Problems 3 and 5. How are they the same? Different?
- Lesson 11 builds understanding that unit fractions can only be compared when they refer to the same whole. In this Debrief, consider laying the foundation for that work by drawing students' attention to the models they drew for Problems 3 and 5. Discussion might include reasoning about why the models they drew facilitated the process of comparison within each problem.

Exit Ticket (3 minutes)

After the Student Debrief, instruct students to complete the Exit Ticket. A review of their work will help with assessing students' understanding of the concepts that were presented in today's lesson and planning more effectively for future lessons. The questions may be read aloud to the students.

A

Number Correct: _______

Multiply and Divide by Eight

1.	2 × 8 =		23.	___ × 8 = 80	
2.	3 × 8 =		24.	___ × 8 = 16	
3.	4 × 8 =		25.	___ × 8 = 24	
4.	5 × 8 =		26.	80 ÷ 8 =	
5.	1 × 8 =		27.	40 ÷ 8 =	
6.	16 ÷ 8 =		28.	8 ÷ 8 =	
7.	24 ÷ 8 =		29.	16 ÷ 8 =	
8.	40 ÷ 8 =		30.	24 ÷ 8 =	
9.	8 ÷ 8 =		31.	___ × 8 = 48	
10.	32 ÷ 8 =		32.	___ × 8 = 56	
11.	6 × 8 =		33.	___ × 8 = 72	
12.	7 × 8 =		34.	___ × 8 = 64	
13.	8 × 8 =		35.	56 ÷ 8 =	
14.	9 × 8 =		36.	72 ÷ 8 =	
15.	10 × 8 =		37.	48 ÷ 8 =	
16.	64 ÷ 8 =		38.	64 ÷ 8 =	
17.	56 ÷ 8 =		39.	11 × 8 =	
18.	72 ÷ 8 =		40.	88 ÷ 8 =	
19.	48 ÷ 8 =		41.	12 × 8 =	
20.	80 ÷ 8 =		42.	96 ÷ 8 =	
21.	___ × 8 = 40		43.	14 × 8 =	
22.	___ × 8 = 8		44.	112 ÷ 8 =	

B

Number Correct: _______

Improvement: _______

Multiply and Divide by Eight

1.	1 × 8 =	
2.	2 × 8 =	
3.	3 × 8 =	
4.	4 × 8 =	
5.	5 × 8 =	
6.	24 ÷ 8 =	
7.	16 ÷ 8 =	
8.	32 ÷ 8 =	
9.	8 ÷ 8 =	
10.	40 ÷ 8 =	
11.	10 × 8 =	
12.	6 × 8 =	
13.	7 × 8 =	
14.	8 × 8 =	
15.	9 × 8 =	
16.	56 ÷ 8 =	
17.	48 ÷ 8 =	
18.	64 ÷ 8 =	
19.	80 ÷ 8 =	
20.	72 ÷ 8 =	
21.	___ × 8 = 8	
22.	___ × 8 = 40	

23.	___ × 8 = 16	
24.	___ × 8 = 80	
25.	___ × 8 = 24	
26.	16 ÷ 8 =	
27.	8 ÷ 8 =	
28.	80 ÷ 8 =	
29.	40 ÷ 8 =	
30.	24 ÷ 8 =	
31.	___ × 8 = 24	
32.	___ × 8 = 32	
33.	___ × 8 = 72	
34.	___ × 8 = 56	
35.	64 ÷ 8 =	
36.	72 ÷ 8 =	
37.	48 ÷ 8 =	
38.	56 ÷ 8 =	
39.	11 × 8 =	
40.	88 ÷ 8 =	
41.	12 × 8 =	
42.	96 ÷ 8 =	
43.	13 × 8 =	
44.	104 ÷ 8 =	

Name ______________________________ Date ____________

1. Each fraction strip is 1 whole. All the fraction strips are equal in length. Color 1 fractional unit in each strip. Then, answer the questions below.

$\frac{1}{2}$

$\frac{1}{4}$

$\frac{1}{8}$

$\frac{1}{3}$

$\frac{1}{6}$

2. Circle *less than* or *greater than*. Whisper the complete sentence.

a. $\frac{1}{2}$ is less than / greater than $\frac{1}{4}$

b. $\frac{1}{6}$ is less than / greater than $\frac{1}{2}$

c. $\frac{1}{3}$ is less than / greater than $\frac{1}{2}$

d. $\frac{1}{3}$ is less than / greater than $\frac{1}{6}$

e. $\frac{1}{8}$ is less than / greater than $\frac{1}{6}$

f. $\frac{1}{8}$ is less than / greater than $\frac{1}{4}$

g. $\frac{1}{2}$ is less than / greater than $\frac{1}{8}$

h. 9 eighths is less than / greater than 2 halves

3. Lily needs $\frac{1}{3}$ cup of oil and $\frac{1}{4}$ cup of water to make muffins. Will Lily use more oil or more water? Explain your answer using pictures, numbers, and words.

4. Use >, <, or = to compare.

a. 1 third ◯ 1 fifth

b. 1 seventh ◯ 1 fourth

c. 1 sixth ◯ $\frac{1}{6}$

d. 1 tenth ◯ $\frac{1}{12}$

e. $\frac{1}{16}$ ◯ 1 eleventh

f. 1 whole ◯ 2 halves

Extension:

g. $\frac{1}{8}$ ◯ 1 eighth ◯ $\frac{1}{6}$ ◯ $\frac{1}{3}$ ◯ 2 halves ◯ 1 whole

5. Your friend Eric says that $\frac{1}{6}$ is greater than $\frac{1}{5}$ because 6 is greater than 5. Is Eric correct? Use words and pictures to explain what happens to the size of a unit fraction when the number of parts gets larger.

Name ___________________________________ Date ________________

1. Each fraction strip is 1 whole. All the fraction strips are equal in length. Color 1 fractional unit in each strip. Then, circle the largest fraction and draw a star to the right of the smallest fraction.

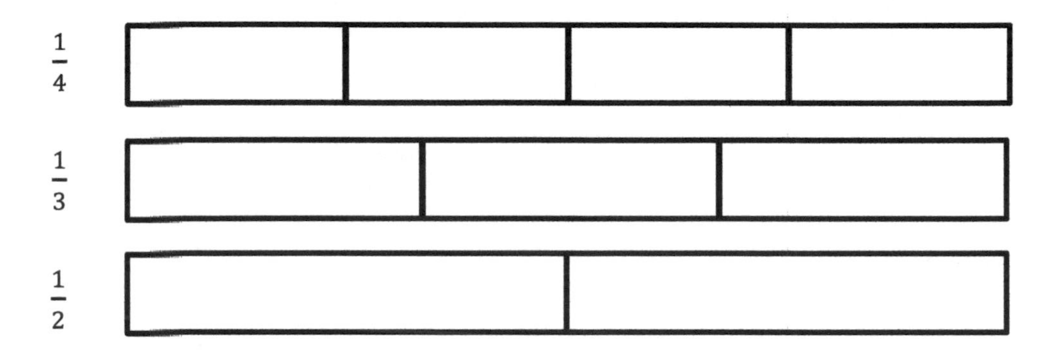

2. Use >, <, or = to compare.

   a. 1 eighth ◯ 1 tenth

   b. 1 whole ◯ 5 fifths

   c. $\frac{1}{7}$ ◯ $\frac{1}{6}$

Name ____________________________________ Date ____________________

1. Each fraction strip is 1 whole. All the fraction strips are equal in length. Color 1 fractional unit in each strip. Then, answer the questions below.

$\frac{1}{2}$

$\frac{1}{3}$

$\frac{1}{5}$

$\frac{1}{4}$

$\frac{1}{9}$

2. Circle *less than* or *greater than*. Whisper the complete sentence.

a. $\frac{1}{2}$ is less than / greater than $\frac{1}{3}$

b. $\frac{1}{9}$ is less than / greater than $\frac{1}{2}$

c. $\frac{1}{4}$ is less than / greater than $\frac{1}{2}$

d. $\frac{1}{4}$ is less than / greater than $\frac{1}{9}$

e. $\frac{1}{5}$ is less than / greater than $\frac{1}{3}$

f. $\frac{1}{5}$ is less than / greater than $\frac{1}{4}$

g. $\frac{1}{2}$ is less than / greater than $\frac{1}{5}$

h. 6 fifths is less than / greater than 3 thirds

3. After his football game, Malik drinks $\frac{1}{2}$ liter of water and $\frac{1}{3}$ liter of juice. Did Malik drink more water or juice? Draw and estimate to partition. Explain your answer.

4. Use >, <, or = to compare.

a.	1 fourth	◯	1 eighth
b.	1 seventh	◯	1 fifth
c.	1 eighth	◯	$\frac{1}{8}$
d.	1 twelfth	◯	$\frac{1}{10}$
e.	$\frac{1}{15}$	◯	1 thirteenth
f.	3 thirds	◯	1 whole

5. Write a word problem about comparing fractions for your friends to solve. Be sure to show the solution so that your friends can check their work.

Lesson 11

Objective: Compare unit fractions with different-sized models representing the whole.

Suggested Lesson Structure

- Fluency Practice (8 minutes)
- Application Problem (6 minutes)
- Concept Development (32 minutes)
- Student Debrief (14 minutes)

Total Time (60 minutes)

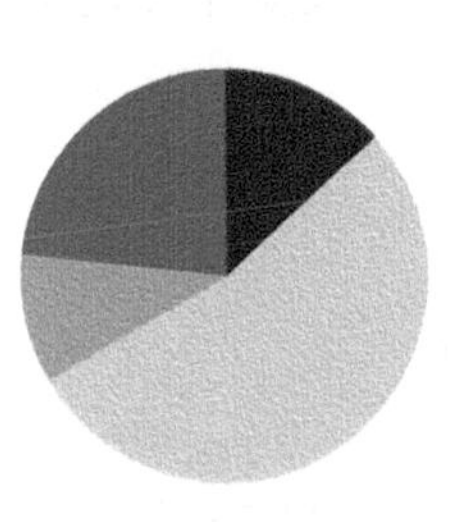

Fluency Practice (8 minutes)

- Skip-Count by Fourths on the Clock **3.G.2, 3.NF.1** (3 minutes)
- Greater or Less Than 1 Whole **3.G.2, 3.NF.2b** (2 minutes)
- Write Fractions Greater Than 1 Whole **3.NF.2b** (3 minutes)

Skip-Count by Fourths on the Clock (3 minutes)

Materials: (T) Clock

Note: This activity reviews counting by fourths on the clock from Module 2.

T: (Hold or project a clock.) Let's skip-count by fourths on the clock, starting with 5 o'clock.

S: 5, 5:15, 5:30, 5:45, 6, 6:15, 6:30, 6:45, 7.

Continue with the following possible sequences:

- 5, 5:15, half past 5, 5:45, 6, 6:15, half past 6, 6:45, 7.
- 5, quarter past 5, half past 5, quarter 'til 6, 6, quarter past 6, half past 6, quarter 'til 7, 7.

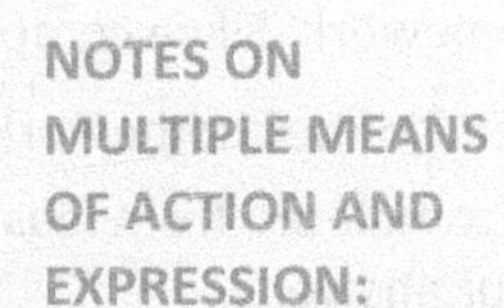

NOTES ON MULTIPLE MEANS OF ACTION AND EXPRESSION:

Skip-Count by Fourths on the Clock is a valuable opportunity for English language learners to practice everyday math language (time on the clock) within the comforts of choral response.

Scaffold this quick oral fluency activity with hand clocks. As students move the minute hand to reflect the count, they are tangibly partitioning fourths of the clock (the whole).

Greater or Less Than 1 Whole (2 minutes)

Note: This activity reviews identifying fractions greater and less than 1 whole.

T: (Write $\frac{1}{2}$.) Is this greater or less than 1 whole?

S: Less!

Continue with the following possible sequence: $\frac{1}{2}, \frac{3}{2}, \frac{1}{3}, \frac{2}{3}, \frac{4}{3}, \frac{5}{3}, \frac{3}{4}, \frac{5}{4}, \frac{11}{10}, \frac{9}{10}, \frac{11}{8}, \frac{5}{8}, \frac{11}{6}, \frac{5}{6}, \frac{11}{12}$, and $\frac{13}{12}$. It may be appropriate for some classes to draw responses on personal white boards for extra support.

Write Fractions Greater Than 1 Whole (3 minutes)

Materials: (S) Personal white board

Note: This activity reviews writing fractions greater than 1 whole from Lesson 9. As students build confidence, omit the first 2 questions.

T: How many halves are in 1 whole?

S: 2 halves.

T: What's 1 more half than 2 halves?

S: 3 halves.

T: Write a fraction on your personal white board that is 1 more half than 1 whole.

S: (Write $\frac{3}{2}$.)

Continue with the following possible sequence: $\frac{1}{3}, \frac{1}{4}, \frac{1}{5}, \frac{1}{10}, \frac{1}{6}$, and $\frac{1}{8}$.

Application Problem (6 minutes)

Rachel, Silvia, and Lola each received the same homework assignment and only completed part of it. Rachel completed $\frac{1}{6}$ of her homework, Silvia completed $\frac{1}{2}$ of her homework, and Lola completed $\frac{1}{4}$ of her homework. Write the amount of homework each girl completed from least to greatest. Draw a picture to prove your answer.

Rachel completed the least, Lola was next, and Silvia completed the most out of the girls.

Note: This problem reviews comparing unit fractions from Lesson 10. If time allows, revisit this problem during today's Debrief. Ask students if they need to adjust their pictures based on what they learned today about comparing fractions.

Concept Development (32 minutes)

Materials: (T) 2 different-sized clear plastic cups, food coloring, water (S) Personal white board

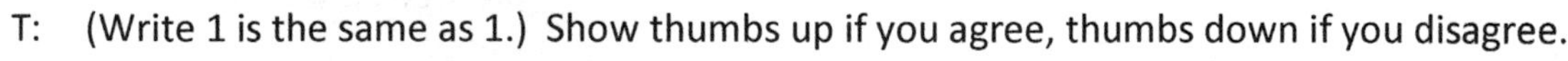

T: (Write 1 is the same as 1.) Show thumbs up if you agree, thumbs down if you disagree.

S: (Show thumbs up or thumbs down.)

T: 1 liter of soda and 1 can of soda. (Draw pictures or show objects.) Is 1 still the same as 1? Turn and talk to your partner.

S: Yes, they're still the same amount. → No, a liter and a can are different. → How *many* stays the same, but a liter is larger than a can, so how *much* in each is different.

MP.6

T: How *many* and how *much* are important to our question. In this case, *what* each thing is changes it, too. Because a liter is larger, it has more soda than a can. Talk to a partner: How does this change your thinking about *1 is the same as 1?*

S: If the thing is larger, then it has more. → Even though the number of things is the same, *what* it is might change how *much* of it there is. → If *what* it is and how *much* it is are different, then 1 and 1 aren't exactly the same.

T: As you compare 1 and 1, I hear you say that the size of the whole *and* how much is in it matters. The same is true when comparing fractions.

T: For breakfast this morning, my brother and I each had a glass of juice. (Present different-sized glasses partitioned into halves and fourths.) What fraction of my glass has juice?

S: 1 fourth.

T: What fraction of my brother's glass has juice?

S: 1 half.

My glass My brother's glass

T: When the wholes are the same, 1 half is greater than 1 fourth. Does this picture prove that? Discuss it with your partner.

S: 1 half is always larger than 1 fourth. → It looks like you might have drunk more, but the wholes aren't the same. → The glasses are different sizes—like the can and the liter of soda. We can't really compare.

T: I'm hearing you say that we have to consider the size of the whole when we compare fractions.

To further illustrate the point, pour each glass of juice into containers that are the same size. It may be helpful to purposefully select your containers so that 1 fourth of the large glass is the larger quantity.

To transition into the pictorial work with wholes that are the same, offer another concrete example. This time use rectangular shaped *wholes* that are different in size, such as those shown to the right.

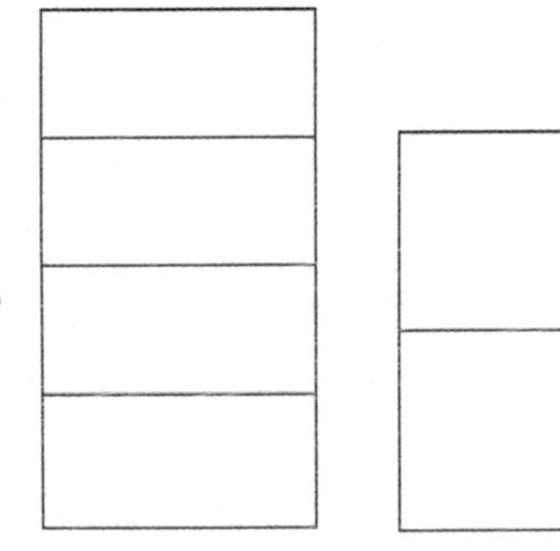

T: Let's see how the comparison changes when our wholes are the same. On your board, draw two rectangles that are the same size. Partition each into thirds.

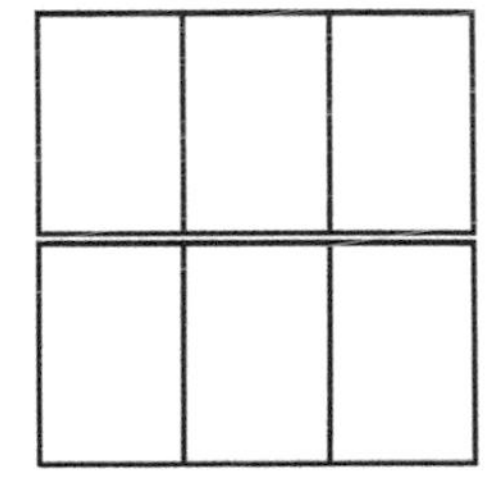

S: (Draw and partition rectangles.)

T: Now, partition the first rectangle into sixths.

S: (Partition the first rectangle from thirds to sixths.)

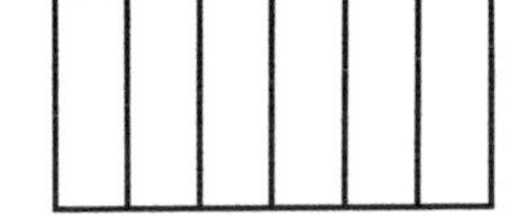

NOTES ON MULTIPLE MEANS OF ENGAGEMENT:

Many students, including those working below grade level, may benefit from having pre-drawn wholes of the same shape and size.

T: Shade the unit fraction in each rectangle. Label your models and use the words *greater than* or *less than* to compare.

S: (Shade, label, and compare models.)

T: Does this picture prove that 1 sixth is less than 1 third? Why or why not? Discuss with your partner.

S: Yes, because the shapes are the same size. → One is just cut into more pieces than the other. → We know the pieces are smaller if there are more of them, as long as the whole is the same.

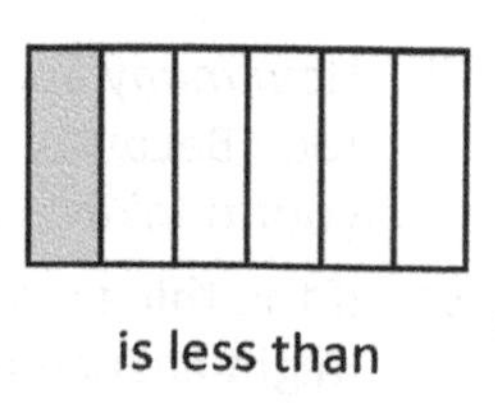

is less than

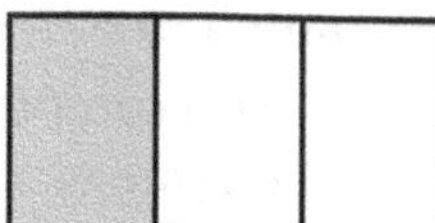

$\frac{1}{6}$ is less than $\frac{1}{3}$

Demonstrate with more examples if necessary, perhaps rotating one of the shapes so it appears different but does not change in size.

Problem Set (10 minutes)

Students should do their personal best to complete the Problem Set within the allotted 10 minutes. For some classes, it may be appropriate to modify the assignment by specifying which problems they work on first. Some problems do not specify a method for solving. Students should solve these problems using the RDW approach used for Application Problems.

NOTES ON MULTIPLE MEANS OF ENGAGEMENT:

The open-ended nature of Problems 1–8 on the Problem Set helps meet the needs of students working above grade level. Encourage creative solutions and maintain high expectations for precision and reasoning.

Student Debrief (14 minutes)

Lesson Objective: Compare unit fractions with different-sized models representing the whole.

The Student Debrief is intended to invite reflection and active processing of the total lesson experience.

Invite students to review their solutions for the Problem Set. They should check work by comparing answers with a partner before going over answers as a class. Look for misconceptions or misunderstandings that can be addressed in the Debrief. Guide students in a conversation to debrief the Problem Set and process the lesson.

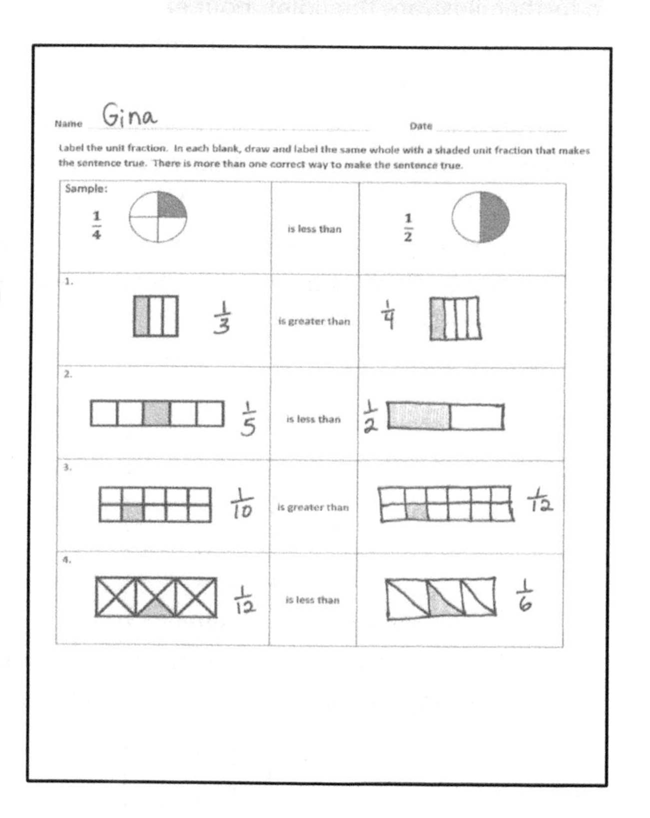

Name Gina Date

Label the unit fraction. In each blank, draw and label the same whole with a shaded unit fraction that makes the sentence true. There is more than one correct way to make the sentence true.

Sample: $\frac{1}{4}$	is less than	$\frac{1}{2}$
1. $\frac{1}{3}$	is greater than	$\frac{1}{4}$
2. $\frac{1}{5}$	is less than	$\frac{1}{2}$
3. $\frac{1}{10}$	is greater than	$\frac{1}{12}$
4. $\frac{1}{12}$	is less than	$\frac{1}{6}$

Any combination of the questions below may be used to lead the discussion.

- Problem 10 presents wholes that are clearly different sizes and also different shapes. Students may already have questioned this as they moved through the Problem Set. If so, consider crediting the student(s) who asked, and then pose the question to the rest of the class for discussion. The question of shape need not be answered today since it will be specifically addressed in Lesson 20. However, allowing the class to grapple with the question now may provide useful information that guides the delivery of Lesson 20.
- Guide a conversation through which students understand that to compare wholes numerically, they must be the same size. Consider closing by having students redraw the diagrams in Problem 9 so that Elizabeth is correct and in Problem 10 so that Manny is correct.

5. $\frac{1}{4}$	is greater than	$\frac{1}{8}$
6. $\frac{1}{10}$	is less than	$\frac{1}{9}$
7. $\frac{1}{8}$	is greater than	$\frac{1}{12}$

8. Fill in the blank with a fraction to make the statement true and draw a matching model.

$\frac{1}{4}$ is less than $\frac{1}{2}$ $\qquad$ $\frac{1}{2}$ is greater than $\frac{1}{4}$

Exit Ticket (3 minutes)

After the Student Debrief, instruct students to complete the Exit Ticket. A review of their work will help with assessing students' understanding of the concepts that were presented in today's lesson and planning more effectively for future lessons. The questions may be read aloud to the students.

9. Robert ate $\frac{1}{2}$ of a small pizza. Elizabeth ate $\frac{1}{4}$ of a large pizza. Elizabeth says, "My piece was larger than yours, so that means $\frac{1}{4} > \frac{1}{2}$." Is Elizabeth correct? Explain your answer.

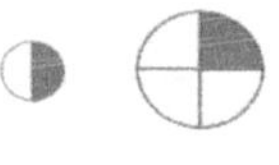

The pizzas are different sizes, which means the wholes aren't the same, so we can't compare. If the wholes were the same, it would show that $\frac{1}{2} > \frac{1}{4}$, but these pictures don't show that.

10. Manny and Daniel each ate $\frac{1}{2}$ of his candy, as shown below. Manny said he ate more candy than Daniel because his half is longer. Is he right? Explain your answer.

Manny's candy bar

Daniel's candy bar

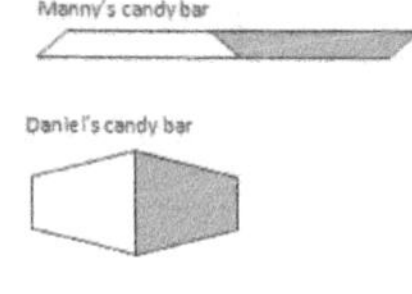

Like above, the wholes are different, so you can't just compare $\frac{1}{2}$ and $\frac{1}{2}$. The wholes are different shapes too, which makes it harder to see if the halves are the same or different. We can't tell if Manny is right.

Name ______________________________ Date ________________

Label the unit fraction. In each blank, draw and label the same whole with a shaded unit fraction that makes the sentence true. There is more than 1 correct way to make the sentence true.

Sample: $\frac{1}{4}$ 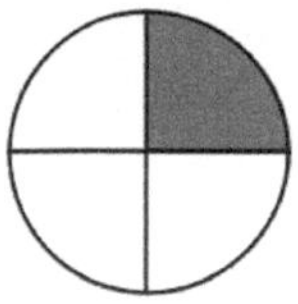	is less than	$\frac{1}{2}$ 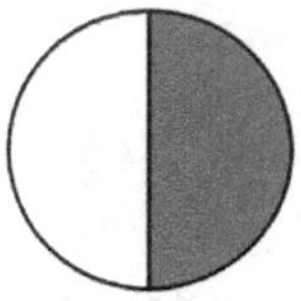
1. 	is greater than	
2. 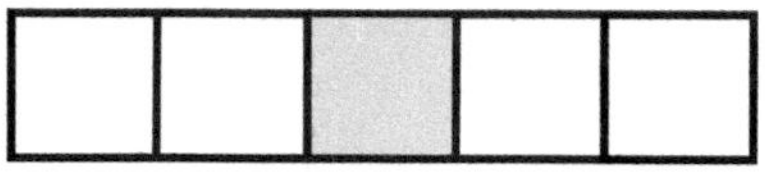	is less than	
3. 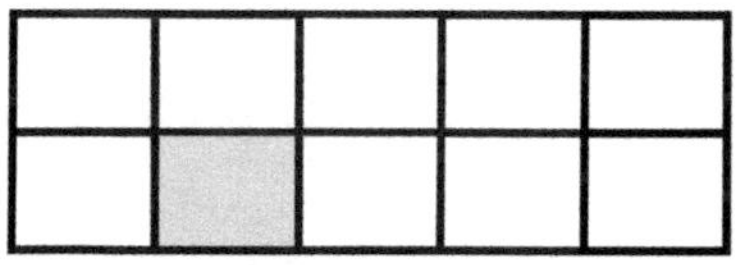	is greater than	
4. 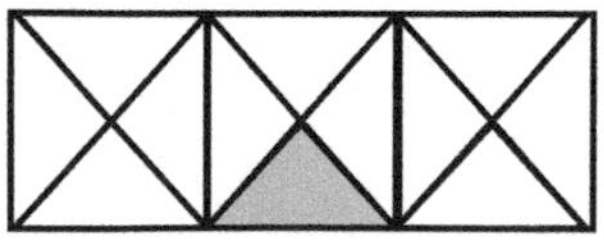	is less than	

5.	is greater than	
6.	is less than	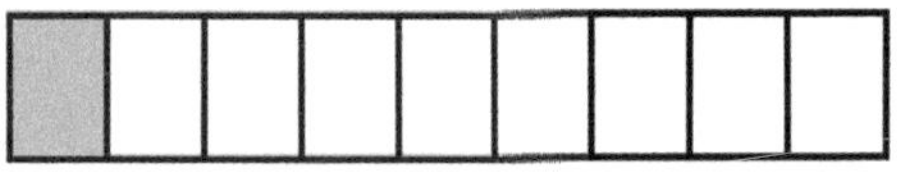
7.	is greater than	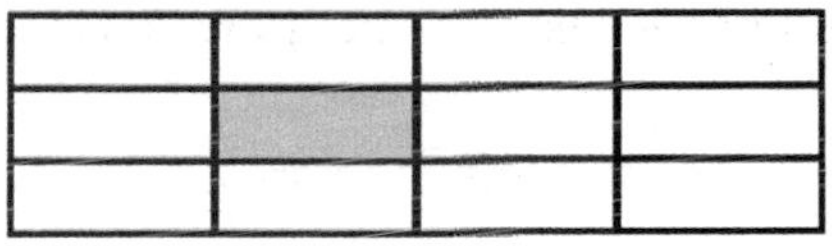

8. Fill in the blank with a fraction to make the statement true, and draw a matching model.

 $\frac{1}{4}$ is less than ☐		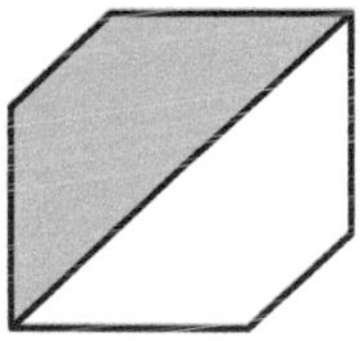 $\frac{1}{2}$ is greater than ☐	

9. Robert ate $\frac{1}{2}$ of a small pizza. Elizabeth ate $\frac{1}{4}$ of a large pizza. Elizabeth says, "My piece was larger than yours, so that means $\frac{1}{4} > \frac{1}{2}$." Is Elizabeth correct? Explain your answer.

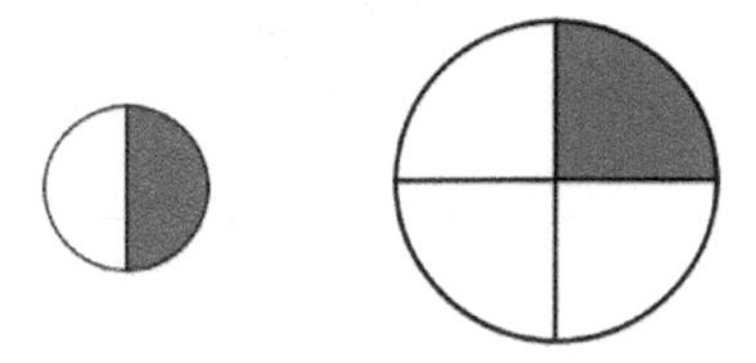

10. Manny and Daniel each ate $\frac{1}{2}$ of his candy, as shown below. Manny said he ate more candy than Daniel because his half is longer. Is he right? Explain your answer.

Manny's Candy Bar

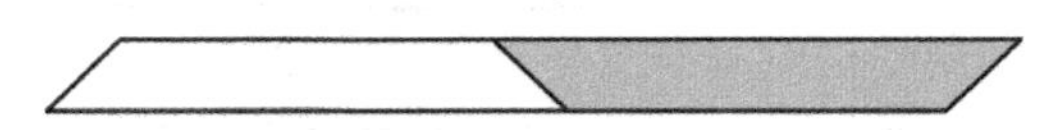

Daniel's Candy Bar

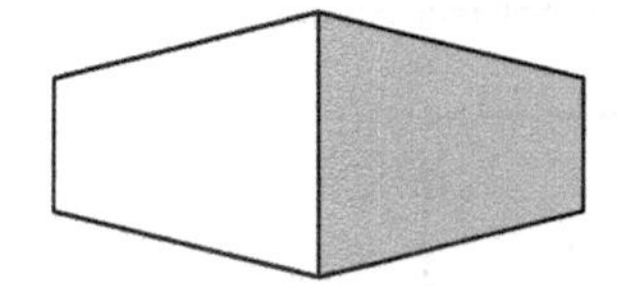

Name ______________________________ Date ______________

1. Fill in the blank with a fraction to make the statement true. Draw a matching model.

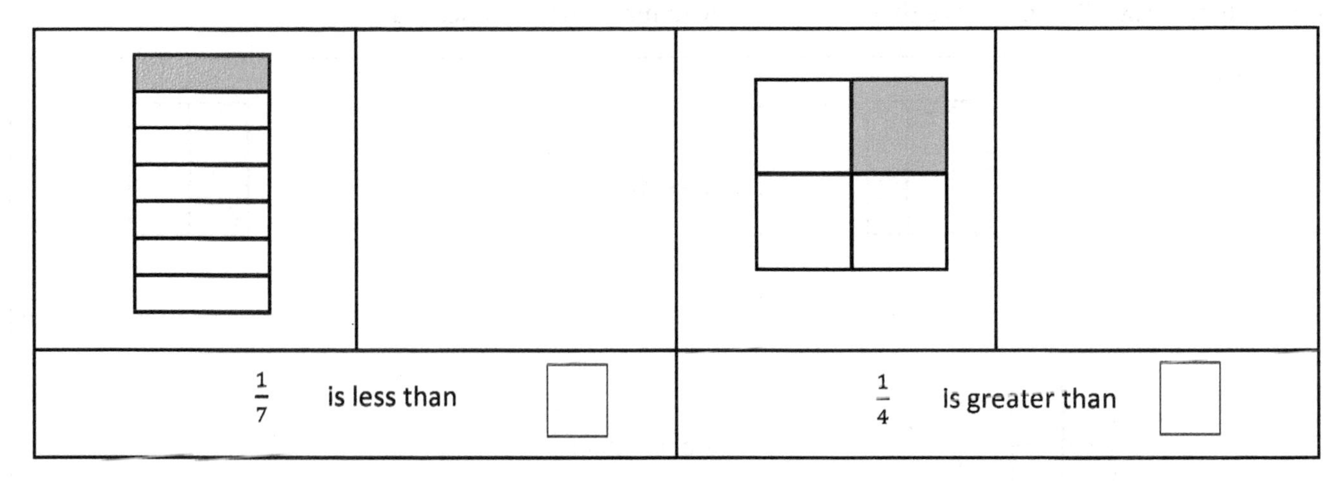

$\frac{1}{7}$ is less than ☐	$\frac{1}{4}$ is greater than ☐

2. Tatiana ate $\frac{1}{2}$ of a small carrot. Louis ate $\frac{1}{4}$ of a large carrot. Who ate more? Use words and pictures to explain your answer.

Name ______________________________ Date ______________

Label the unit fraction. In each blank, draw and label the same whole with a shaded unit fraction that makes the sentence true. There is more than 1 correct way to make the sentence true.

Sample: $\frac{1}{3}$ 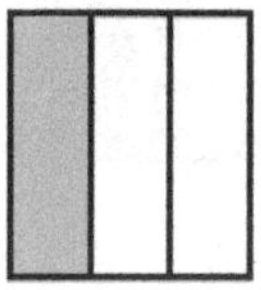	is less than	$\frac{1}{2}$ 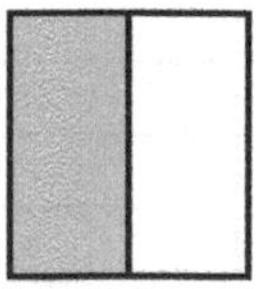
1. 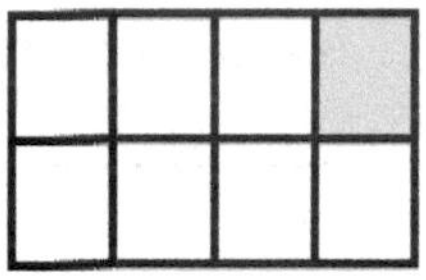	is greater than	
2. 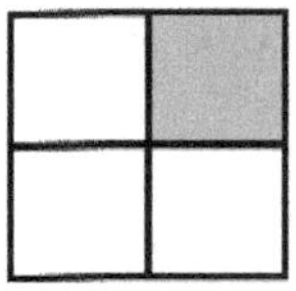	is less than	
3. 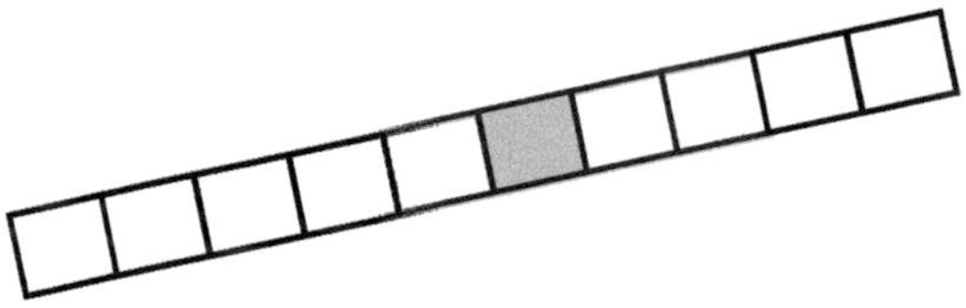	is greater than	
4. 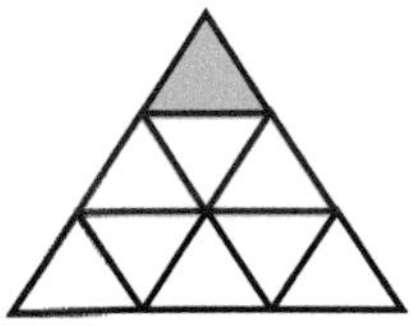	is less than	

5.	is greater than	
6.	is less than	
7.	is greater than	

8. Fill in the blank with a fraction to make the statement true. Draw a matching model.

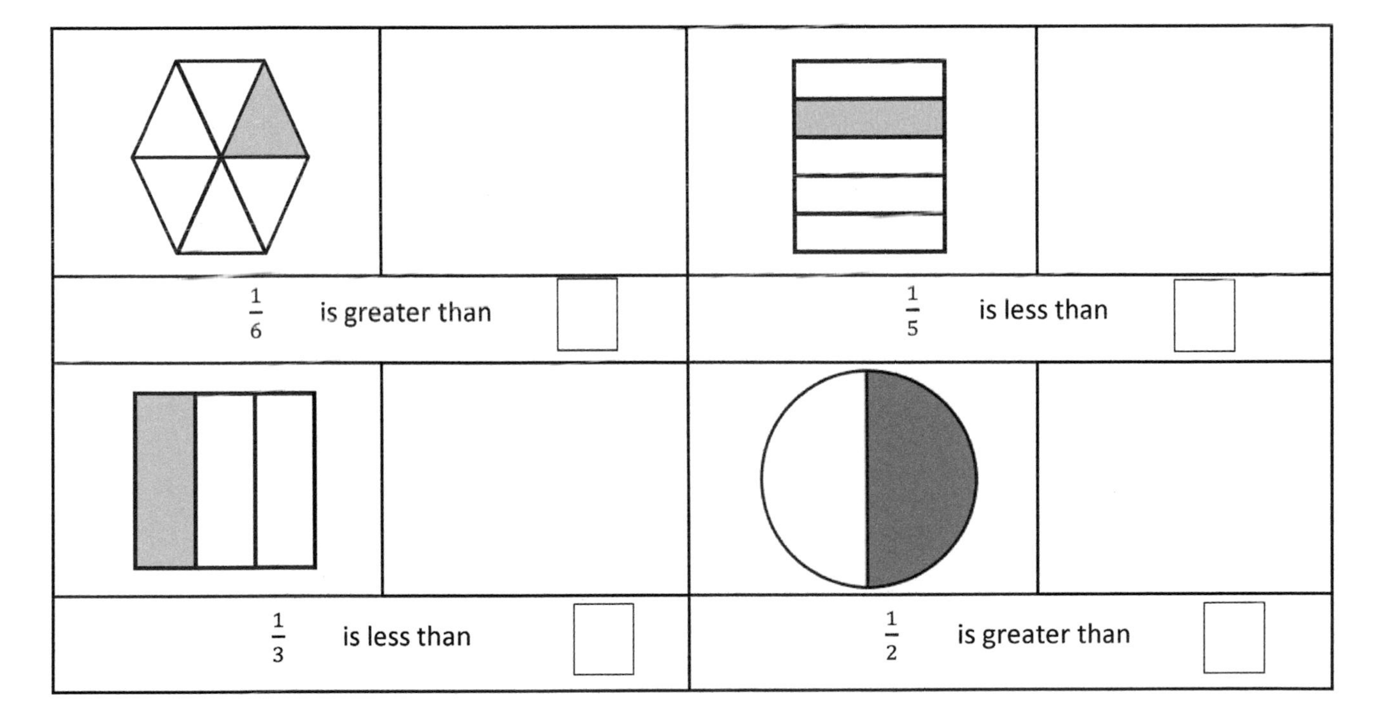

$\frac{1}{6}$ is greater than ☐	$\frac{1}{5}$ is less than ☐
$\frac{1}{3}$ is less than ☐	$\frac{1}{2}$ is greater than ☐

9. Debbie ate $\frac{1}{8}$ of a large brownie. Julian ate $\frac{1}{2}$ of a small brownie. Julian says, "I ate more than you because $\frac{1}{2} > \frac{1}{8}$."

   a. Use pictures and words to explain Julian's mistake.

   b. How could you change the problem so that Julian is correct? Use pictures and words to explain.

Lesson 12

Objective: Specify the corresponding whole when presented with one equal part.

Suggested Lesson Structure

■ Fluency Practice	(12 minutes)
■ Application Problem	(8 minutes)
■ Concept Development	(32 minutes)
■ Student Debrief	(8 minutes)
Total Time	**(60 minutes)**

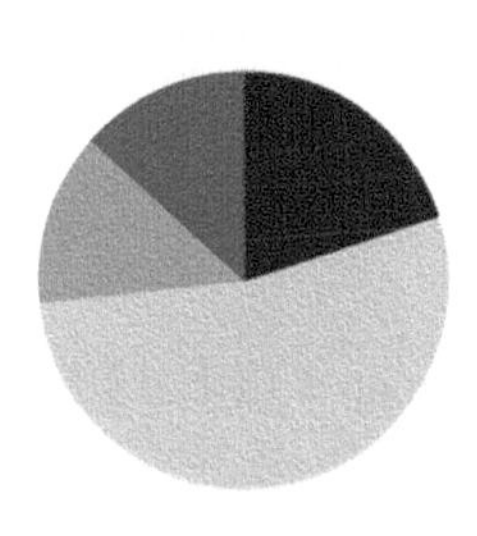

Fluency Practice (12 minutes)

- Sprint: Multiply with Nine **3.OA.4** (6 minutes)
- Unit and Non-Unit Fractions of 1 Whole **3.G.2, 3.NF.2** (3 minutes)
- More Units Than 1 Whole **3.NF.2b** (3 minutes)

Sprint: Multiply with Nine (6 minutes)

Materials: (S) Multiply with Nine Sprint

Note: This Sprint supports fluency with multiplication using units of 9.

Unit and Non-Unit Fractions of 1 Whole (3 minutes)

Materials: (S) Personal white board

Note: This activity reviews naming the shaded and unshaded equal parts of a whole, as well as drawing number bonds to represent the fractional parts of 1 whole.

T: (Draw a shape partitioned in halves with 1 half shaded.) Write the fraction that is shaded.
S: (Write $\frac{1}{2}$.)
T: Write the fraction that is not shaded.
S: (Write $\frac{1}{2}$.)
T: Draw the number bond.
S: (Draw a number bond showing that 1 half and 1 half equal 2 halves.)

Continue with the following possible sequence: $\frac{2}{3}$ and $\frac{1}{3}$, $\frac{4}{5}$ and $\frac{1}{5}$, $\frac{9}{10}$ and $\frac{1}{10}$, and $\frac{7}{8}$ and $\frac{1}{8}$.

More Units Than 1 Whole (3 minutes)

Materials: (S) Personal white board (optional)

Note: This activity reviews naming fractions greater than 1 whole from Lesson 9. It may be appropriate for some classes to draw responses on personal white boards for extra support.

T: What's 1 more fifth than 1 whole?
S: 6 fifths.
T: 2 more fifths than 1 whole?
S: 7 fifths.

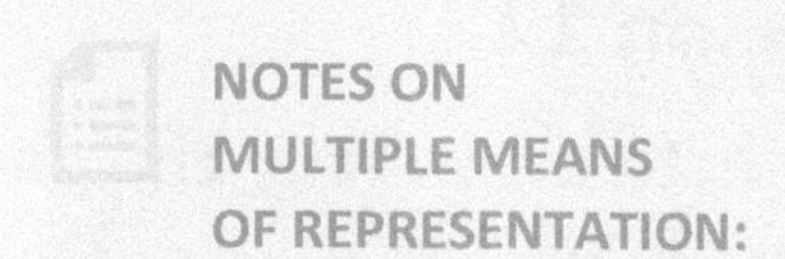

NOTES ON MULTIPLE MEANS OF REPRESENTATION:

In *More Units Than 1 Whole,* support students working below grade level with pictorial models drawn on personal white boards by either the teacher or the students. Alternatively, begin with halves, thirds, and fourths, gradually progressing to tenths.

Continue with the following possible sequence: 4 fifths, 3 fifths, 1 tenth, 7 tenths, 1 third, 2 thirds, 1 eighth, 5 eighths, 1 sixth, and 5 sixths.

Application Problem (8 minutes)

Jennifer hid half of her birthday money in the dresser drawer. The other half she put in her jewelry box. If she hid \$8 in the drawer, how much money did she get for her birthday?

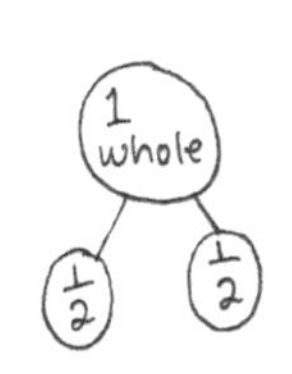

Note: This problem reviews the concept that 2 halves make 1 whole. Students may solve by adding or multiplying. They might draw a tape diagram or a number bond to model the problem. Invite students to share their pictures with a partner.

Concept Development (32 minutes)

Materials: (S) Use similar materials to those used in Lesson 4 (at least 75 copies of each), 10-centimeter length of yarn, 4" × 1" rectangular piece of yellow construction paper, 3" × 1" brown paper, 1" × 1" orange square, water, small plastic cups, clay

Exploration: Designate the following stations for groups of 3 (more than 3 not suggested).

Station A: 1 half and 1 fourth
Station B: 1 half and 1 third
Station C: 1 third and 1 fourth
Station D: 1 third and 1 sixth
Station E: 1 fourth and 1 sixth
Station F: 1 fourth and 1 eighth
Station G: 1 fifth and 1 tenth
Station H: 1 fifth and 1 sixth

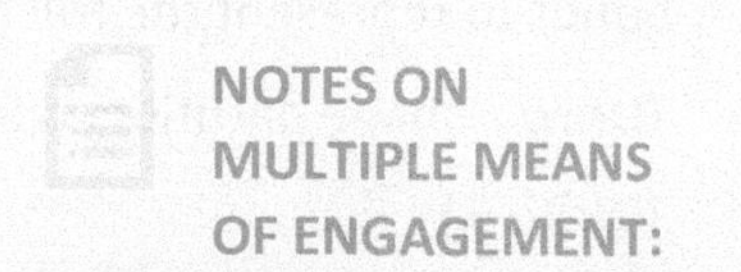

NOTES ON MULTIPLE MEANS OF ENGAGEMENT:

Organize students working below grade level at the stations with the easier fractional units and students working above grade level at the stations with the most challenging fractional units.

The students represent 1 whole using the materials at their stations.

Notes:

- Each item at the station represents the indicated unit fractions.
- Students show 1 whole corresponding to the given unit fraction. Each station includes 2 objects representing unit fractions, and therefore 2 different whole amounts.

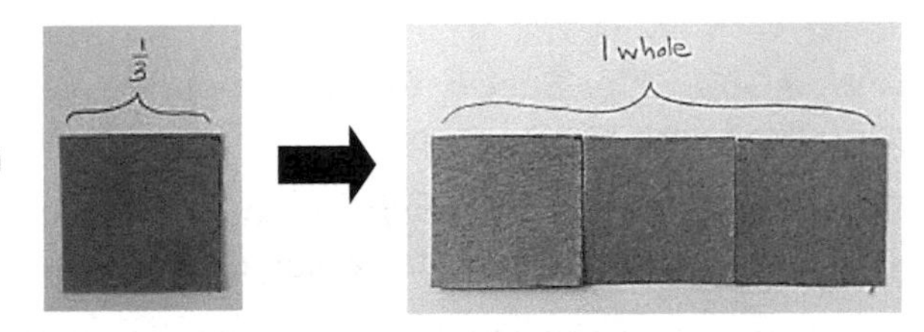

- The entire quantity of each item must be used as the fraction indicated. For example, if showing 1 third with the orange square, the whole must use 3 thirds or 3 of the orange squares (pictured above to the right).

T: (Hold up the same size ball of clay—200 g—from Lesson 4.) This piece of clay represents 1 third. What might 1 whole look like? Discuss with your partner.

S: (Discuss.)

T: (After discussion, model the whole as 3 equal lumps of clay weighing 600 g.)

T: (Hold up a 12-inch by 1-inch yellow strip.) This strip represents 1 fourth. What might 1 whole look like?

S: (Discuss.)

T: (After discussion, model the whole using 4 equal strips laid end-to-end for a length of 48 inches.)

T: (Show a 12-ounce cup of water.) The water in this cup represents 1 fifth. What might the whole look like? What if the water represents 1 fourth? (Measure the 2 quantities into 2 separate containers.)

> **NOTES ON MULTIPLE MEANS OF REPRESENTATION:**
>
> Give English language learners a little more time to respond, either in writing or in their first language.

Give the students 15 minutes to create their display. Next, conduct a museum walk where they tour the work of the other stations. During the tour, students should identify the fractions and think about their relationships. Use the following points to guide the students:

- Identify the unit fraction.
- Think about how the whole amount relates to your own and to other whole amounts.
- Compare the yarn to the yellow strip.
- Compare the yellow strip to the brown paper.

> **NOTES ON MULTIPLE MEANS OF ACTION AND EXPRESSION:**
>
> The museum walk is a rich opportunity for students to practice language. Pair students and give them sentence frames or prompts to use at each station to help them discuss what they see with their partner.

Problem Set (10 minutes)

Students should do their personal best to complete the Problem Set within the allotted 10 minutes. For some classes, it may be appropriate to modify the assignment by specifying which problems they work on first. Some problems do not specify a method for solving. Students should solve these problems using the RDW approach used for Application Problems.

Student Debrief (8 minutes)

Lesson Objective: Specify the corresponding whole when presented with one equal part.

The Student Debrief is intended to invite reflection and active processing of the total lesson experience.

Invite students to review their solutions for the Problem Set. They should check work by comparing answers with a partner before going over answers as a class. Look for misconceptions or misunderstandings that can be addressed in the Debrief. Guide students in a conversation to debrief the Problem Set and process the lesson.

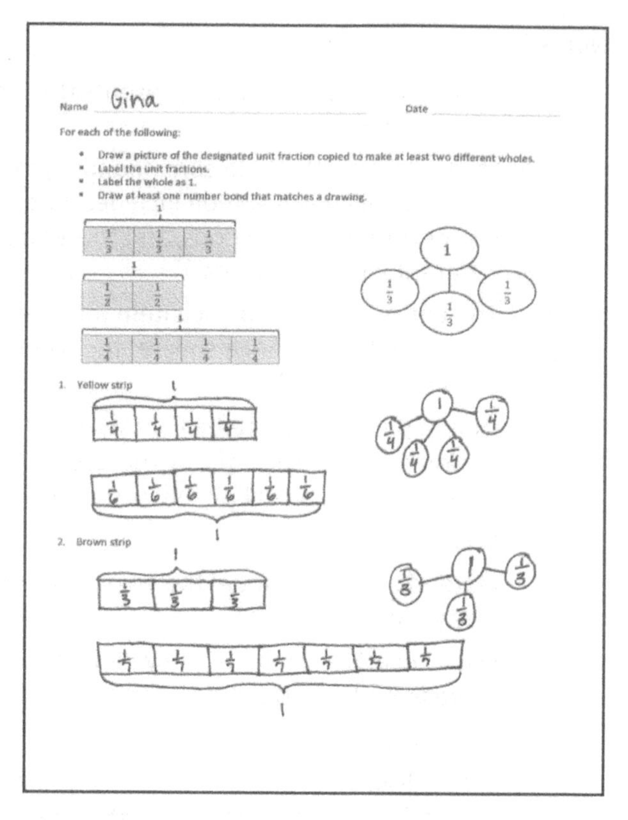

Any combination of the questions below may be used to lead the discussion.

MP.2

- What were the different wholes we saw at each station that were the same?
- What different unit fractions did you see as you went from station to station?
- What did you notice about different unit fractions at the stations?
- Which wholes had the most equal parts?
- Which wholes had the least equal parts?
- What surprised you about the different representations of thirds or any other fraction?
- How does the water compare to the clay? The clay to the yarn?
- What if all the wholes were the same size? What would happen to the equal parts?
- Does the picture in Problem 2 show that $\frac{1}{3}$ equals $\frac{1}{7}$? Why or why not? How would you need to change your picture to compare $\frac{1}{3}$ and $\frac{1}{7}$?

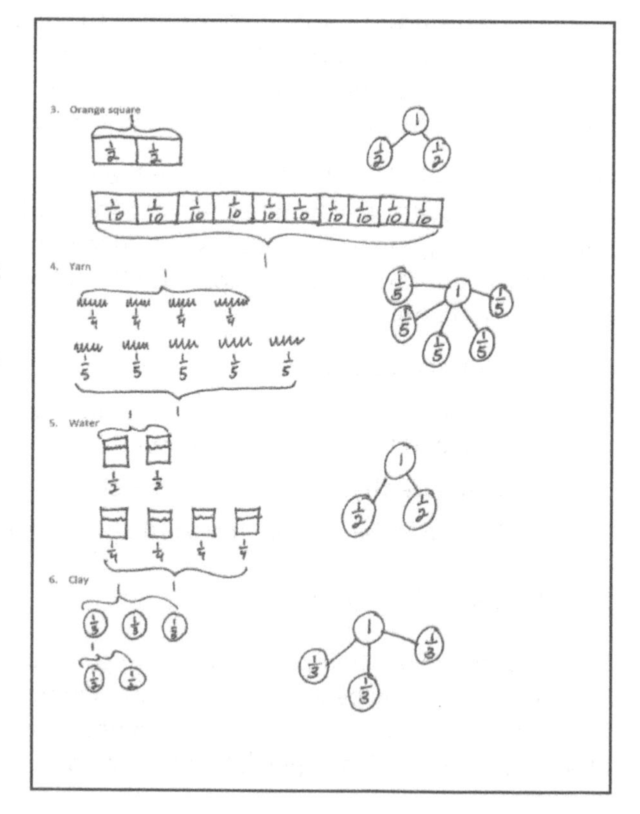

Exit Ticket (3 minutes)

After the Student Debrief, instruct students to complete the Exit Ticket. A review of their work will help with assessing students' understanding of the concepts that were presented in today's lesson and planning more effectively for future lessons. The questions may be read aloud to the students.

A

Number Correct: _______

Multiply with Nine

1.	9 × 1 =	
2.	1 × 9 =	
3.	9 × 2 =	
4.	2 × 9 =	
5.	9 × 3 =	
6.	3 × 9 =	
7.	9 × 4 =	
8.	4 × 9 =	
9.	9 × 5 =	
10.	5 × 9 =	
11.	9 × 6 =	
12.	6 × 9 =	
13.	9 × 7 =	
14.	7 × 9 =	
15.	9 × 8 =	
16.	8 × 9 =	
17.	9 × 9 =	
18.	9 × 10 =	
19.	10 × 9 =	
20.	1 × 9 =	
21.	10 × 9 =	
22.	2 × 9 =	

23.	9 × 9 =	
24.	3 × 9 =	
25.	8 × 9 =	
26.	4 × 9 =	
27.	7 × 9 =	
28.	5 × 9 =	
29.	6 × 9 =	
30.	9 × 5 =	
31.	9 × 10 =	
32.	9 × 1 =	
33.	9 × 6 =	
34.	9 × 4 =	
35.	9 × 9 =	
36.	9 × 2 =	
37.	9 × 7 =	
38.	9 × 3 =	
39.	9 × 8 =	
40.	11 × 9 =	
41.	9 × 11 =	
42.	12 × 9 =	
43.	9 × 12 =	
44.	13 × 9 =	

B

Number Correct: _______

Improvement: _______

Multiply with Nine

1.	$1 \times 9 =$		23.	$10 \times 9 =$	
2.	$9 \times 1 =$		24.	$9 \times 9 =$	
3.	$2 \times 9 =$		25.	$4 \times 9 =$	
4.	$9 \times 2 =$		26.	$8 \times 9 =$	
5.	$3 \times 9 =$		27.	$3 \times 9 =$	
6.	$9 \times 3 =$		28.	$7 \times 9 =$	
7.	$4 \times 9 =$		29.	$6 \times 9 =$	
8.	$9 \times 4 =$		30.	$9 \times 10 =$	
9.	$5 \times 9 =$		31.	$9 \times 5 =$	
10.	$9 \times 5 =$		32.	$9 \times 6 =$	
11.	$6 \times 9 =$		33.	$9 \times 1 =$	
12.	$9 \times 6 =$		34.	$9 \times 9 =$	
13.	$7 \times 9 =$		35.	$9 \times 4 =$	
14.	$9 \times 7 =$		36.	$9 \times 3 =$	
15.	$8 \times 9 =$		37.	$9 \times 2 =$	
16.	$9 \times 8 =$		38.	$9 \times 7 =$	
17.	$9 \times 9 =$		39.	$9 \times 8 =$	
18.	$10 \times 9 =$		40.	$11 \times 9 =$	
19.	$9 \times 10 =$		41.	$9 \times 11 =$	
20.	$9 \times 3 =$		42.	$12 \times 9 =$	
21.	$1 \times 9 =$		43.	$9 \times 12 =$	
22.	$2 \times 9 =$		44.	$13 \times 9 =$	

Name ______________________________ Date ______________

For each of the following:

- Draw a picture of the designated unit fraction copied to make at least two different wholes.
- Label the unit fractions.
- Label the whole as 1.
- Draw at least one number bond that matches a drawing.

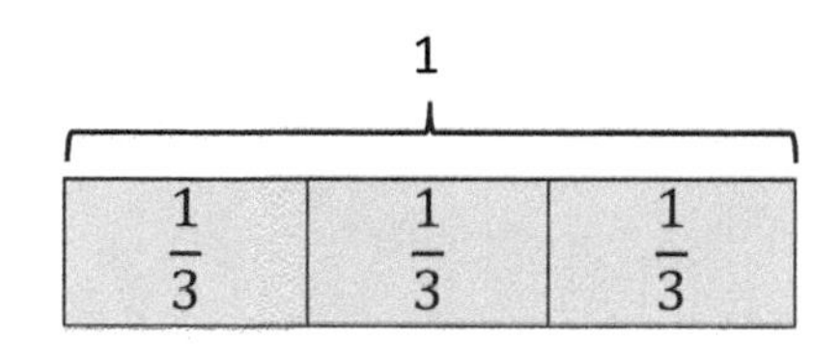

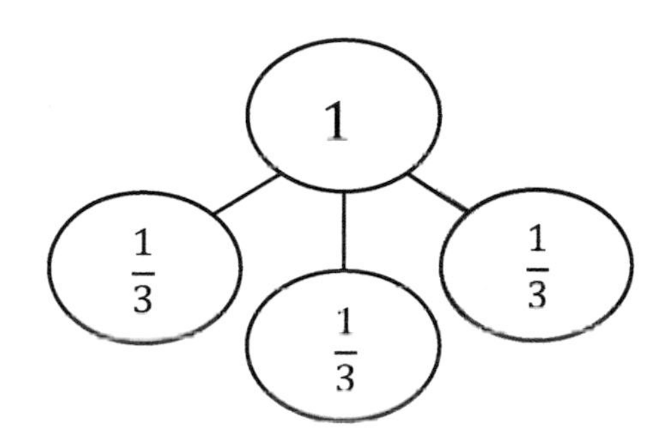

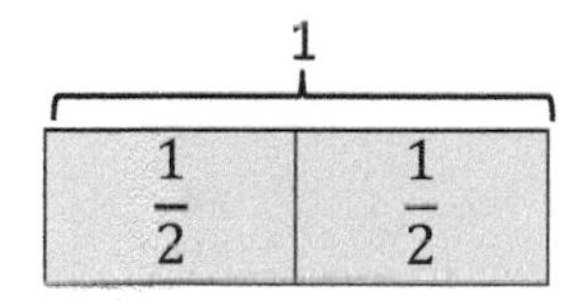

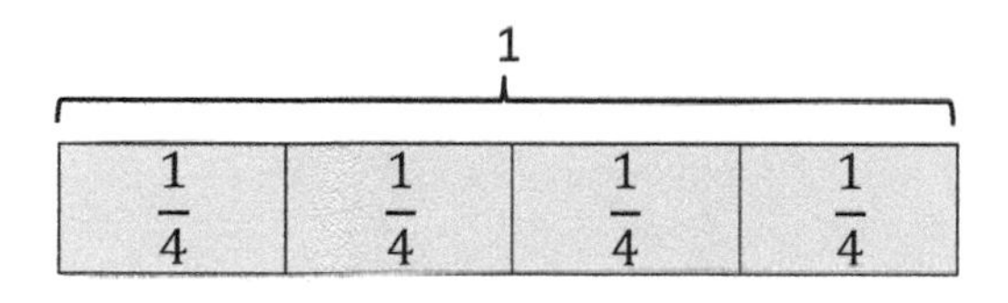

1. Yellow strip

2. Brown strip

3. Orange square

4. Yarn

5. Water

6. Clay

Name ______________________________ Date ____________

Each shape represents the unit fraction. Draw a picture representing a possible whole.

1. 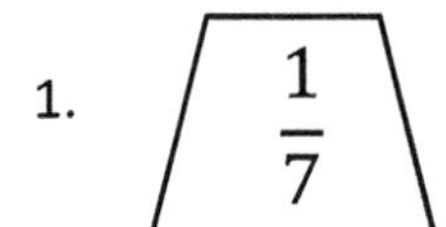

2. $\frac{1}{9}$

3. Aileen and Jack used the same triangle representing the unit fraction $\frac{1}{4}$ to create 1 whole. Who did it correctly? Explain your answer.

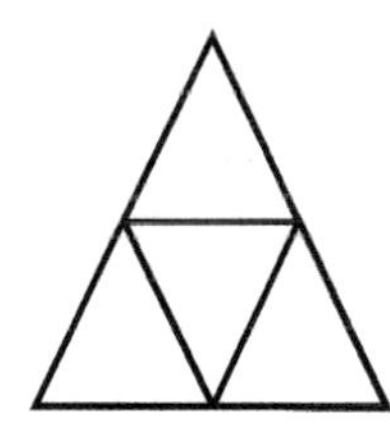

Aileen's Drawing

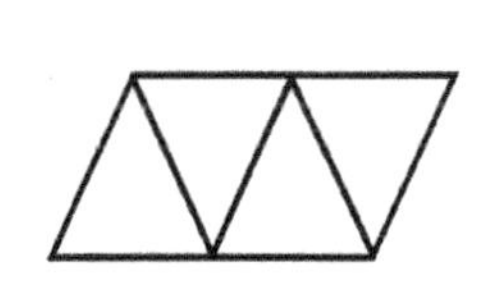

Jack's Drawing

Name ______________________________ Date ________________

Each shape represents the given unit fraction. Estimate to draw a possible whole.

1. 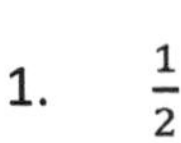 $\frac{1}{2}$

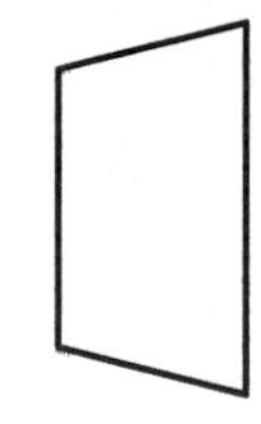

2. $\frac{1}{6}$

3. 1 third

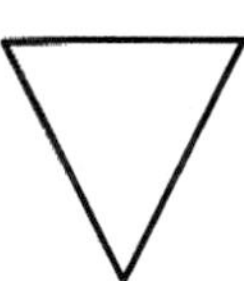

4. 1 fourth

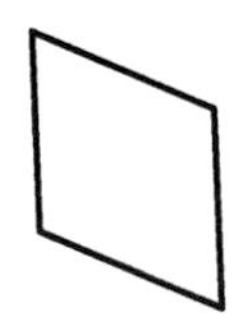

Each shape represents the given unit fraction. Estimate to draw a possible whole, label the unit fractions, and draw a number bond that matches the drawing. The first one is done for you.

5. $\frac{1}{3}$

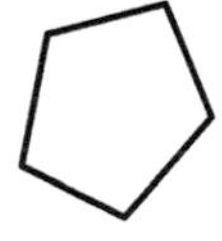

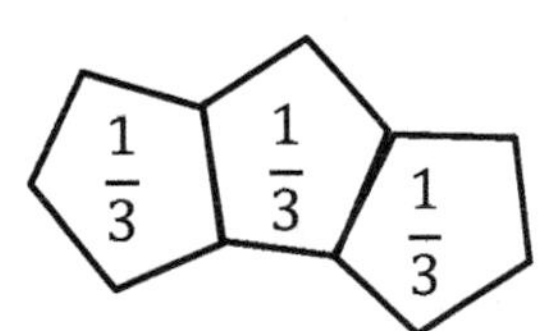

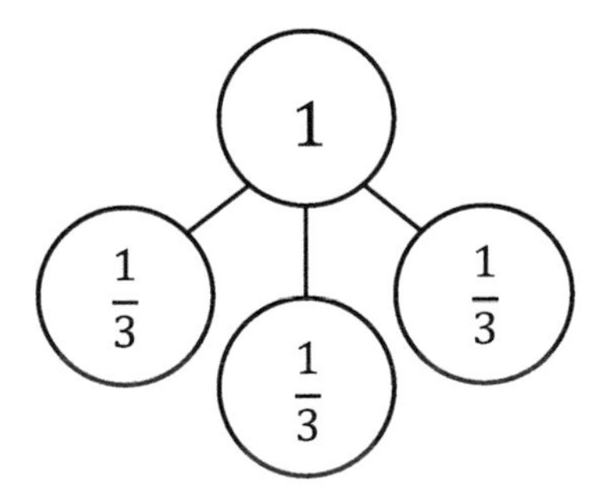

6. $\frac{1}{2}$

7. $\frac{1}{5}$

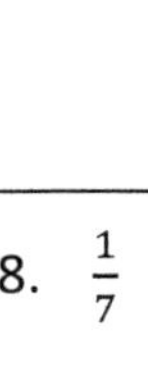

8. $\frac{1}{7}$

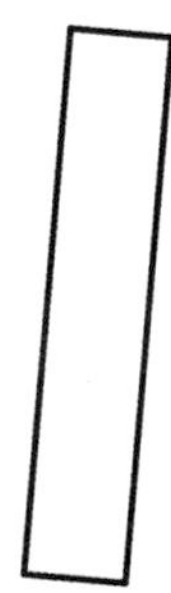

9. Evan and Yong used this shape, representing the unit fraction $\frac{1}{3}$, to draw 1 whole. Shania thinks both of them did it correctly. Do you agree with her? Explain your answer.

Evan's Shape

Yong's Shape

Lesson 13

Objective: Identify a shaded fractional part in different ways depending on the designation of the whole.

Suggested Lesson Structure

Fluency Practice	(9 minutes)
Application Problem	(5 minutes)
Concept Development	(35 minutes)
Student Debrief	(11 minutes)
Total Time	**(60 minutes)**

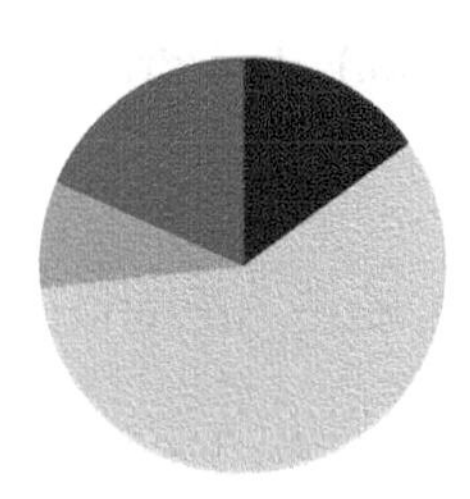

Fluency Practice (9 minutes)

- Skip-Count by Fourths on the Clock **3.G.2, 3.NF.1** (3 minutes)
- Division **3.OA.2** (3 minutes)
- Draw a Whole **3.NF.3c** (3 minutes)

Skip-Count by Fourths on the Clock (3 minutes)

Materials: (T) Clock

Note: This activity reviews counting by fourths on the clock from Module 2.

T: (Hold or project a clock.) Let's skip-count by fourths on the clock starting with 1 o'clock.

S: 1, quarter past 1, half past 1, quarter 'til 2, 2, quarter past 2, half past 2, quarter 'til 3, 3.

T: Stop. From 3:00, skip-count by fourths backward.

S: 3, quarter 'til 3, half past 2, quarter past 2, 2, quarter 'til 2, half past 1, quarter past 1, 1.

Continue counting up and down by fourths.

Division (3 minutes)

Note: This activity reviews division from Modules 1 and 3.

T: (Write $4 \div 2 =$ ____.) Say the number sentence and the answer.

S: 4 divided by 2 equals 2.

Continue with the following possible sequence: $6 \div 2$, $6 \div 3$, $8 \div 2$, $8 \div 4$, $10 \div 2$, $10 \div 5$, $12 \div 2$, $12 \div 6$, $12 \div 4$, and $12 \div 3$.

Draw a Whole (3 minutes)

Materials: (S) Personal white board

Note: This activity reviews representing the whole when given 1 equal part from Lesson 12.

T: Draw 1 unit on your personal white board.

S: (Draw 1 unit.)

T: Label the unit $\frac{1}{3}$. (After students label.) Now draw a possible whole that corresponds to your unit of $\frac{1}{3}$.

Continue with the following possible sequence: $\frac{1}{5}$, $\frac{1}{6}$, $\frac{1}{4}$, and $\frac{1}{2}$.

Application Problem (5 minutes)

Davis wants to make a picture using 9 square tiles. What fraction of the picture does 1 tile represent? Draw 3 different ways Davis could make his picture.

Note: This problem reviews identifying the unit fraction from Topic B. Invite students to share their pictures and discuss why their unit fraction is the same, even though their pictures are different.

Concept Development (35 minutes)

Materials: (S) 1 index card (or per pair), black marker, fraction strips, personal white board

T: Fold your index card to make 4 equal units. Shade and label the first unit. Each part is equal to what fraction of the whole?

$\frac{1}{4}$

1 card is 1 whole

S: 1 fourth.

T: What is the whole?

S: The index card.

T: With a black marker, trace the outside of your card to show the whole.

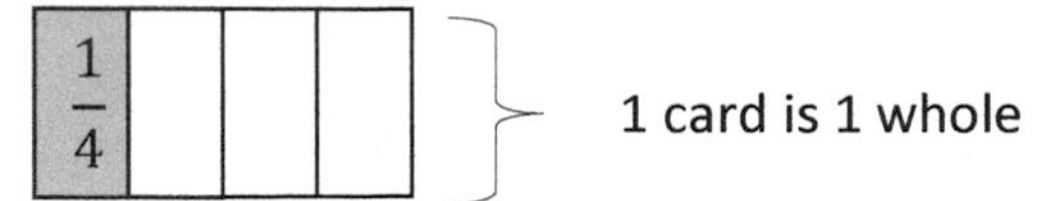

Have students working above grade level answer an open-ended question such as, "What number patterns (or relationships) do you notice?"

T: Flip your index card over so you cannot see the fraction you wrote. The new whole is half of the card. Outline it with a marker. (After students outline.) Use your pencil to shade the same amount of space you shaded on the other side. (After students shade.) Talk with your partner about how to label the shaded amount on this side of the card.

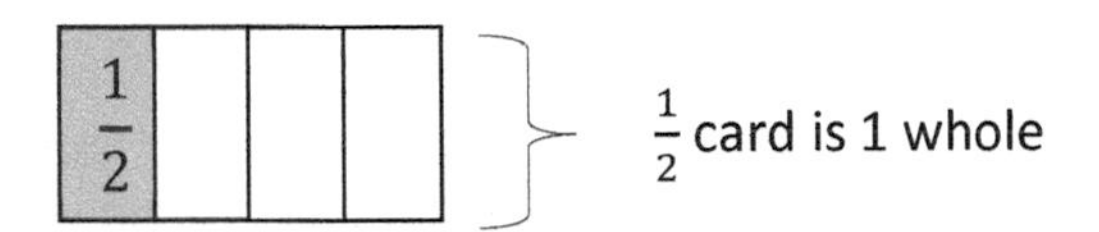

S: The shaded part is $\frac{1}{2}$ because the new whole is different. I see the whole. The shaded part is just half of that.

T: Changing the whole changed the unit fraction that we use to describe the shaded part. What *was* 1 fourth of the whole card is 1 half of the new, smaller whole.

Display the figure to the right, and give students a fraction strip of the same length.

T: This time, the whole is the entire rectangle. Trace the outline of your fraction strip, and then shade to draw the model on your board.

S: (Draw the model.)

T: Tell your partner how you can figure out what fraction is shaded.

S: I can estimate and draw lines to partition the rectangle. → I can fold my fraction strip to figure out the unit fraction. → Either way, 2 thirds are shaded.

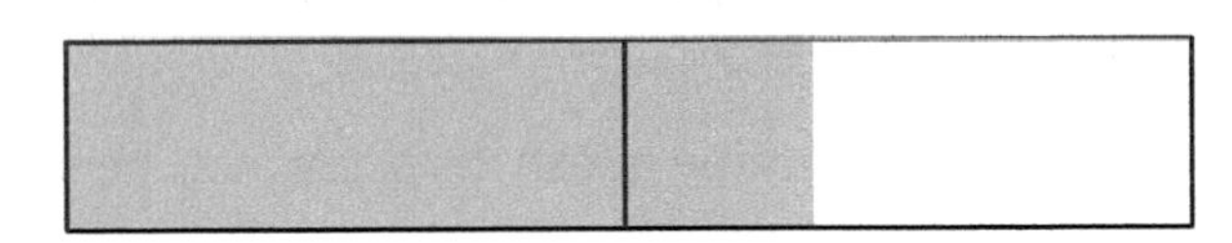

MP.3

T: Now, use your fraction strip to measure, partition, and label.

T: (Show the figure to the right, and have students draw it on their boards using fraction strips for accuracy.) If each of the outlined rectangles represents 1 whole, then what fraction is shaded? Discuss with your partner.

S: I can fold my fraction strip to measure the parts. → I can estimate to draw lines inside the small rectangles and partition each into 3 equal pieces. → Then 1 whole rectangle and 1 third are shaded, or $\frac{4}{3}$.

T: Talk with your partner about why it's important to know the whole.

S: (Discuss.)

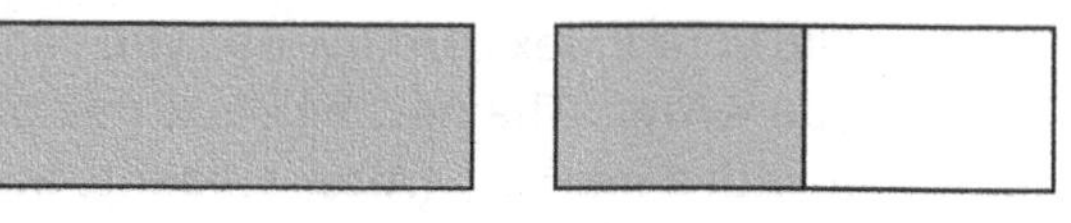

MP.3

Display the picture to the right.

T: For his birthday, Kyle's mom brought in cake to share with the class. When she picked up the 2 cake pans at the end of the day, she said, "Wow, your friends ate $\frac{3}{4}$ of the cake." Kyle said, "No, Mom, we ate $\frac{6}{4}$ cakes." Talk to a partner: Who is right? Use pictures, words, or numbers on your boards to help prove your answer.

S: (Discuss.)

NOTES ON MULTIPLE MEANS OF ACTION AND EXPRESSION:

Support English language learners as they construct their written response to Problem 7. Read the prompt aloud, or have students read chorally. Provide sentence starters and a word bank.

Sentence starters may include:

- "I agree with...because..."
- "I think...is right because..."

Below are some possible words for the word bank:

specify	shaded	rectangle
whole	fourths	halves

Problem Set (10 minutes)

Students should do their personal best to complete the Problem Set within the allotted 10 minutes. For some classes, it may be appropriate to modify the assignment by specifying which problems they work on first. Some problems do not specify a method for solving. Students should solve these problems using the RDW approach used for Application Problems.

Student Debrief (11 minutes)

Lesson Objective: Identify a shaded fractional part in different ways depending on the designation of the whole.

The Student Debrief is intended to invite reflection and active processing of the total lesson experience.

Invite students to review their solutions for the Problem Set. They should check work by comparing answers with a partner before going over answers as a class. Look for misconceptions or misunderstandings that can be addressed in the Debrief. Guide students in a conversation to debrief the Problem Set and process the lesson.

Any combination of the questions below may be used to lead the discussion.

- In Problems 6(a)–6(d), box the rope that represents the whole. Circle the rope that represents the part.
- Compare Problems 6(e) and 6(f) to illustrate the part–whole relationship.
- Compare Rope C in Problems 6(a) and 6(d).
- Compare Rope B in Problems 6(a) and 6(d).

NOTES ON MULTIPLE MEANS OF ENGAGEMENT:

Model a few examples of the Problem Set activities to support students working below grade level. Make sure they can specify the whole. Say, "Trace the whole with your finger."

To aid partitioning in Part B, cover all but the shaded part.

For Problem 6, have students organize the data with a chart or table to facilitate comparisons, if needed.

Exit Ticket (3 minutes)

After the Student Debrief, instruct students to complete the Exit Ticket. A review of their work will help with assessing students' understanding of the concepts that were presented in today's lesson and planning more effectively for future lessons. The questions may be read aloud to the students.

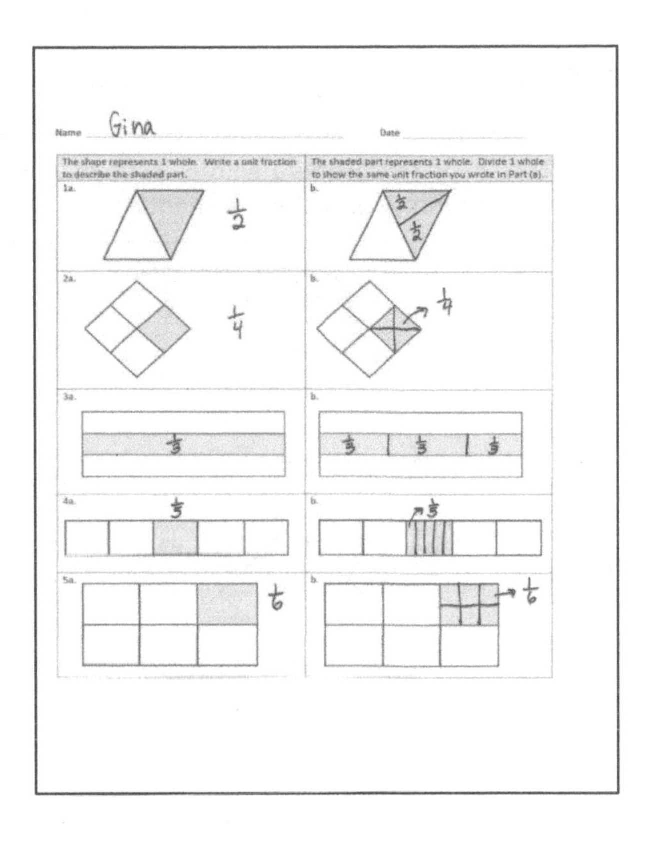

6. Use the diagram below to complete the following statements.

Rope A (1 m)

Rope B

Rope C

a. Rope C is $\frac{1}{2}$ the length of Rope B.

b. Rope B is $\frac{1}{2}$ the length of Rope A.

c. Rope C is $\frac{1}{4}$ the length of Rope A.

d. If Rope B measures 1 m long, then Rope A is 2 m long, and Rope C is $\frac{1}{2}$ m long.

e. If Rope A measures 1 m long, Rope B is $\frac{1}{2}$ m long, and Rope C is $\frac{1}{4}$ m long.

7. Ms. Fan drew the figure below on the board. She asked the class to name the shaded fraction. Charlie answered $\frac{3}{4}$. Janice answered $\frac{3}{2}$. Jenna thinks they're both right. With whom do you agree? Explain your thinking.

It depends on what the whole is! If it's the large rectangle, then $\frac{3}{4}$ is right. But, if it's the smaller rectangle, $\frac{3}{2}$ is right. So, I can't agree with anyone because I don't know what the whole is!

Name ______________________________ Date ________________

The shape represents 1 whole. Write a unit fraction to describe the shaded part.	The shaded part represents 1 whole. Divide 1 whole to show the same unit fraction you wrote in Part (a).
1. a. 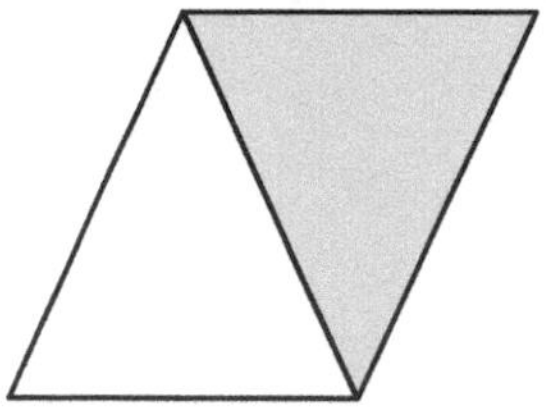	b. 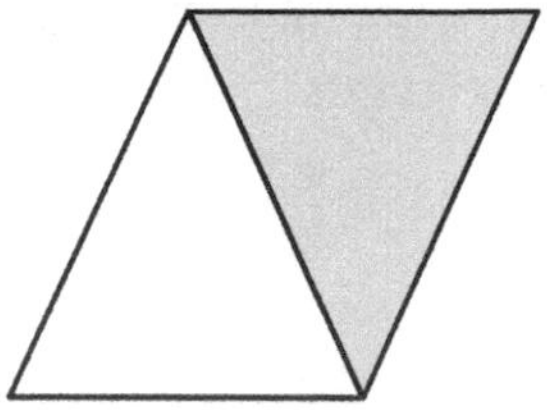
2. a. 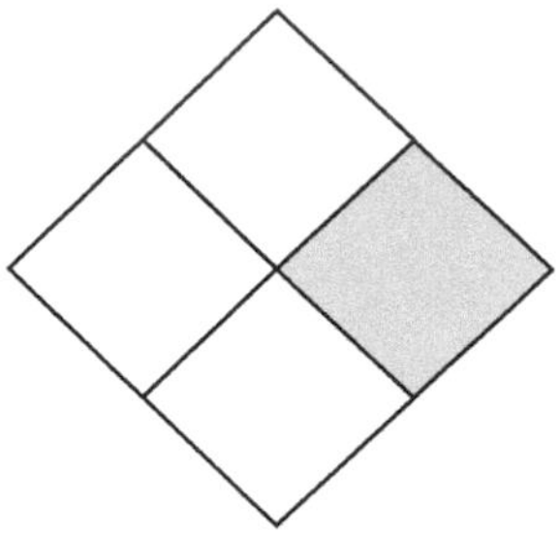	b. 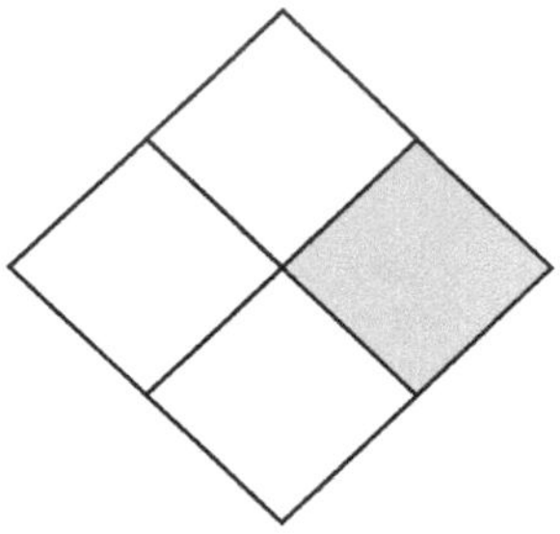
3. a. 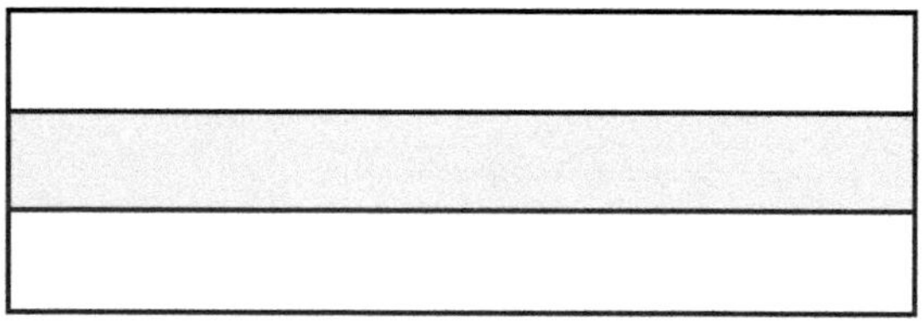	b. 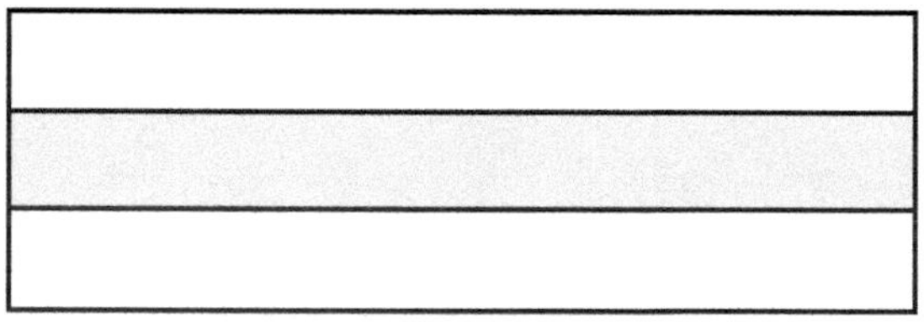
4. a. 	b.
5. a. 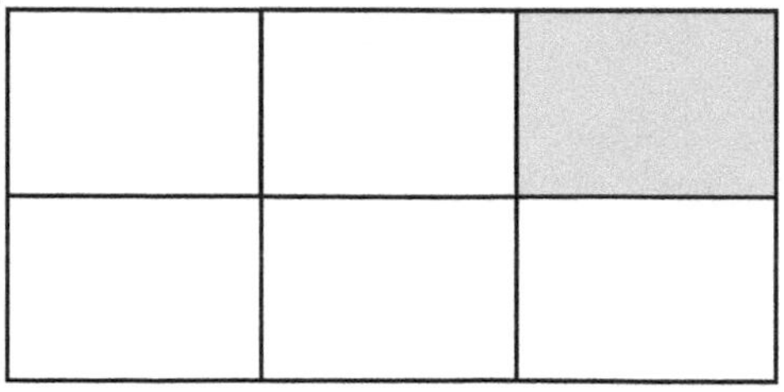	b. 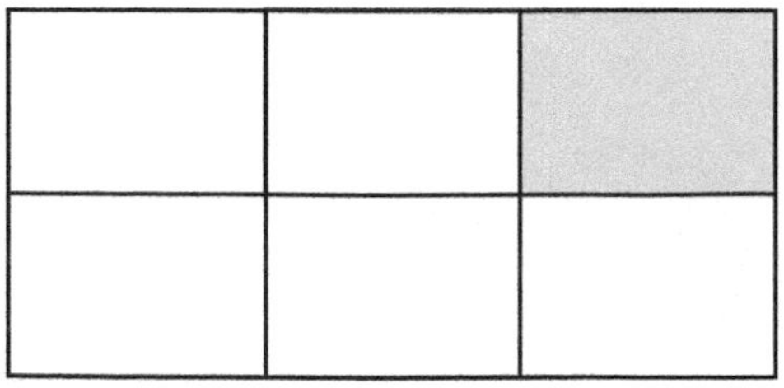

6. Use the diagram below to complete the following statements.

Rope A

Rope B

Rope C

a. Rope ____________ is $\frac{1}{2}$ the length of Rope B.

b. Rope ____________ is $\frac{1}{2}$ the length of Rope A.

c. Rope C is $\frac{1}{4}$ the length of Rope ____________.

d. If Rope B measures 1 m long, then Rope A is ____________ m long, and Rope C is ____________ m long.

e. If Rope A measures 1 m long, Rope B is ____________ m long, and Rope C is ____________ m long.

7. Ms. Fan drew the figure below on the board. She asked the class to name the shaded fraction. Charlie answered $\frac{3}{4}$. Janice answered $\frac{3}{2}$. Jenna thinks they're both right. With whom do you agree? Explain your thinking.

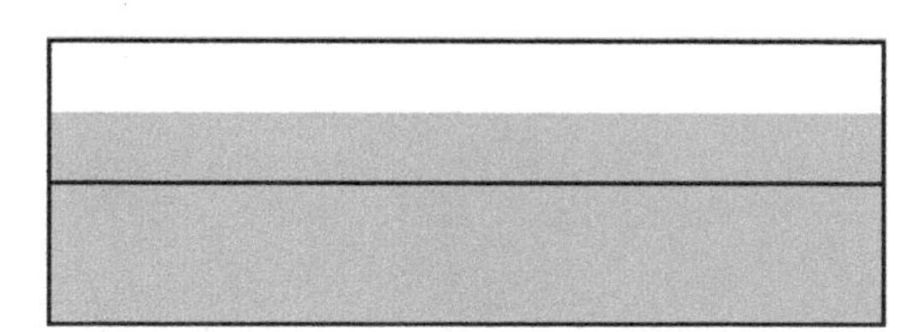

Name ______________________________ Date ________________

Ms. Silverstein asked the class to draw a model showing $\frac{2}{3}$ shaded. Karol and Deb drew the models below. Whose model is correct? Explain how you know.

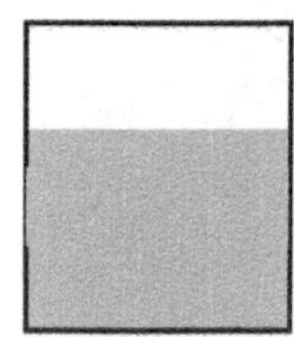

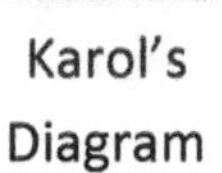

Karol's Diagram

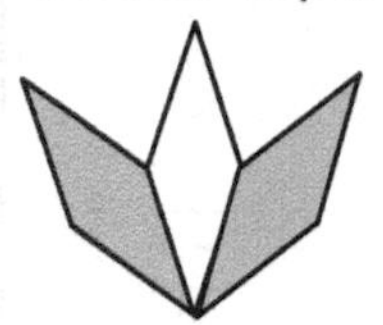

Deb's Diagram

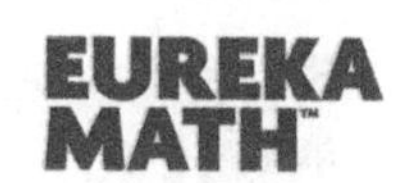

Name ______________________________ Date ____________

The shape represents 1 whole. Write a fraction to describe the shaded part.	The shaded part represents 1 whole. Divide 1 whole to show the same unit fraction you wrote in Part (a).
1. a.	b.
2. a.	b.
3. a.	b.
4. a.	b.

5. Use the pictures below to complete the following statements.

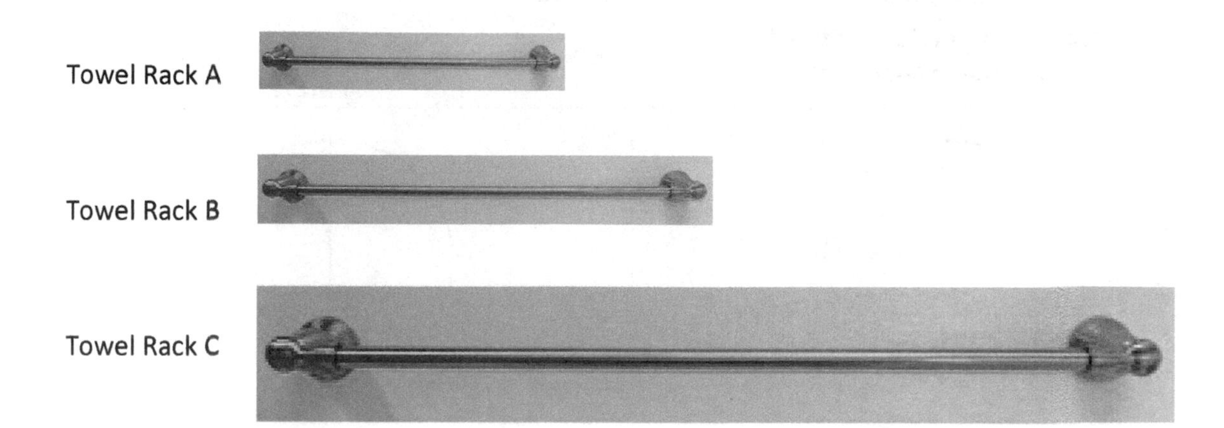

a. Towel Rack ____________ is about $\frac{1}{2}$ the length of Towel Rack C.

b. Towel Rack ____________ is about $\frac{1}{3}$ the length of Towel Rack C.

c. If Towel Rack C measures 6 ft long, then Towel Rack B is about ____________ ft long, and Towel Rack A is about ____________ ft long.

d. About how many copies of Towel Rack A equal the length of Towel Rack C? Draw number bonds to help you.

e. About how many copies of Towel Rack B equal the length of Towel Rack C? Draw number bonds to help you.

6. Draw 3 strings—B, C, and D—by following the directions below. String A is already drawn for you.

   - String B is $\frac{1}{3}$ of String A.
   - String C is $\frac{1}{2}$ of String B.
   - String D is $\frac{1}{3}$ of String C.

Extension: String E is 5 times the length of String D.

String A

Name ______________________________ Date ______________

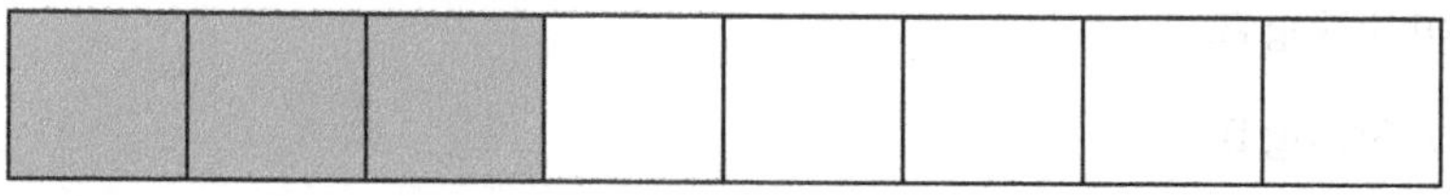

1. Natalie folded 1 whole fraction strip as pictured above.

   a. How many equal parts did she divide the whole into?

   b. Label each equal part with a unit fraction.

   c. Identify the fraction of the strip she shaded.

   d. Identify the fraction of the strip she did not shade.

2. Draw 2 rectangles the same size. Each rectangle represents 1 whole.

   a. Partition each rectangle into 3 equal parts. Shade and label a fraction greater than 1.

   b. Draw a number bond that shows 1 whole rectangle as 3 unit fractions.

3. The bakery had a chocolate cake and a vanilla cake that were exactly the same size. Mr. Chu bought 1 fourth of the chocolate cake. Mrs. Ramirez bought 1 sixth of the vanilla cake. Who bought a larger piece of cake? Explain your answer using words, pictures, and numbers.

4. Natalie explained, "My drawing shows a picture of $\frac{3}{2}$." Kosmo says, "It looks like a picture of $\frac{3}{4}$ to me."

   a. Show and explain how they could both be correct by choosing different wholes. Use words, pictures, and numbers.

   b. Natalie said to Kosmo, "One part can represent either 1 half or 1 fourth. That must mean $\frac{1}{2} = \frac{1}{4}$." Do you agree with Natalie? Use words, pictures, and numbers to explain your reasoning.

Mid-Module Assessment Task Standards Addressed		Topics A–C
Develop understanding of fractions as numbers.		
3.NF.1	Understand a fraction 1/*b* as the quantity formed by 1 part when a whole is partitioned into *b* equal parts; understand a fraction *a*/*b* as the quantity formed by *a* parts of size 1/*b*.	
3.NF.3	Explain equivalence of fractions in special cases, and compare fractions by reasoning about their size. c. Express whole numbers as fractions, and recognize fractions that are equivalent to whole numbers. *Examples: Express 3 in the form 3 = 3/1; recognize that 6/1 = 6; locate 4/4 and 1 at the same point of a number line.* d. Compare two fractions with the same numerator or the same denominator by reasoning about their size. Recognize that comparisons are valid only when the two fractions refer to the same whole. Record the results of comparisons with the symbols >, =, or <, and justify the conclusions, e.g., by using a visual fraction model.	
Reason with shapes and their attributes.		
3.G.2	Partition shapes into parts with equal areas. Express the area of each part as a unit fraction of the whole. *For example, partition a shape into 4 parts with equal area, and describe the area of each part as 1/4 of the area of the shape.*	

Evaluating Student Learning Outcomes

A Progression Toward Mastery is provided to describe steps that illuminate the gradually increasing understandings that students develop *on their way to proficiency.* In this chart, this progress is presented from left (Step 1) to right (Step 4). The learning goal for students is to achieve Step 4 mastery. These steps are meant to help teachers and students identify and celebrate what the students CAN do now and what they need to work on next.

A Progression Toward Mastery				
Assessment Task Item and Standards Assessed	**STEP 1 Little evidence of reasoning without a correct answer. (1 Point)**	**STEP 2 Evidence of some reasoning without a correct answer. (2 Points)**	**STEP 3 Evidence of some reasoning with a correct answer or evidence of solid reasoning with an incorrect answer. (3 Points)**	**STEP 4 Evidence of solid reasoning with a correct answer. (4 Points)**
1 **3.NF.1**	The student has one answer correct.	The student has two answers correct.	The student answers Parts (b) through (d) correctly but answers Part (a) with a fractional answer or has answered one of the four questions incorrectly or incompletely.	The student correctly: a. Identifies how many parts the whole is divided into—8. b. Labels each unit fraction as $\frac{1}{8}$. c. Identifies the fraction shaded—$\frac{3}{8}$. d. Identifies the fraction unshaded—$\frac{5}{8}$.
2 **3.NF.3c** **3.G.2**	The student is unable to answer either question correctly.	The student is unable to shade a fraction greater than 1 but answers Part (b) correctly.	The student answers Part (a) correctly but does not seem to understand Part (b).	The student correctly: a. Shows two rectangles divided into thirds with a fraction greater than $\frac{3}{3}$ shaded. b. Writes a number bond with the whole as 1 or $\frac{3}{3}$ and $\frac{1}{3}$, $\frac{1}{3}$, and $\frac{1}{3}$ as the parts.
3 **3.NF.3d** **3.G.2**	The student's work shows no evidence of being able to partition the cakes into fractional units to make sense of the problem.	The student has poorly represented the cakes, making it difficult to compare the fractions. The student incorrectly states that Mrs. Ramirez bought the larger piece.	The student draws two equivalent cakes and realizes Mr. Chu has the larger piece, but the explanation is not clear, perhaps poorly labeled, lacking a statement of the solution.	The student clearly: Explains that Mr. Chu bought the larger piece of cake using words, pictures, and numbers.

A Progression Toward Mastery				
4 **3.NF.1** **3.NF.3d** **3.G.2**	The student is unable to recognize or show that he recognizes either fraction in the model.	The student recognizes 3 fourths but is unable to recognize 3 halves within the picture or vice versa.	The student is able to recognize 3 fourths and 3 halves within the same picture, which is clear perhaps by markings on the strip, but the explanation lacks clarity.	The student clearly: a. Uses words, pictures, and numbers to explain how the picture can be interpreted either as 4 halves with $\frac{3}{2}$ shaded, with the whole being defined by the middle line of the strip, *or* as 4 fourths with $\frac{3}{4}$ shaded, with the whole being defined by the entire strip. b. Uses word, pictures, and numbers to explain that Natalie is not correct because the whole is different for each fractional unit.

Name Gina Date ____________

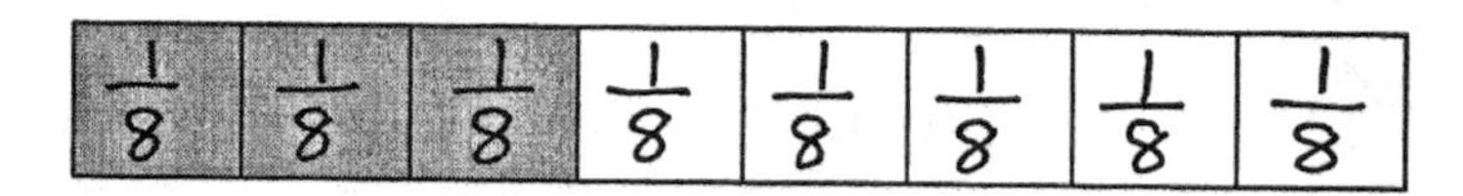

1. Natalie folded 1 whole fraction strip as pictured above.

   a. How many equal parts did she divide the whole into? 8 equal parts

   b. Label each equal part with a unit fraction.

   c. Identify the fraction of the strip she shaded. $\frac{3}{8}$

   d. Identify the fraction of the strip she did not shade. $\frac{5}{8}$

2. Draw 2 rectangles the same size. Each rectangle represents 1 whole.

   a. Partition each rectangle into 3 equal parts. Shade and label a fraction greater than 1.

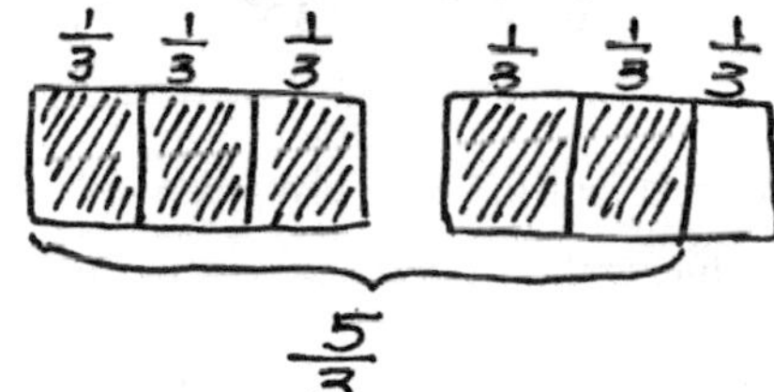

   b. Draw a number bond that shows 1 whole rectangle as 3 unit fractions.

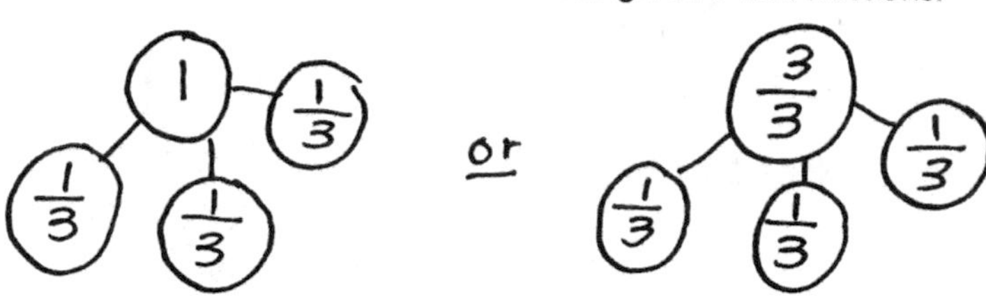

3. The bakery had a chocolate cake and a vanilla cake that were exactly the same size. Mr. Chu bought 1 fourth of the chocolate cake. Mrs. Ramirez bought 1 sixth of the vanilla cake. Who bought a larger piece of cake? Explain your answer using words, pictures, and numbers.

Mr. Chu bought a larger piece of cake because $\frac{1}{4} > \frac{1}{6}$. Fourths have fewer equal parts, so each piece is bigger.

4. Natalie explained, "My drawing shows a picture of $\frac{3}{2}$." Kosmo says, "It looks like a picture of $\frac{3}{4}$ to me."

   a. Show and explain how they could both be correct by choosing different wholes. Use words, pictures, and numbers.

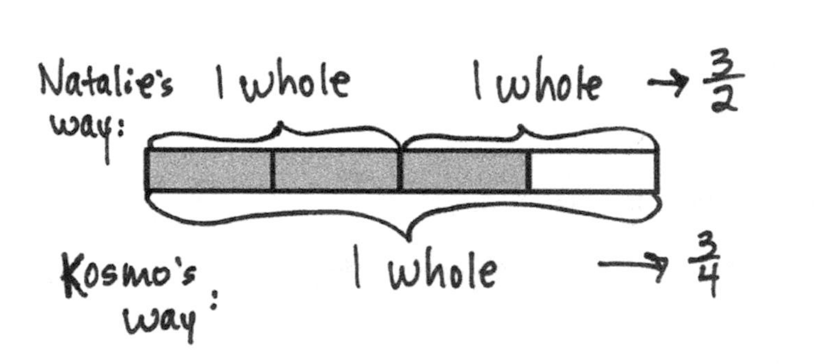

They can both be right. It depends on the whole and they don't know what it is.

   b. Natalie said to Kosmo, "One part can represent either 1 half or 1 fourth. That must mean $\frac{1}{2} = \frac{1}{4}$." Do you agree with Natalie? Use words, pictures, and numbers to explain your reasoning.

Natalie is wrong because the wholes are not the same size. The wholes have to be the same size to compare fractions.

Natalie's whole:

$\frac{1}{2}$	$\frac{1}{2}$

Kosmo's whole:

$\frac{1}{4}$	$\frac{1}{4}$	$\frac{1}{4}$	$\frac{1}{4}$

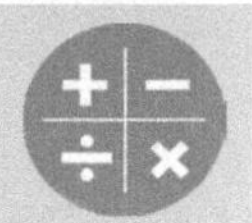

Topic D

Fractions on the Number Line

3.NF.2ab, 3.NF.3cd

Focus Standards:		3.NF.2	Understand a fraction as a number on the number line; represent fractions on a number line diagram. a. Represent a fraction 1/*b* on a number line diagram by defining the interval from 0 to 1 as the whole and partitioning it into *b* equal parts. Recognize that each part has size 1/*b* and that the endpoint of the part based at 0 locates the number 1/*b* on the number line. b. Represent a fraction *a*/*b* on a number line diagram by marking off *a* lengths 1/*b* from 0. Recognize that the resulting interval has size *a*/*b* and that its endpoint locates the number *a*/*b* on the number line.
		3.NF.3	Explain equivalence of fractions in special cases, and compare fractions by reasoning about their size. c. Express whole numbers as fractions, and recognize fractions that are equivalent to whole numbers. *Examples: Express 3 in the form 3 = 3/1; recognize that 6/1 = 6; locate 4/4 and 1 at the same point of a number line diagram.* d. Compare two fractions with the same numerator or the same denominator by reasoning about their size. Recognize that comparisons are valid only when the two fractions refer to the same whole. Record the results of comparisons with the symbols >, =, <, and justify the conclusions, e.g., by using a visual fraction model.
Instructional Days:		6	
Coherence	**-Links from:**	G2–M8	Time, Shapes, and Fractions as Equal Parts of Shapes
	-Links to:	G4–M5	Fraction Equivalence, Ordering, and Operations

In Topic C, students compared unit fractions and explored the importance of specifying the whole when doing so. In Topic D, they apply their learning to the number line. Number bonds and fraction strips serve as bridges into this work. Students see intervals on the number line as wholes. They initially measure equal lengths between 0 and 1 with their fraction strips. They then work with number lines that have endpoints other than 0 and 1 or include multiple whole number intervals. This naturally transitions into comparing fractions with the same denominator, as well as fractional numbers and whole numbers on the number line. As students compare, they reason about the size of fractions and contextualize their learning within real-world applications.

A Teaching Sequence Toward Mastery of Fractions on the Number Line

Objective 1: Place fractions on a number line with endpoints 0 and 1.
(Lesson 14)

Objective 2: Place any fraction on a number line with endpoints 0 and 1.
(Lesson 15)

Objective 3: Place whole number fractions and fractions between whole numbers on the number line.
(Lesson 16)

Objective 4: Practice placing various fractions on the number line.
(Lesson 17)

Objective 5: Compare fractions and whole numbers on the number line by reasoning about their distance from 0.
(Lesson 18)

Objective 6: Understand distance and position on the number line as strategies for comparing fractions.
(Optional)
(Lesson 19)

Lesson 14

Objective: Place fractions on a number line with endpoints 0 and 1.

Suggested Lesson Structure

- Fluency Practice (12 minutes)
- Application Problem (7 minutes)
- Concept Development (33 minutes)
- Student Debrief (8 minutes)

Total Time **(60 minutes)**

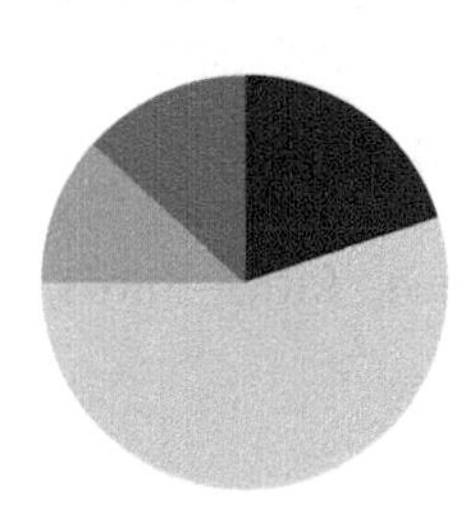

Fluency Practice (12 minutes)

- Division **3.OA.2** (8 minutes)
- Counting by Fractional Units **3.NF.1, 3.NF.3** (3 minutes)
- Unit Fractions in 1 Whole **3.NF.1** (1 minute)

Division (8 minutes)

Materials: (T) Timer (S) Personal white board or paper

Note: This activity supports fluency with division facts.

T: Write as many different division facts as you can in the next three minutes. Take your mark, get set, go!

S: (Work independently.)

T: (At three minutes.) Share your work with your partner. Check to see if your partner's problems are correct.

S: (Work with a partner.)

T: Try again for three minutes. Take your mark, get set, go!

S: (Work independently.)

T: (At three minutes.) Check your work with your partner. Tell your partner what division facts are easy for you.

S: (Work with a partner.)

T: Who improved? How did you improve? What helped you do more problems correctly?

Counting by Fractional Units (3 minutes)

Note: This activity reviews counting by fractional units and supports students as they work with fractions on the number line in Topic D.

T: Count by eighths from 1 eighth to 8 eighths and back to 0.

S: $\frac{1}{8}, \frac{2}{8}, \frac{3}{8}, \frac{4}{8}, \frac{5}{8}, \frac{6}{8}, \frac{7}{8}, \frac{8}{8}, \frac{7}{8}, \frac{6}{8}, \frac{5}{8}, \frac{4}{8}, \frac{3}{8}, \frac{2}{8}, \frac{1}{8}$, 0.

Continue with the following possible sequence: fifths, thirds, and fourths.

NOTES ON MULTIPLE MEANS OF ENGAGEMENT:

- Change directions so that the sequence stays unpredictable.
- React to misunderstandings by repeating transitions until mastery.
- Support by recording on a number line as students count.
- Extend by having students say "1" or "1 whole" instead of a fraction (e.g., "..., 6 eighths, 7 eighths, 1, 7 eighths, 6 eighths, ...").

Unit Fractions in 1 Whole (1 minute)

Note: This activity reviews how many unit fractions are in 1 whole, which is a skill that the students use during the Concept Development.

T: I'll say a unit fraction. You say how many there are in 1 whole. 1 fifth.

S: 5. It takes 5 copies of 1 fifth to make 1 whole.

Continue with the following possible sequence: 1 tenth, 1 fourth, 1 third, 1 eighth, and 1 half.

Application Problem (7 minutes)

Mr. Ray is knitting a scarf. He says that he has completed 1 fifth of the total length of the scarf.

Draw a picture of the final scarf. Label what he has finished and what he still has to make. Draw a number bond with 2 parts to show the fraction he has made and the fraction he has not made.

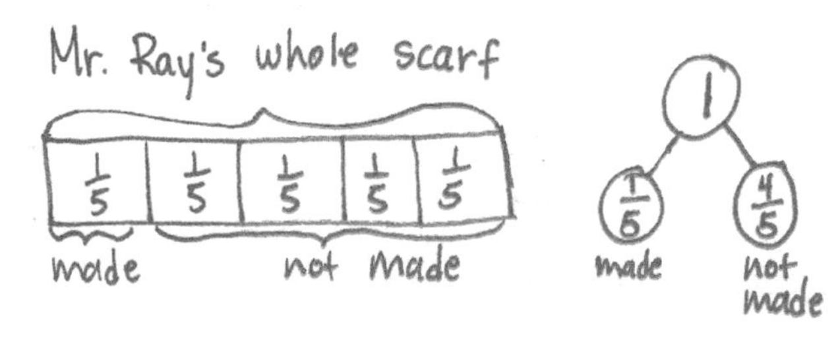

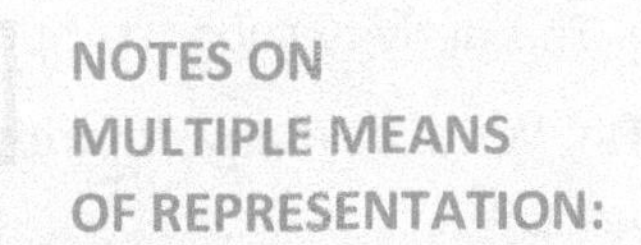

NOTES ON MULTIPLE MEANS OF REPRESENTATION:

Empower English language learners to solve word problems by activating prior knowledge. Guide students to make personal connections. Discuss their own experiences with knitting and scarves.

Note: This problem reviews the concept from Lesson 12 of representing the whole when given one equal part.

Concept Development (33 minutes)

Materials: (T) Board space, yardstick, large fraction strip for modeling (S) Fraction strips, blank paper, ruler

MP.7

Part 1: Measure a line of length 1 whole.

T: (Model the steps below as students follow along on their personal white boards.)

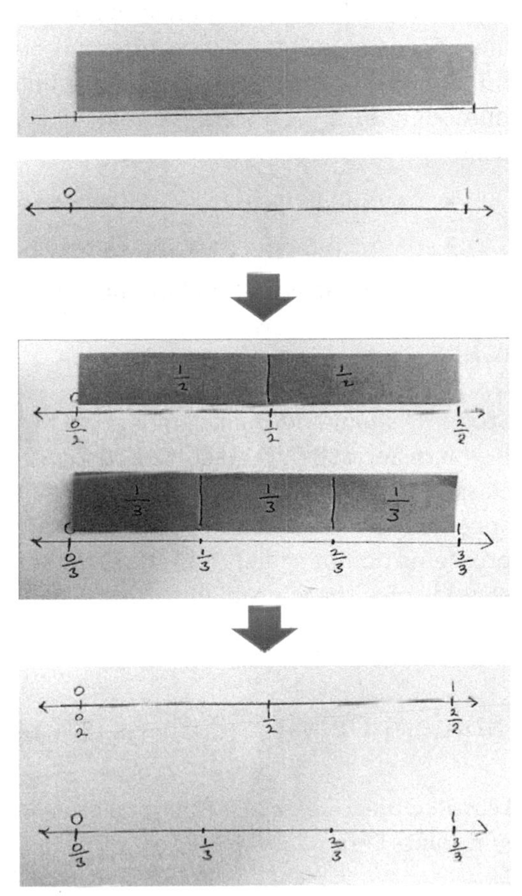

1. Draw a horizontal line with your ruler that is a bit longer than one of your fraction strips.
2. Place a whole fraction strip just above the line you drew.
3. Make a small mark on your line that is even with the left end of your strip.
4. Label that mark 0 above the line. This is where we start measuring the length of the strip.
5. Make a small mark on your line that is even with the right end of your strip.
6. Label that mark 1 above the line. If we start at 0, the 1 tells us when we've travelled 1 whole length of the strip.

Part 2: Measure the fractions.

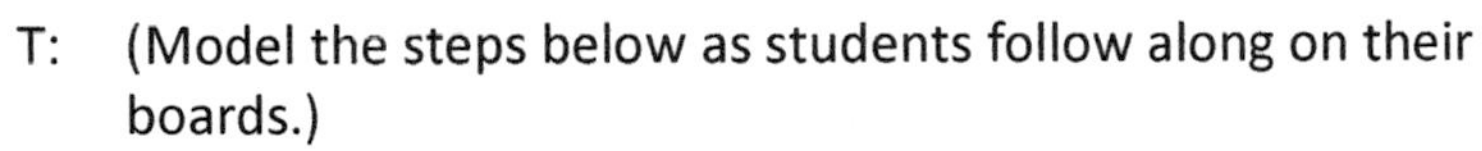

T: (Model the steps below as students follow along on their boards.)

1. Place your fraction strip with halves above the line.
2. Make a mark on the number line at the right end of 1 half. This is the length of 1 half of the fraction strip.
3. Label that mark $\frac{1}{2}$. Label 0 halves and 2 halves.
4. Repeat the process to measure and label other fractional numbers on a number line.

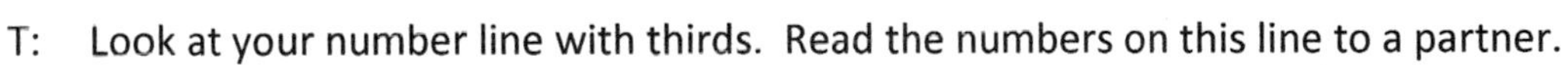

T: Look at your number line with thirds. Read the numbers on this line to a partner.

S: 0, 1. → I think it's 0, $\frac{1}{3}$, $\frac{2}{3}$, 1. → What about $\frac{0}{3}$, $\frac{1}{3}$, $\frac{2}{3}$, $\frac{3}{3}$? → Are fractions numbers?

T: Some of you read the whole numbers, and others read whole numbers and fractions. Fractions are numbers. Let's read the numbers from least to greatest, and let's say 0 thirds and 3 thirds for now rather than zero and one.

S: (Read numbers, $\frac{0}{3}$, $\frac{1}{3}$, $\frac{2}{3}$, $\frac{3}{3}$.)

T: Let's read again and this time say zero and 1 rather than 0 thirds and 3 thirds.

S: (Read numbers, 0, $\frac{1}{3}$, $\frac{2}{3}$, 1 .)

Part 3: Draw number bonds to correspond with the number lines.

Once students have become excellent at making and labeling fractions on number lines using strips to measure, have them draw number bonds to correspond. Use questioning while circulating to help them see similarities and differences between the bonds, fraction strips, and fractions on the number line. Guide students to recognize that placing fractions on the number line is analogous to placing whole numbers on the number line. If preferred, the following suggestions can be used:

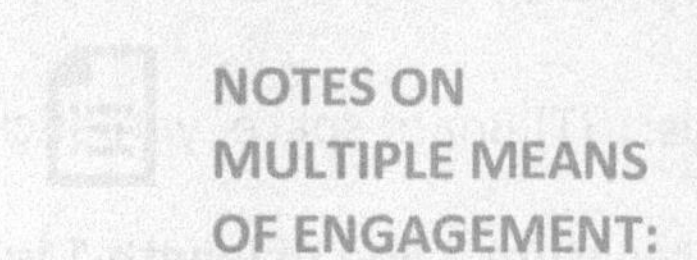

NOTES ON MULTIPLE MEANS OF ENGAGEMENT:

This lesson gradually leads students from the concrete level (fraction strips) to the pictorial level (number lines).

- What do both the number bond and number line show?
- Which model best shows how big the unit fraction is in relation to the whole? Explain how.
- How do your number lines help you make number bonds?

Problem Set (10 minutes)

Students should do their personal best to complete the Problem Set within the allotted 10 minutes. For some classes, it may be appropriate to modify the assignment by specifying which problems they work on first. Some problems do not specify a method for solving. Students should solve these problems using the RDW approach used for Application Problems.

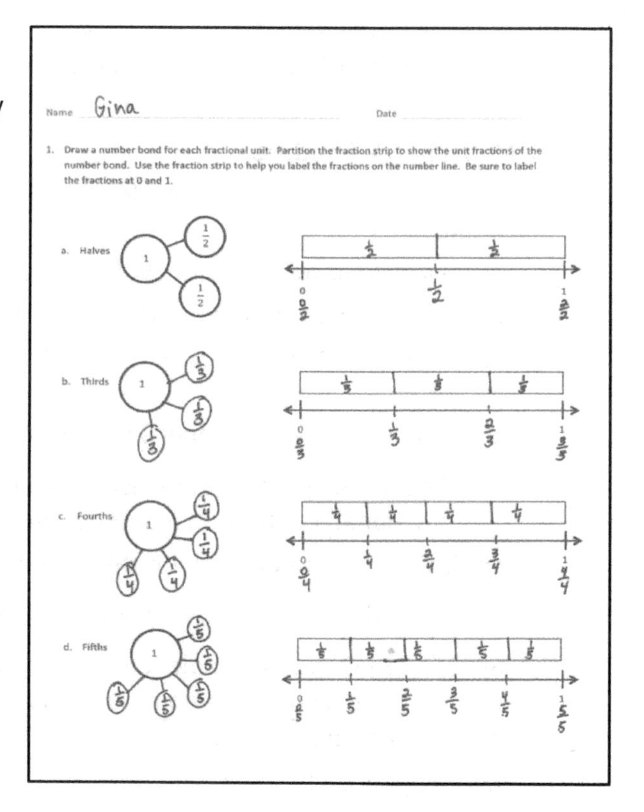

Name Gina Date

1. Draw a number bond for each fractional unit. Partition the fraction strip to show the unit fractions of the number bond. Use the fraction strip to help you label the fractions on the number line. Be sure to label the fractions at 0 and 1.

a. Halves

b. Thirds

c. Fourths

d. Fifths

Student Debrief (8 minutes)

Lesson Objective: Place fractions on a number line with endpoints 0 and 1.

The Student Debrief is intended to invite reflection and active processing of the total lesson experience.

Invite students to review their solutions for the Problem Set. They should check work by comparing answers with a partner before going over answers as a class. Look for misconceptions or misunderstandings that can be addressed in the Student Debrief. Guide students in a conversation to debrief the Problem Set and process the lesson.

Any combination of the questions below may be used to lead the discussion.

- Look at the number line you made for Problem 3. What does each point on the number line mean? (The following response is possible: "$\frac{1}{5}$ marks the distance from 0—the end of the ribbon—to where Mrs. Lee sews on the first bead." "It tells us what number that point represents.")
- In Problem 2, the point is a point in time, not the whole length. In Problem 3, the point indicates the location of a bead. Let students have fun with the difference between these two problems. The puppy is in one location, which is like the mark on the line. The ribbon is the entire length. If preferred, the following suggestion can be used to guide the discussion:
  - Think about the units of measure in Problems 2 and 3. How are they the same? How are they different?
- What unit do we use to make intervals when we measure and mark 2 inches on a number line? (Inches.) How many times we do we mark off 1 inch to get to 2 inches? (2 times.) What unit do we use to make intervals when we measure and mark 2 hours? (Hours.) How many times do we mark off 1 hour to get to 2 hours? (2 times.) What unit do we use to make intervals when we measure and mark 2 halves? (Halves.) How many times do we mark off 1 half to get to 2 halves? (2 times.) Fractions are numbers that are measured and marked on the number line the same way as whole numbers. They're just another type of unit.
- Describe the process for labeling fractions on the number line.
- Why is the fraction strip an important tool to use when labeling fractions on a number line?
- What does the fraction strip help you measure?

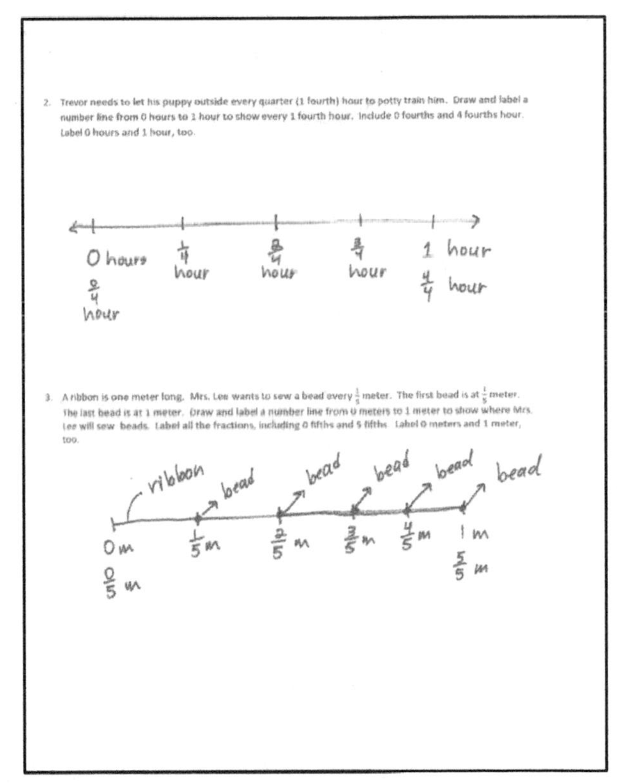
2. Trevor needs to let his puppy outside every quarter (1 fourth) hour to potty train him. Draw and label a number line from 0 hours to 1 hour to show every 1 fourth hour. Include 0 fourths and 4 fourths hour. Label 0 hours and 1 hour, too.

3. A ribbon is one meter long. Mrs. Lee wants to sew a bead every $\frac{1}{5}$ meter. The first bead is at $\frac{1}{5}$ meter. The last bead is at 1 meter. Draw and label a number line from 0 meters to 1 meter to show where Mrs. Lee will sew beads. Label all the fractions, including 0 fifths and 5 fifths. Label 0 meters and 1 meter, too.

Exit Ticket (3 minutes)

After the Student Debrief, instruct students to complete the Exit Ticket. A review of their work will help with assessing students' understanding of the concepts that were presented in today's lesson and planning more effectively for future lessons. The questions may be read aloud to the students.

Name ______________________________ Date ______________

1. Draw a number bond for each fractional unit. Partition the fraction strip to show the unit fractions of the number bond. Use the fraction strip to help you label the fractions on the number line. Be sure to label the fractions at 0 and 1.

 a. Halves

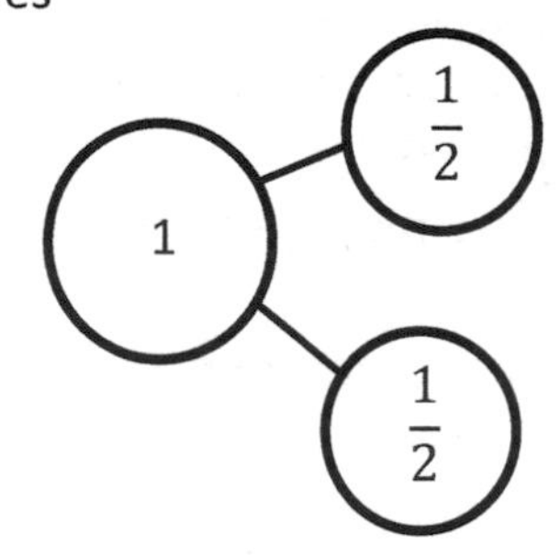

 b. Thirds

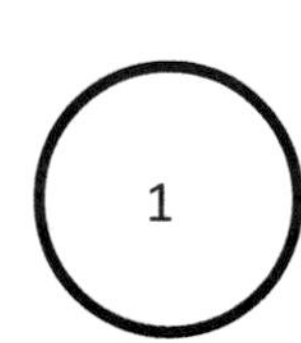

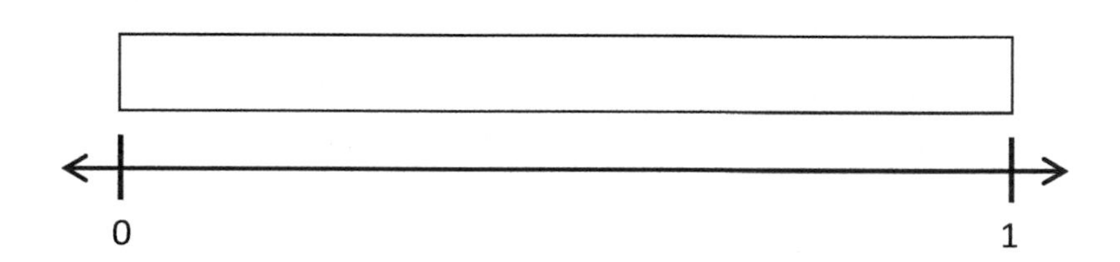

 c. Fourths

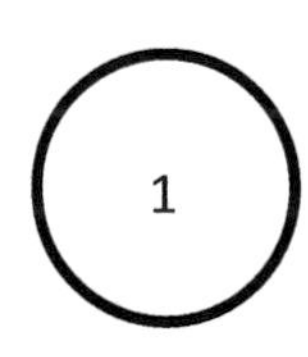

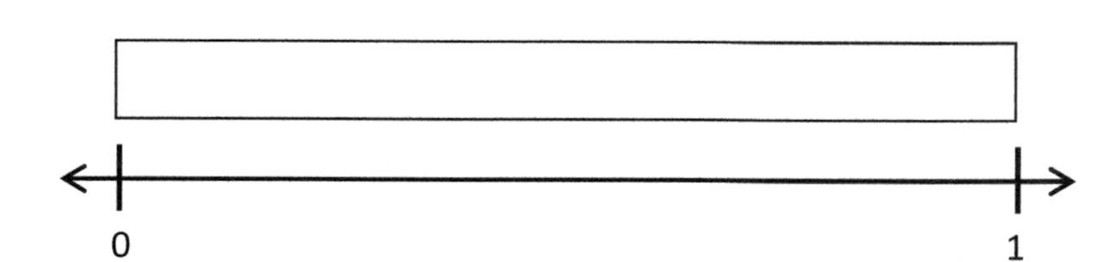

 d. Fifths

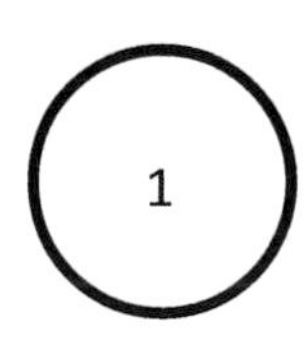

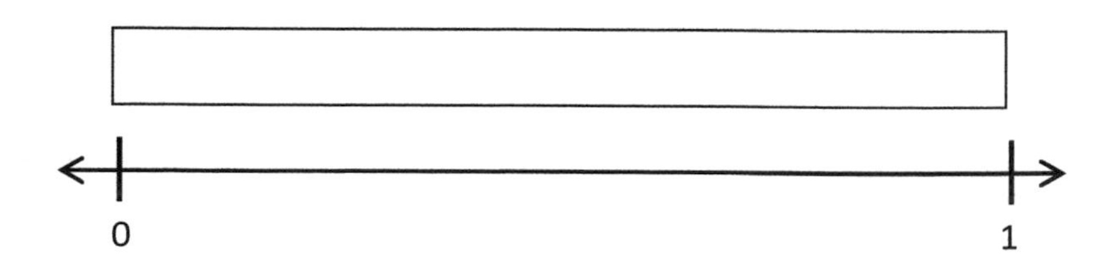

2. Trevor needs to let his puppy outside every quarter (1 fourth) hour to potty train him. Draw and label a number line from 0 hours to 1 hour to show every 1 fourth hour. Include 0 fourths and 4 fourths hour. Label 0 hours and 1 hour, too.

3. A ribbon is 1 meter long. Mrs. Lee wants to sew a bead every $\frac{1}{5}$ meter. The first bead is at $\frac{1}{5}$ meter. The last bead is at 1 meter. Draw and label a number line from 0 meters to 1 meter to show where Mrs. Lee will sew beads. Label all the fractions, including 0 fifths and 5 fifths. Label 0 meters and 1 meter, too.

Name ______________________________ Date ______________

1. Draw a number bond for the fractional unit. Partition the fraction strip, and draw and label the fractions on the number line. Be sure to label the fractions at 0 and 1.

Sixths (1)

0 1

2. Ms. Metcalf wants to share $1 equally among 5 students. Draw a number bond and a number line to help explain your answer.

a. What fraction of a dollar will each student get?

b. How much money will each student get?

Name ______________________________ Date ________________

1. Draw a number bond for each fractional unit. Partition the fraction strip to show the unit fractions of the number bond. Use the fraction strip to help you label the fractions on the number line. Be sure to label the fractions at 0 and 1.

   a. Halves

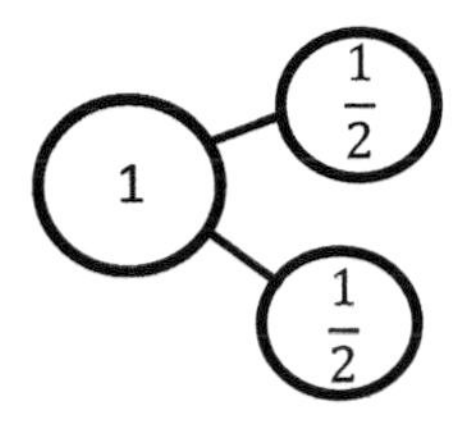

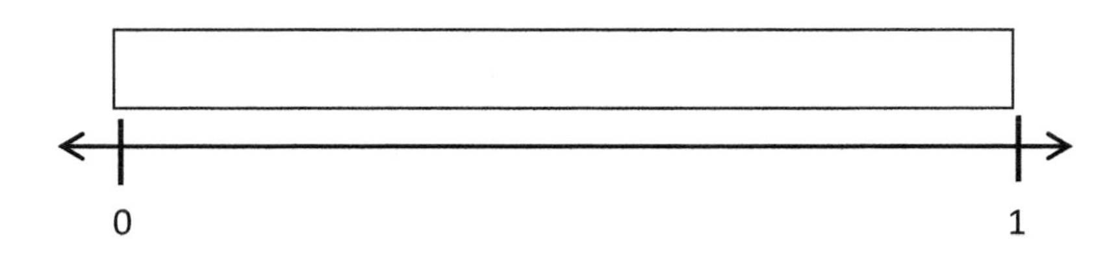

   b. Eighths

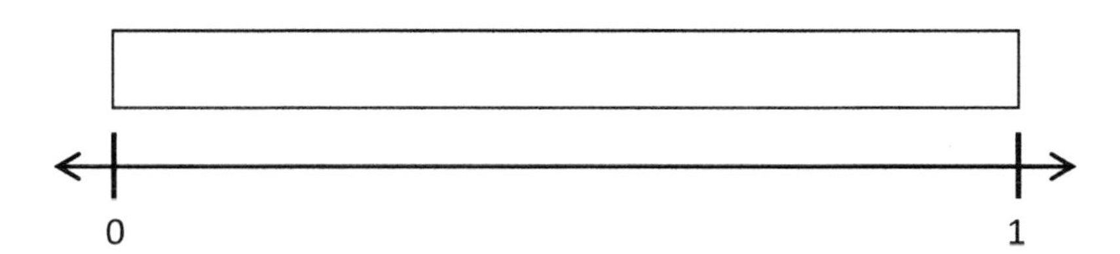

   c. Fifths

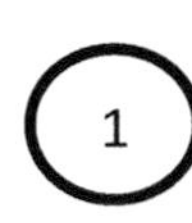

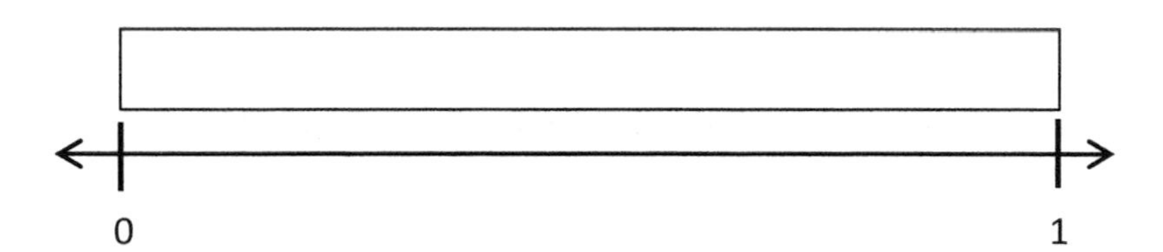

2. Carter needs to wrap 7 presents. He lays the ribbon out flat and says, "If I make 6 equally spaced cuts, I'll have just enough pieces. I can use 1 piece for each package, and I won't have any pieces left over." Does he have enough pieces to wrap all the presents?

3. Mrs. Rivera is planting flowers in her 1-meter long rectangular plant box. She divides the plant box into sections $\frac{1}{9}$ meter in length, and plants 1 seed in each section. Draw and label a fraction strip representing the plant box from 0 meters to 1 meter. Represent each section where Mrs. Rivera will plant a seed. Label all the fractions.

   a. How many seeds will she be able to plant in 1 plant box?

   b. How many seeds will she be able to plant in 4 plant boxes?

   c. Draw a number line below your fraction strip and mark all the fractions.

Lesson 15

Objective: Place any fraction on a number line with endpoints 0 and 1.

Suggested Lesson Structure

- Fluency Practice (9 minutes)
- Application Problem (7 minutes)
- Concept Development (35 minutes)
- Student Debrief (9 minutes)

Total Time **(60 minutes)**

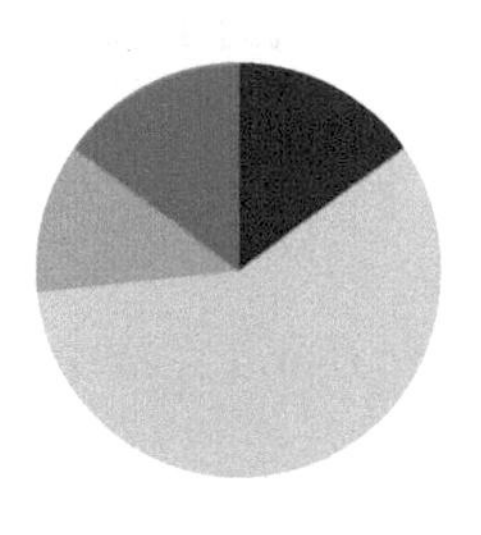

Fluency Practice (9 minutes)

- Counting by Fractional Units **3.NF.1, 3.NF.3c** (3 minutes)
- Division **3.OA.2** (3 minutes)
- Place Unit Fractions on a Number Line Between 0 and 1 **3.NF.2a** (3 minutes)

Counting by Fractional Units (3 minutes)

Note: This activity reviews counting by fractional units and supports students as they work with fractions on the number line in Topic D.

T: Count by fourths from 1 fourth to 8 fourths and back to 0.

S: $\frac{1}{4}, \frac{2}{4}, \frac{3}{4}, \frac{4}{4}, \frac{5}{4}, \frac{6}{4}, \frac{7}{4}, \frac{8}{4}, \frac{7}{4}, \frac{6}{4}, \frac{5}{4}, \frac{4}{4}, \frac{3}{4}, \frac{2}{4}, \frac{1}{4}, 0.$

Continue with the following possible sequence: thirds, halves, and fifths.

Division (3 minutes)

Note: This activity supports fluency with division facts.

T: (Write 4 ÷ 4 = _____.) Say the number sentence and answer.

S: 4 divided by 4 equals 1.

Continue with the following possible sequence: 4 ÷ 2, 4 ÷ 1, 10 ÷ 10, 10 ÷ 5, 10 ÷ 2, 10 ÷ 1, 6 ÷ 6, 6 ÷ 3, 6 ÷ 1, 8 ÷ 8, 8 ÷ 4, 8 ÷ 2, 8 ÷ 1, 15 ÷ 15, 15 ÷ 5, 15 ÷ 3, 15 ÷ 1, 12 ÷ 12, 12 ÷ 6, 12 ÷ 4, 12 ÷ 3, 12 ÷ 2, 12 ÷ 1, 16 ÷ 16, 16 ÷ 8, 16 ÷ 4, 16 ÷ 2, and 16 ÷ 1.

Place Unit Fractions on a Number Line Between 0 and 1 (3 minutes)

Materials: (S) Personal white board

Note: This activity reviews the concept of placing unit fractions on a number line from Lesson 14.

> **NOTES ON MULTIPLE MEANS OF REPRESENTATION:**
>
> As students estimate to equally partition fourths and eighths on the number line, guide them to begin by finding the midpoint—first by drawing 2 equal parts and then continuing *halving* until the desired unit fraction is created.

T: (Draw a number line with endpoints 0 and 1.) Draw my number line on your personal white board.

S: (Draw.)

T: Estimate to show and label 1 half.

S: (Estimate the halfway point between 0 and 1, and write $\frac{1}{2}$.)

Continue with the following possible sequence: $\frac{1}{10}$, $\frac{1}{4}$, $\frac{1}{8}$, $\frac{1}{3}$, $\frac{1}{5}$, and $\frac{1}{6}$.

Application Problem (7 minutes)

In baseball, it is about 30 yards from home plate to first base. The batter got tagged out about halfway to first base. About how many yards from home plate was he when he got tagged out? Draw a number line to show the point where he was when he got tagged out.

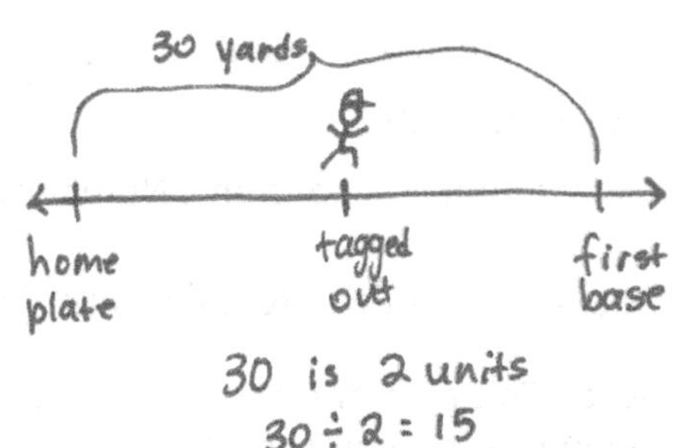

Note: This problem reviews the concept of placing fractions on a number line from Lesson 14. It also reviews division by units of 2. Invite students to share their strategies for dividing 30 by 2.

Concept Development (35 minutes)

Materials: (S) Personal white board

Problem 1: Locate the point 2 thirds on a number line.

T: 2 thirds. How many equal parts are in the whole?

S: Three.

T: How many of those equal parts have been counted?

S: Two.

T: Count up to 2 thirds, starting at 1 third.

S: 1 third, 2 thirds.

T: Draw a 2-part number bond of 1 whole with 1 part as 2 thirds.

S: (Draw a number bond.)

T: What is the unknown part?

S: 1 third.

T: Draw a number line with endpoints of 0 and 1—with 0 thirds and 3 thirds—to match your number bond.

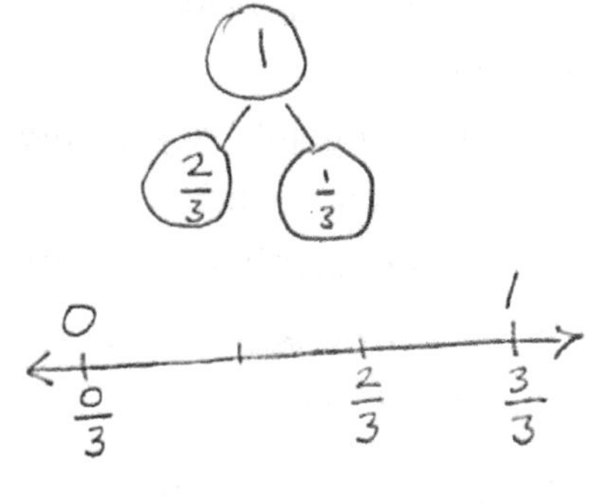

S: (Draw a number line, and label the endpoints.)

T: Mark off your thirds without labeling the fractions.

S: (Mark the thirds.)

T: Slide your finger along the length of the first part of your number bond. Speak the fraction as you do.

S: 2 thirds (sliding up to the point 2 thirds).

T: Label that point as 2 thirds.

S: (Label 2 thirds.)

T: Put your finger back on 2 thirds. Slide and speak the next part.

S: 1 third.

T: At what point are you now?

S: 3 thirds or 1 whole.

T: Our number bond is complete.

Problem 2: Locate the point 3 fifths on a number line.

T: 3 fifths. How many equal parts are in the whole?

S: Five.

T: How many of those equal parts have been counted?

S: Three.

T: Count up to 3 fifths, starting at 1 fifth.

S: 1 fifth, 2 fifths, 3 fifths.

T: Draw a 2-part number bond of 1 whole with 1 part as 3 fifths.

S: (Draw a number bond.)

T: What is the unknown part?

S: 2 fifths.

T: Draw a number line with endpoints of 0 and 1—with 0 fifths and 5 fifths—to match your number bond.

S: (Draw a number line, and label the endpoints.)

T: Mark off your fifths without labeling the fractions.

S: (Mark the fifths.)

T: Slide your finger along the length of the first part of your number. Speak the fraction as you do.

S: 3 fifths (sliding up to the point 3 fifths).

T: Label that point as 3 fifths.
S: (Label 3 fifths.)
T: Put your finger back on 3 fifths. Slide and speak the next part.
S: 2 fifths.
T: At what point are you now?
S: 5 fifths or 1 whole.
T: Our number bond is complete.

Repeat the process with other fractions such as 3 fourths, 6 eighths, 2 sixths, and 1 seventh. Release the students to work independently as they demonstrate their skills and understanding.

Problem Set (10 minutes)

Students should do their personal best to complete the Problem Set within the allotted 10 minutes. For some classes, it may be appropriate to modify the assignment by specifying which problems they work on first. Some problems do not specify a method for solving. Students should solve these problems using the RDW approach used for Application Problems.

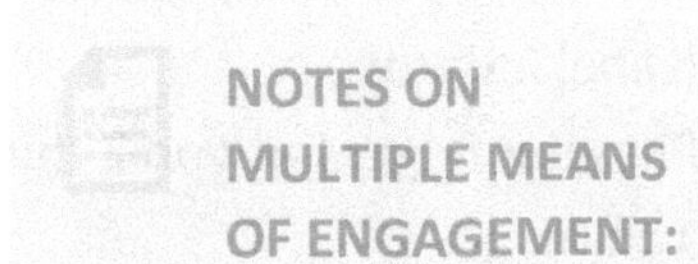

NOTES ON MULTIPLE MEANS OF ENGAGEMENT:

The Problem Set offers practice of increasing difficulty. Expect and coach students working above grade level to complete the entire Problem Set with excellence.

Student Debrief (9 minutes)

Lesson Objective: Place any fraction on a number line with endpoints 0 and 1.

The Student Debrief is intended to invite reflection and active processing of the total lesson experience.

Invite students to review their solutions for the Problem Set. They should check work by comparing answers with a partner before going over answers as a class. Look for misconceptions or misunderstandings that can be addressed in the Student Debrief. Guide students in a conversation to debrief the Problem Set and process the lesson.

Any combination of the questions below may be used to lead the discussion.

- How does the number bond relate to the number line?
- How do the number bond and number line with fractions relate to the number bond and number line with whole numbers?

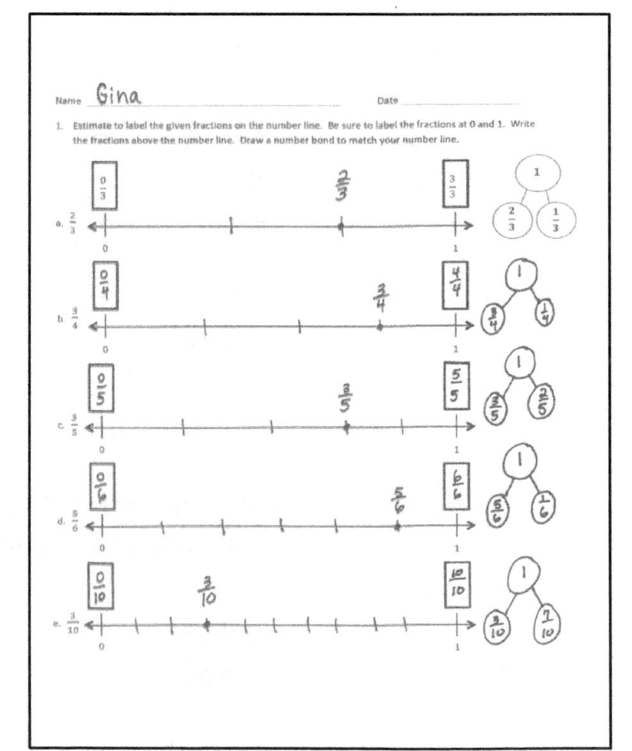

Name Gina Date

1. Estimate to label the given fractions on the number line. Be sure to label the fractions at 0 and 1. Write the fractions above the number line. Draw a number bond to match your number line.

a. $\frac{2}{3}$

b. $\frac{3}{4}$

c. $\frac{3}{5}$

d. $\frac{5}{6}$

e. $\frac{3}{10}$

- Part–part–whole thinking has been in your life since kindergarten. When might a kindergartener draw a number bond? A first grader? Second grader? Third grader?
- When you think of a number bond, do you usually think of chunks of things? To you, does using it with the number line give it a new meaning? It does for me. Now, I see it can also be about distances on a line, too.

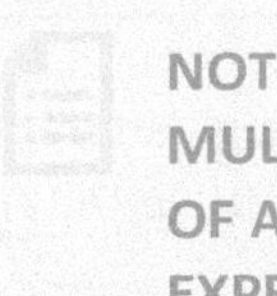

NOTES ON MULTIPLE MEANS OF ACTION AND EXPRESSION:

Facilitate math meaning for English language learners through discussion. The daily debriefs and frequent Turn and Talks in each lesson enhance the English language learners' understanding of math concepts and language, build confidence and comfort, and communicate high expectations for English language learners' participation.

Exit Ticket (3 minutes)

After the Student Debrief, instruct students to complete the Exit Ticket. A review of their work will help with assessing students' understanding of the concepts that were presented in today's lesson and planning more effectively for future lessons. The questions may be read aloud to the students.

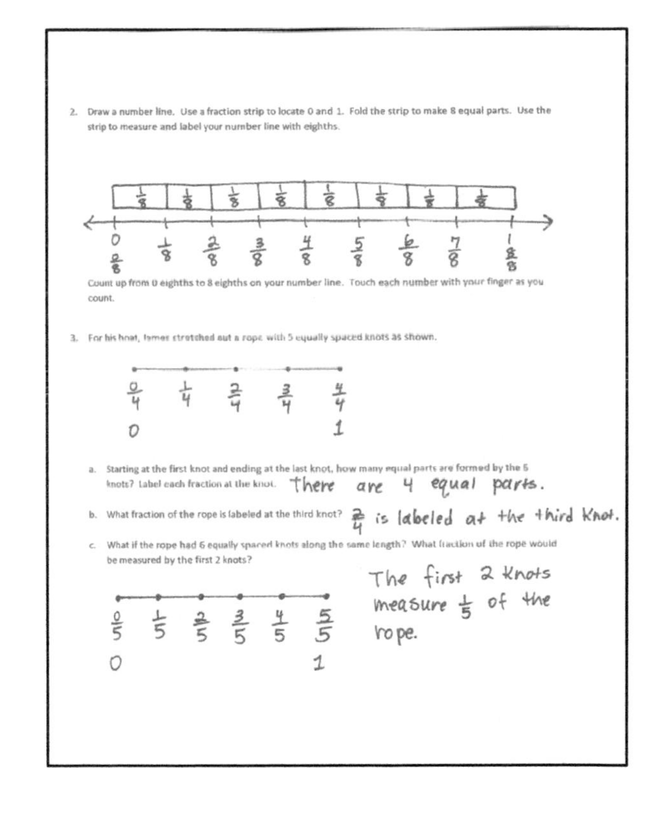

2. Draw a number line. Use a fraction strip to locate 0 and 1. Fold the strip to make 8 equal parts. Use the strip to measure and label your number line with eighths.

$\frac{1}{8}$ $\frac{1}{8}$ $\frac{1}{8}$ $\frac{1}{8}$ $\frac{1}{8}$ $\frac{1}{8}$ $\frac{1}{8}$ $\frac{1}{8}$

0, $\frac{0}{8}$ $\frac{1}{8}$ $\frac{2}{8}$ $\frac{3}{8}$ $\frac{4}{8}$ $\frac{5}{8}$ $\frac{6}{8}$ $\frac{7}{8}$ 1, $\frac{8}{8}$

Count up from 0 eighths to 8 eighths on your number line. Touch each number with your finger as you count.

3. For his boat, James stretched out a rope with 5 equally spaced knots as shown.

$\frac{0}{4}$ $\frac{1}{4}$ $\frac{2}{4}$ $\frac{3}{4}$ $\frac{4}{4}$

0 1

a. Starting at the first knot and ending at the last knot, how many equal parts are formed by the 5 knots? Label each fraction at the knot. There are 4 equal parts.

b. What fraction of the rope is labeled at the third knot? $\frac{2}{4}$ is labeled at the third Knot.

c. What if the rope had 6 equally spaced knots along the same length? What fraction of the rope would be measured by the first 2 knots?

$\frac{0}{5}$ $\frac{1}{5}$ $\frac{2}{5}$ $\frac{3}{5}$ $\frac{4}{5}$ $\frac{5}{5}$

0 1

The first 2 Knots measure $\frac{1}{5}$ of the rope.

Name ______________________________ Date ______________

1. Estimate to label the given fractions on the number line. Be sure to label the fractions at 0 and 1. Write the fractions above the number line. Draw a number bond to match your number line.

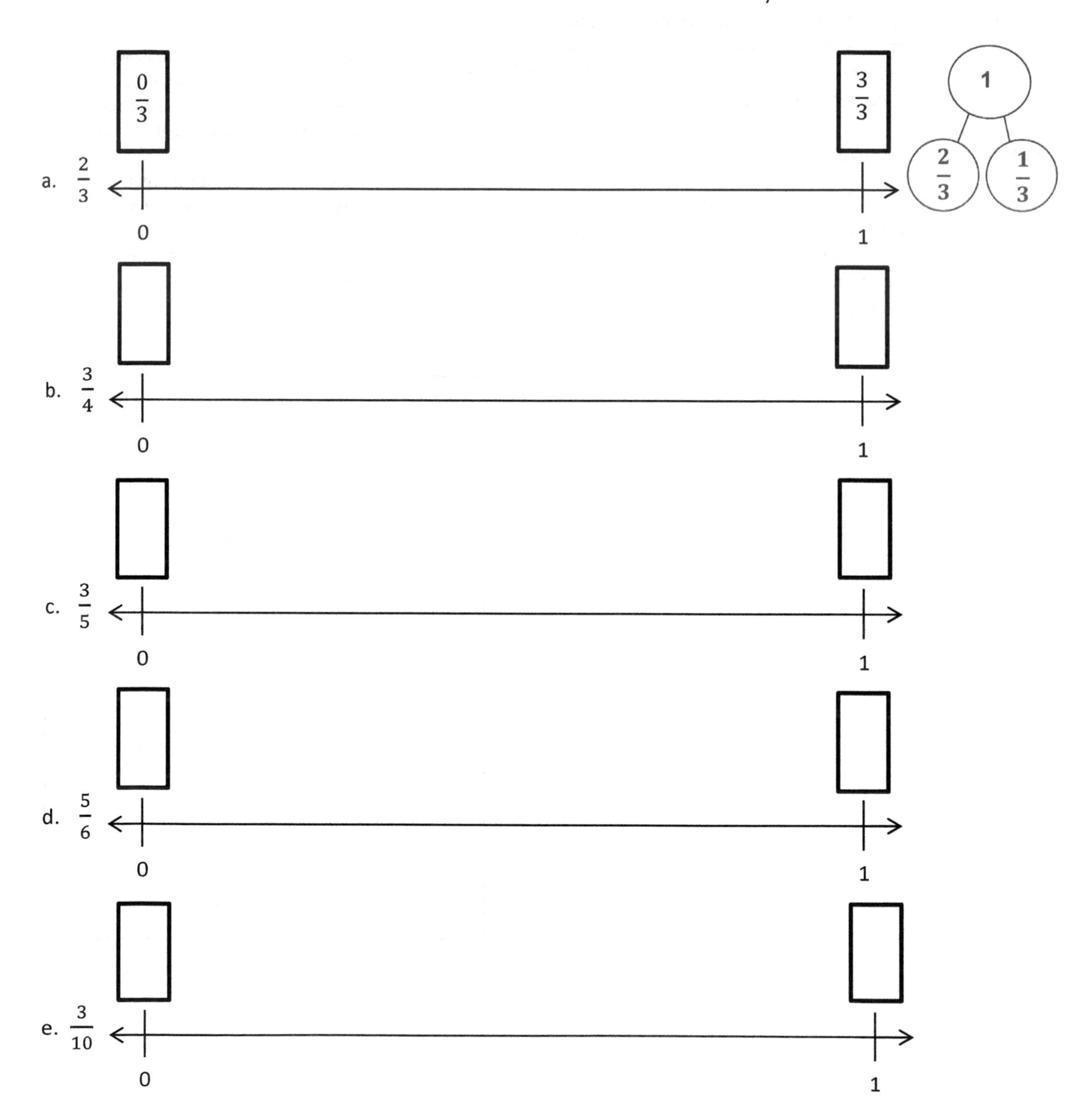

2. Draw a number line. Use a fraction strip to locate 0 and 1. Fold the strip to make 8 equal parts. Use the strip to measure and label your number line with eighths.

   Count up from 0 eighths to 8 eighths on your number line. Touch each number with your finger as you count.

3. For his boat, James stretched out a rope with 5 equally spaced knots as shown.

   a. Starting at the first knot and ending at the last knot, how many equal parts are formed by the 5 knots? Label each fraction at the knot.

   b. What fraction of the rope is labeled at the third knot?

   c. What if the rope had 6 equally spaced knots along the same length? What fraction of the rope would be measured by the first 2 knots?

Name ______________________________ Date ______________

1. Estimate to label the given fraction on the number line. Be sure to label the fractions at 0 and 1. Write the fractions above the number line. Draw a number bond to match your number line.

2. Partition the number line. Then, place each fraction on the number line: $\frac{3}{6}$, $\frac{1}{6}$, and $\frac{5}{6}$.

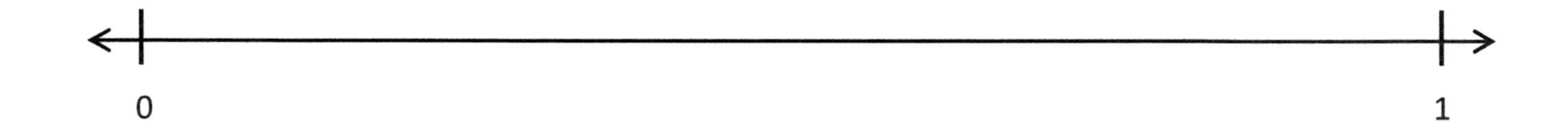

Name ______________________________ Date ____________

1. Estimate to label the given fractions on the number line. Be sure to label the fractions at 0 and 1. Write the fractions above the number line. Draw a number bond to match your number line. The first one is done for you.

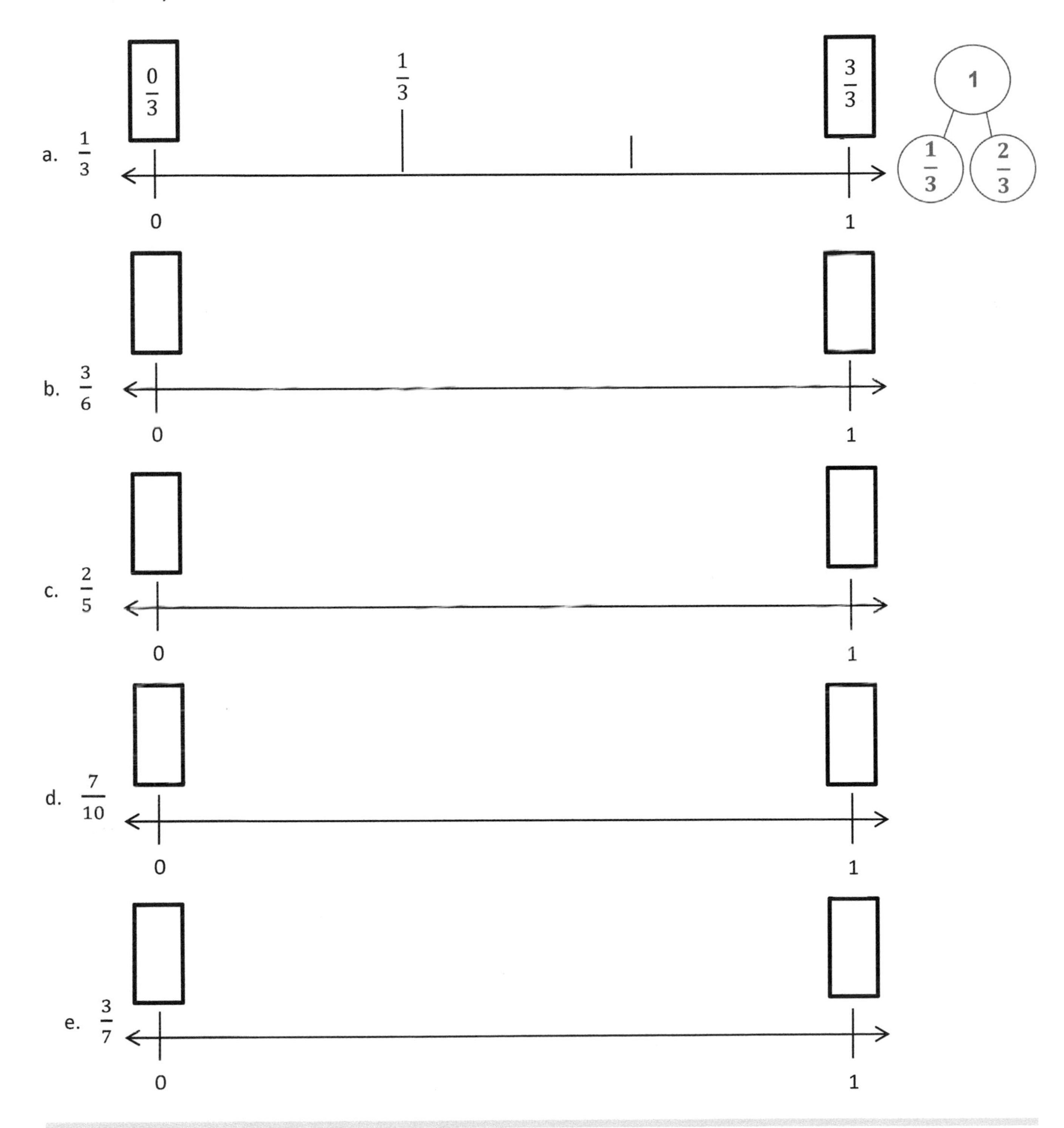

2. Henry has 5 dimes. Ben has 9 dimes. Tina has 2 dimes.

   a. Write the value of each person's money as a fraction of a dollar:

      Henry:

      Ben:

      Tina:

   b. Estimate to place each fraction on the number line.

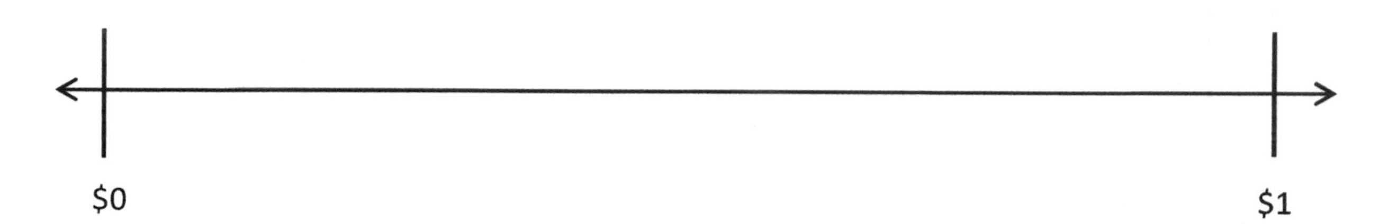

3. Draw a number line. Use a fraction strip to locate 0 and 1. Fold the strip to make 8 equal parts.

   a. Use the strip to measure and label your number line with eighths.

   b. Count up from 0 eighths to 8 eighths on your number line. Touch each number with your finger as you count.

Lesson 16

Objective: Place whole number fractions and fractions between whole numbers on the number line.

Suggested Lesson Structure

- Fluency Practice (12 minutes)
- Application Problem (7 minutes)
- Concept Development (31 minutes)
- Student Debrief (10 minutes)

Total Time **(60 minutes)**

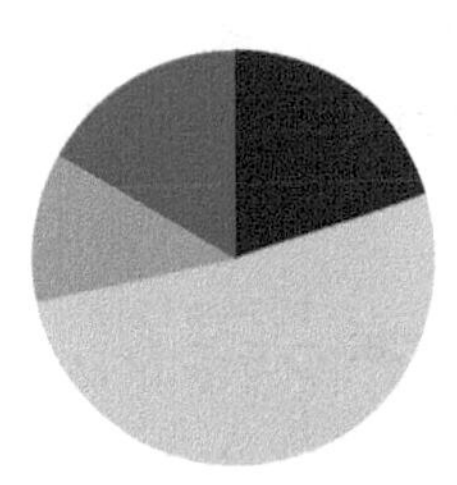

Fluency Practice (12 minutes)

- Sprint: Multiply and Divide by Nine **3.OA.4** (7 minutes)
- Counting by Fractional Units **3.NF.1, 3.NF.3c** (2 minutes)
- Place Fractions on a Number Line Between 0 and 1 **3.NF.2a** (3 minutes)

Sprint: Multiply and Divide by Nine (7 minutes)

Materials: (S) Multiply and Divide by Nine Sprint

Note: This Sprint supports fluency with multiplication and division using units of 9.

Counting by Fractional Units (2 minutes)

Note: This activity reviews counting by fractional units and supports students as they work with fractions on the number line in Topic D.

T: Count by halves from 1 half to 6 halves and back to 0.

S: $\frac{1}{2}, \frac{2}{2}, \frac{3}{2}, \frac{4}{2}, \frac{5}{2}, \frac{6}{2}, \frac{5}{2}, \frac{4}{2}, \frac{3}{2}, \frac{2}{2}, \frac{1}{2}$, 0.

Continue with the following possible sequence: thirds, fifths, and fourths.

Place Fractions on a Number Line Between 0 and 1 (3 minutes)

Materials: (S) Personal white board

Note: This activity reviews the concept of placing fractions on a number line from Lesson 15.

T: (Draw a number line with endpoints 0 and 1.) Draw my number line on your personal white board.

S: (Draw.)

T: Estimate to mark and label 1 fifth.

S: (Estimate 1 fifth of the distance between 0 and 1, and write $\frac{1}{5}$.)

T: Estimate to mark and label 4 fifths.

S: (Estimate 4 fifths of the distance between 0 and 1, and write $\frac{4}{5}$.)

Continue with the following possible sequence: $\frac{1}{8}$, $\frac{7}{8}$, $\frac{3}{8}$, $\frac{5}{8}$, $\frac{3}{4}$, and $\frac{1}{4}$.

NOTES ON MULTIPLE MEANS OF REPRESENTATION:

Check English language learners' listening comprehension of math language during the fluency activity *Place Fractions on a Number Line Between 0 and 1.* Celebrate improvement. "You heard 1 fifth and showed 1 fifth. Great job!"

Application Problem (7 minutes)

Hannah bought 1 yard of ribbon to wrap 4 small presents. She wants to cut the ribbon into equal parts. Draw and label a number line from 0 yards to 1 yard to show where Hannah will cut the ribbon. Label all the fractions, including 0 fourths and 4 fourths. Also, label 0 yards and 1 yard.

Note: This problem reviews the concept of placing fractions on a number line from Lessons 14 and 15.

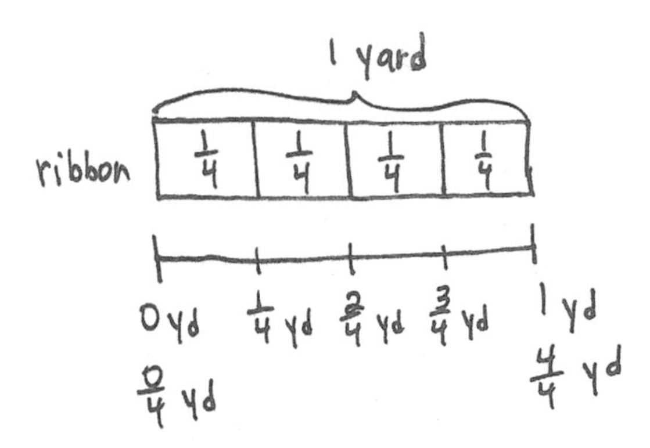

Concept Development (31 minutes)

Materials: (S) Personal white board

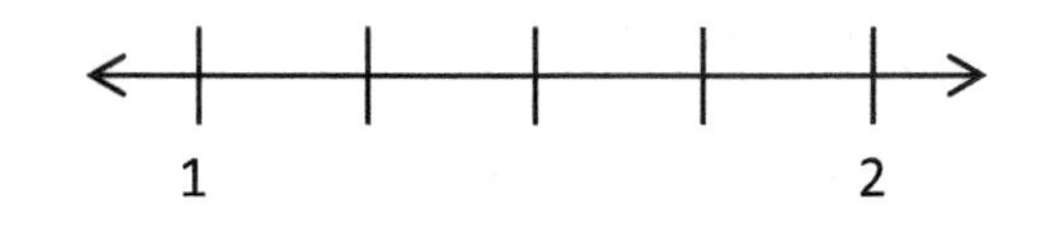

T: Draw a number line on your board with the endpoints 1 and 2. The last few days, our left endpoint was 0. Talk to a partner: Where has 0 gone?

S: It didn't disappear; it is to the left of the 1. → The arrow on the number line tells us that there are more numbers, but we just didn't show them.

T: It's as if we took a picture of a piece of the number line, but those missing numbers still exist. Partition your whole into 4 equal lengths. (Model.)

T: Our number line doesn't start at 0, so we can't start at 0 fourths. How many fourths are in 1 whole?

S: 4 fourths.

NOTES ON MULTIPLE MEANS OF ENGAGEMENT:

If gauging that students working below grade level need it, build understanding with pictures or concrete materials. Extend the number line back to 0. Have students shade in fourths as they count. Use fraction strips as in Lesson 14, if needed.

T: We will label 4 fourths at whole number 1. Label the rest of the fractions up to 2. Check your work with a partner. (Allow work time.) What are the whole number fractions—the fractions equal to 1 and 2?

S: 4 fourths and 8 fourths.

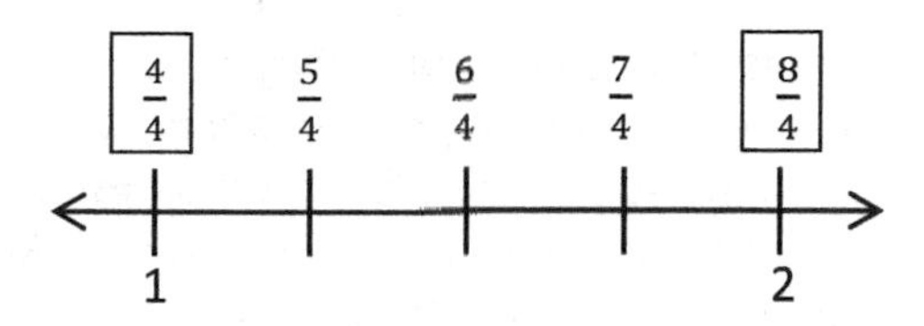

T: Draw boxes around those fractions. (Model.)

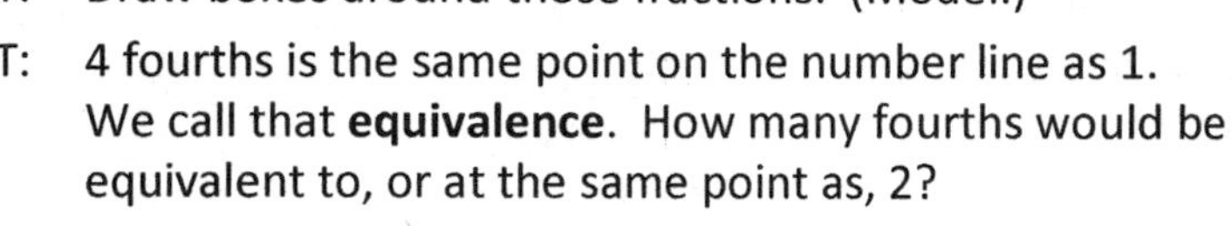

MP.7 T: 4 fourths is the same point on the number line as 1. We call that **equivalence**. How many fourths would be equivalent to, or at the same point as, 2?

S: 8 fourths.

T: Talk to a partner: What fraction is equivalent to, at the same point as, 3?

S: (After discussion.) 12 fourths.

T: Draw a number line with the endpoints 2 and 4. What whole number is missing from this number line?

S: The number 3.

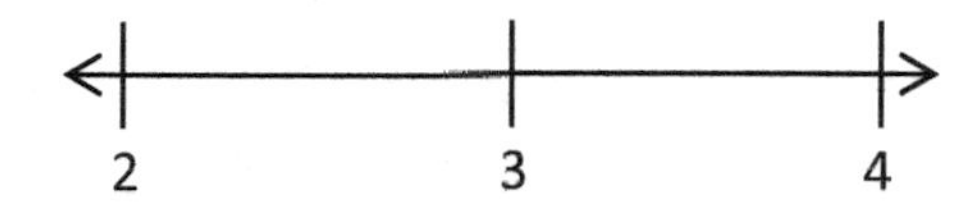

T: Let's place the number 3. It should be equally spaced between 2 and 4. Draw that in. (Model.)

T: We will partition each whole number interval into 3 equal lengths. Tell your partner what your number line will look like.

S: (Discuss.)

T: To label the number line that starts at 2, we have to know how many thirds are equivalent to 2 wholes. Discuss with your partner how to find the number of thirds in 2 wholes.

S: 3 thirds made 1 whole. So, 6 units of thirds make 2 wholes. → 6 thirds are equivalent to 2 wholes.

T: Fill in the rest of your number line.

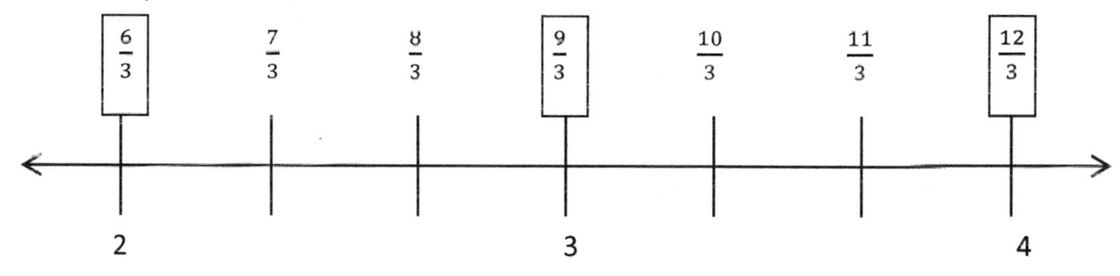

Follow with an example using endpoints 3 and 6 so students place 2 whole numbers on the number line, and then partition into halves.

Close the guided practice by having students work in pairs. Partner A names a number line with endpoints between 0 and 5 and a unit fraction. Partners begin with halves and thirds. When they have demonstrated that they have done 2 number lines correctly, they may try fourths and fifths, etc. Partner B draws, and Partner A assesses. Then, partners switch roles.

NOTES ON MULTIPLE MEANS OF ENGAGEMENT:

Students working above grade level may solve quickly using mental math. Push students to notice and articulate patterns and relationships. As they work in pairs to partition number lines, have students make and analyze their predictions.

Problem Set (10 minutes)

Students should do their personal best to complete the Problem Set within the allotted 10 minutes. For some classes, it may be appropriate to modify the assignment by specifying which problems they work on first. Some problems do not specify a method for solving. Students should solve these problems using the RDW approach used for Application Problems.

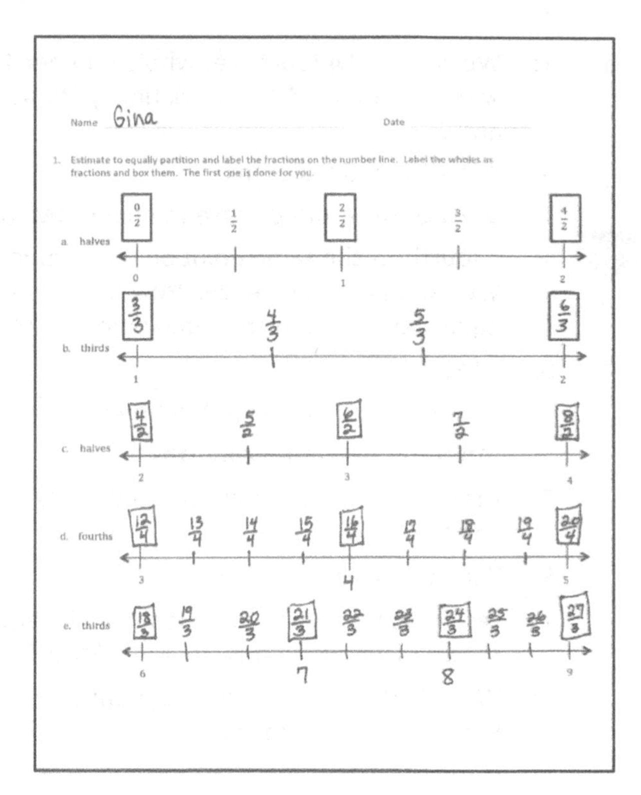

Name Gina Date

1. Estimate to equally partition and label the fractions on the number line. Label the wholes as fractions and box them. The first one is done for you.

a. halves

b. thirds

c. halves

d. fourths

e. thirds

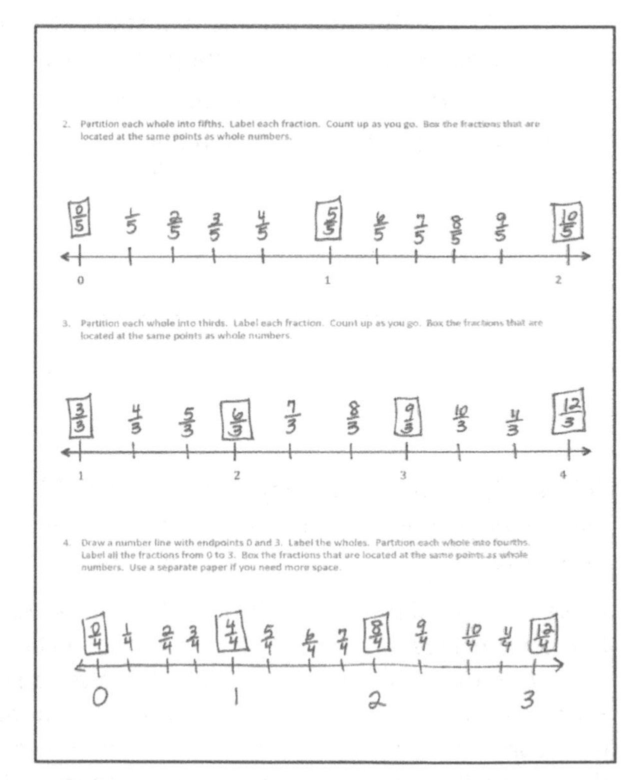

2. Partition each whole into fifths. Label each fraction. Count up as you go. Box the fractions that are located at the same points as whole numbers.

3. Partition each whole into thirds. Label each fraction. Count up as you go. Box the fractions that are located at the same points as whole numbers.

4. Draw a number line with endpoints 0 and 3. Label the wholes. Partition each whole into fourths. Label all the fractions from 0 to 3. Box the fractions that are located at the same points as whole numbers. Use a separate paper if you need more space.

Student Debrief (10 minutes)

Lesson Objective: Place whole number fractions and fractions between whole numbers on the number line.

The Student Debrief is intended to invite reflection and active processing of the total lesson experience.

Invite students to review their solutions for the Problem Set. They should check work by comparing answers with a partner before going over answers as a class. Look for misconceptions or misunderstandings that can be addressed in the Debrief. Guide students in a conversation to debrief the Problem Set and process the lesson.

Any combination of the questions below may be used to lead the discussion.

- In Problem 1, what fractions are **equivalent** to, or at the exact same point as, 3 on the number line?
- What number is equivalent to, or at the exact same point as, 12 fourths in Problem 1?
- Point out Problem 3, which counts 3 thirds, 6 thirds, 9 thirds, and 12 thirds:
  - Look at the fractions you boxed in Problem 3. What pattern do you notice?
  - What is the connection between multiplication and fractions equal to whole numbers?
  - How do you think that strategy might help you find other whole number fractions?

Exit Ticket (3 minutes)

After the Student Debrief, instruct students to complete the Exit Ticket. A review of their work will help with assessing students' understanding of the concepts that were presented in today's lesson and planning more effectively for future lessons. The questions may be read aloud to the students.

A

Number Correct: _______

Multiply and Divide by Nine

1.	2 × 9 =		23.	___ × 9 = 90	
2.	3 × 9 =		24.	___ × 9 = 18	
3.	4 × 9 =		25.	___ × 9 = 27	
4.	5 × 9 =		26.	90 ÷ 9 =	
5.	1 × 9 =		27.	45 ÷ 9 =	
6.	18 ÷ 9 =		28.	9 ÷ 9 =	
7.	27 ÷ 9 =		29.	18 ÷ 9 =	
8.	45 ÷ 9 =		30.	27 ÷ 9 =	
9.	9 ÷ 9 =		31.	___ × 9 = 54	
10.	36 ÷ 9 =		32.	___ × 9 = 63	
11.	6 × 9 =		33.	___ × 9 = 81	
12.	7 × 9 =		34.	___ × 9 = 72	
13.	8 × 9 =		35.	63 ÷ 9 =	
14.	9 × 9 =		36.	81 ÷ 9 =	
15.	10 × 9 =		37.	54 ÷ 9 =	
16.	72 ÷ 9 =		38.	72 ÷ 9 =	
17.	63 ÷ 9 =		39.	11 × 9 =	
18.	81 ÷ 9 =		40.	99 ÷ 9 =	
19.	54 ÷ 9 =		41.	12 × 9 =	
20.	90 ÷ 9 =		42.	108 ÷ 9 =	
21.	___ × 9 = 45		43.	14 × 9 =	
22.	___ × 9 = 9		44.	126 ÷ 9 =	

B

Number Correct: _______

Improvement: _______

Multiply and Divide by Nine

1.	1 × 9 =	
2.	2 × 9 =	
3.	3 × 9 =	
4.	4 × 9 =	
5.	5 × 9 =	
6.	27 ÷ 9 =	
7.	18 ÷ 9 =	
8.	36 ÷ 9 =	
9.	9 ÷ 9 =	
10.	45 ÷ 9 =	
11.	10 × 9 =	
12.	6 × 9 =	
13.	7 × 9 =	
14.	8 × 9 =	
15.	9 × 9 =	
16.	63 ÷ 9 =	
17.	54 ÷ 9 =	
18.	72 ÷ 9 =	
19.	90 ÷ 9 =	
20.	81 ÷ 9 =	
21.	____ × 9 = 9	
22.	____ × 9 = 45	

23.	____ × 9 = 18	
24.	____ × 9 = 90	
25.	____ × 9 = 27	
26.	18 ÷ 9 =	
27.	9 ÷ 9 =	
28.	90 ÷ 9 =	
29.	45 ÷ 9 =	
30.	27 ÷ 9 =	
31.	____ × 9 = 27	
32.	____ × 9 = 36	
33.	____ × 9 = 81	
34.	____ × 9 = 63	
35.	72 ÷ 9 =	
36.	81 ÷ 9 =	
37.	54 ÷ 9 =	
38.	63 ÷ 9 =	
39.	11 × 9 =	
40.	99 ÷ 9 =	
41.	12 × 9 =	
42.	108 ÷ 9 =	
43.	13 × 9 =	
44.	117 ÷ 9 =	

Name ______________________________ Date ________________

1. Estimate to equally partition and label the fractions on the number line. Label the wholes as fractions, and box them. The first one is done for you.

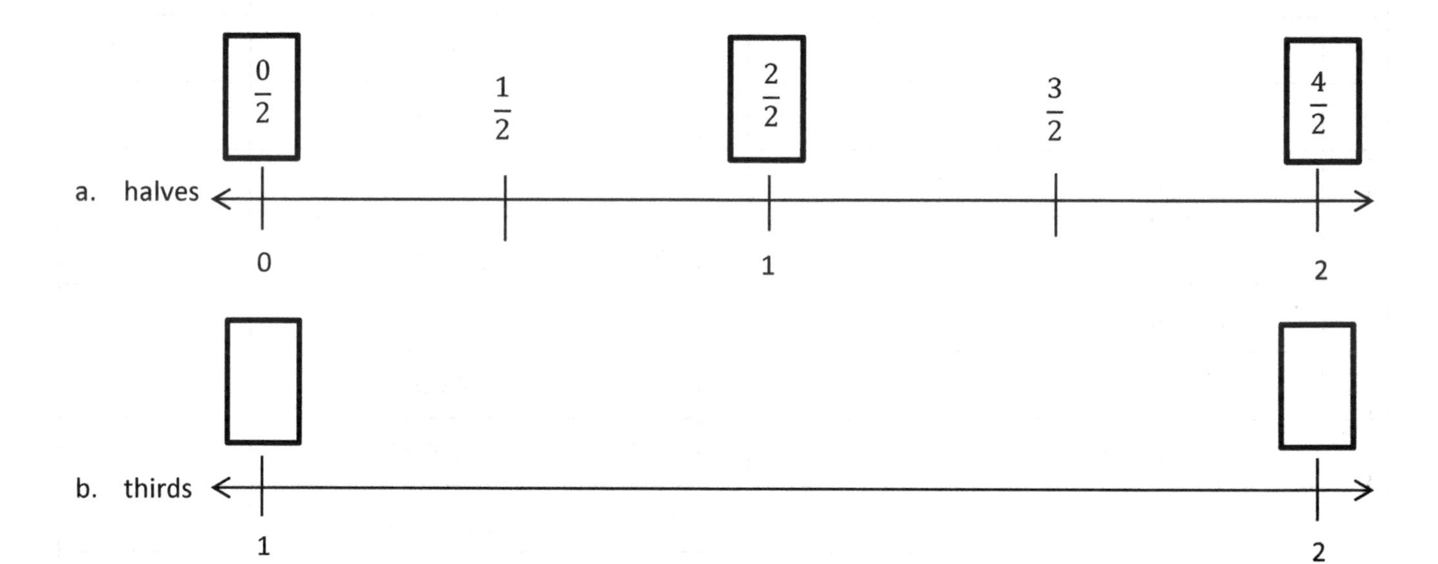

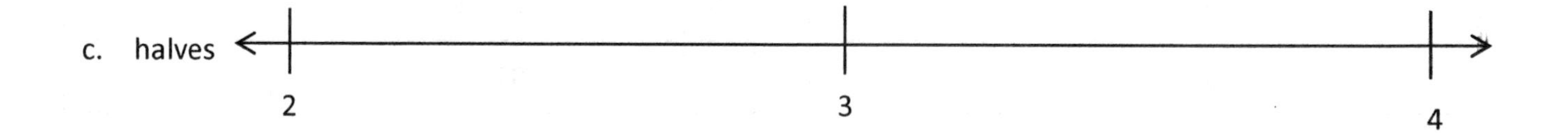

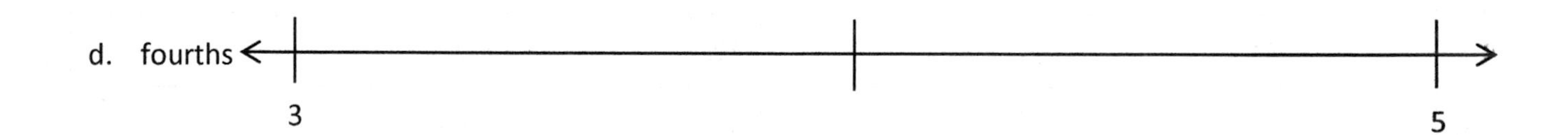

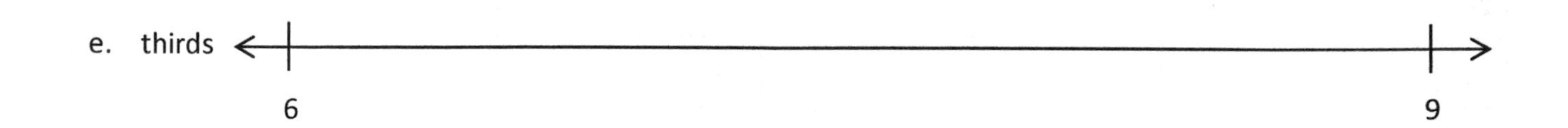

2. Partition each whole into fifths. Label each fraction. Count up as you go. Box the fractions that are located at the same points as whole numbers.

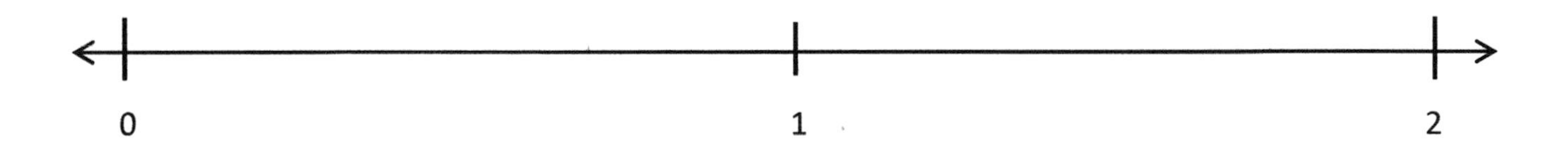

3. Partition each whole into thirds. Label each fraction. Count up as you go. Box the fractions that are located at the same points as whole numbers.

4. Draw a number line with endpoints 0 and 3. Label the wholes. Partition each whole into fourths. Label all the fractions from 0 to 3. Box the fractions that are located at the same points as whole numbers. Use a separate paper if you need more space.

Name ______________________________ Date ________________

1. Estimate to equally partition and label the fractions on the number line. Label the wholes as fractions, and box them.

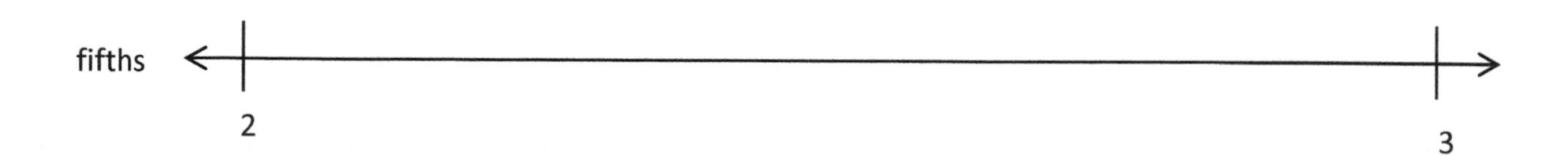

2. Draw a number line with endpoints 0 and 2. Label the wholes. Estimate to partition each whole into sixths, and label them. Box the fractions that are located at the same points as whole numbers.

Name ______________________________ Date ________________

1. Estimate to equally partition and label the fractions on the number line. Label the wholes as fractions, and box them. The first one is done for you.

a. thirds

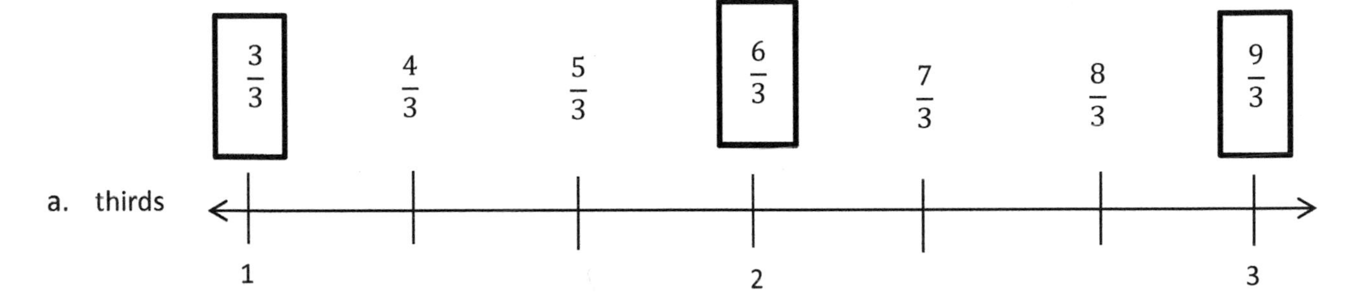

b. eighths

c. fourths

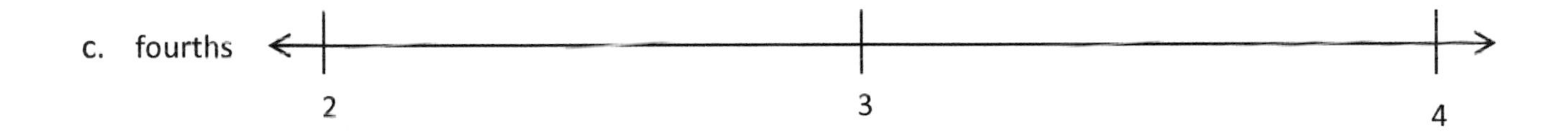

d. halves

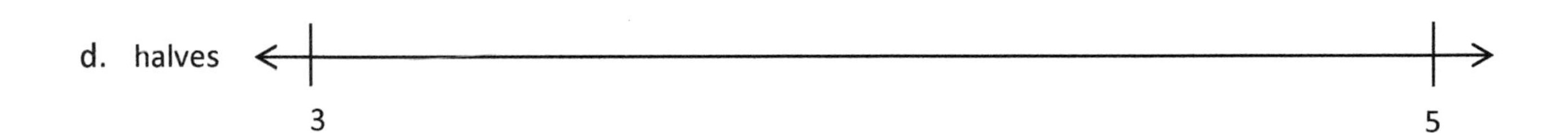

e. fifths

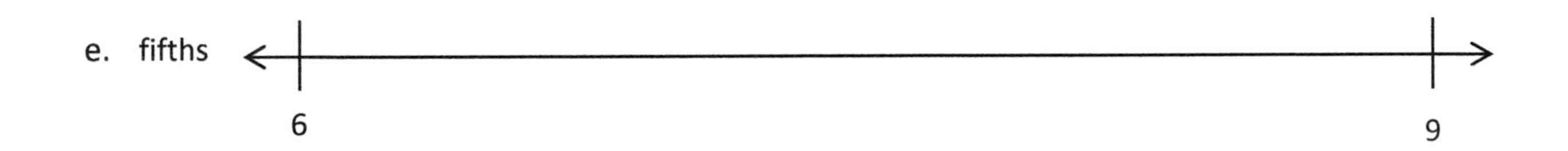

2. Partition each whole into sixths. Label each fraction. Count up as you go. Box the fractions that are located at the same points as whole numbers.

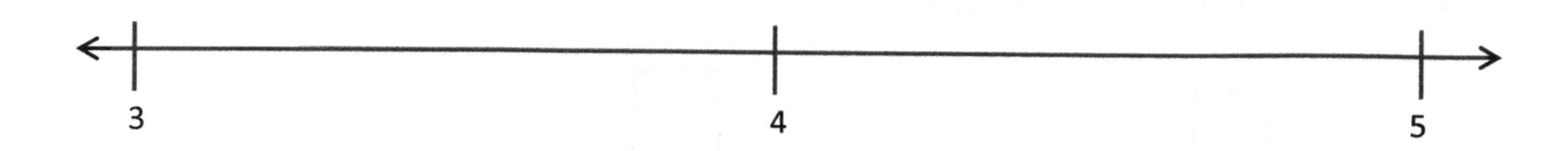

3. Partition each whole into halves. Label each fraction. Count up as you go. Box the fractions that are located at the same points as whole numbers.

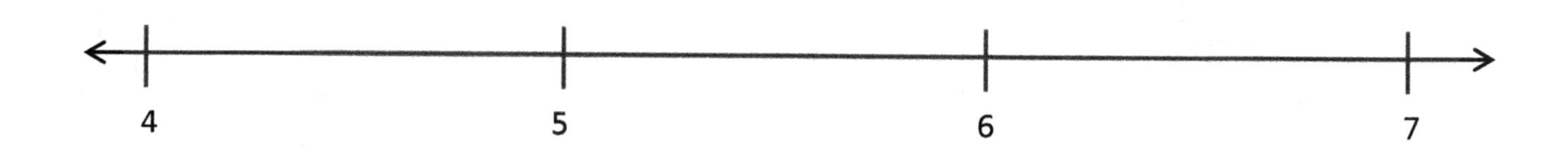

4. Draw a number line with endpoints 0 and 3. Label the wholes. Partition each whole into fifths. Label all the fractions from 0 to 3. Box the fractions that are located at the same points as whole numbers. Use a separate paper if you need more space.

Lesson 17

Objective: Practice placing various fractions on the number line.

Suggested Lesson Structure

- Fluency Practice (12 minutes)
- Application Problem (6 minutes)
- Concept Development (32 minutes)
- Student Debrief (10 minutes)

Total Time (60 minutes)

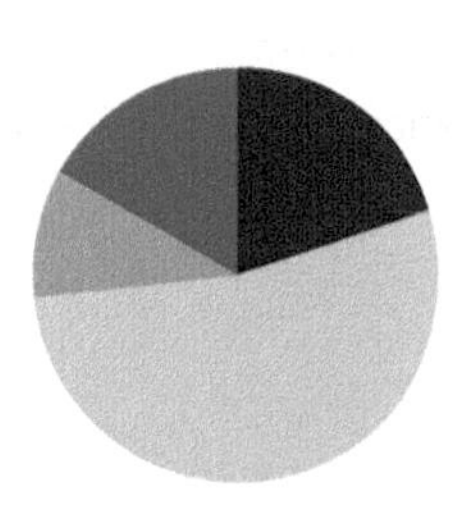

Fluency Practice (12 minutes)

- Sprint: Division **3.OA.2** (8 minutes)
- Place Fractions on a Number Line **3.NF.2b** (3 minutes)
- Compare Unit Fractions **3.NF.3d** (1 minutes)

Sprint: Division (8 minutes)

Materials: (S) Division Sprint

Note: This Sprint supports fluency with division using various units.

Place Fractions on a Number Line (3 minutes)

Materials: (S) Personal white board

Note: This activity reviews the concept of placing fractions on a number line from Lesson 16.

T: (Draw a number line marked at 0, 1, 2, and 3.) Draw my number line on your personal white board.

S: (Draw.)

T: Estimate to mark and label 1 half within the interval 0 to 1.

S: (Estimate the halfway point between 0 and 1 and write $\frac{1}{2}$.)

T: Estimate to mark 2 halves. Label 2 halves as a fraction.

S: (Write $\frac{2}{2}$ above the 1 on the number line.)

Continue with the following possible sequence, drawing a new number line for the different fractional units: $\frac{4}{2}, \frac{6}{2}, \frac{1}{5}, \frac{5}{5}, \frac{10}{5}, \frac{15}{5}, \frac{1}{3}, \frac{3}{3}, \frac{9}{3}, \frac{6}{3}, \frac{1}{4}, \frac{8}{4}, \frac{12}{4}$, and $\frac{4}{4}$.

Compare Unit Fractions (1 minute)

Note: This activity reviews the concept of comparing unit fractions from Topic C.

T: (Write $\frac{1}{2}$ and $\frac{1}{10}$.) Both fractions refer to the same whole. Say the largest fraction.

S: 1 half.

Continue with the following possible sequence: $\frac{1}{2}$ and $\frac{1}{3}$, $\frac{1}{3}$ and $\frac{1}{4}$, $\frac{1}{4}$ and $\frac{1}{6}$, $\frac{1}{4}$ and $\frac{1}{2}$, $\frac{1}{6}$ and $\frac{1}{8}$, $\frac{1}{6}$ and $\frac{1}{5}$, and $\frac{1}{5}$ and $\frac{1}{10}$.

Application Problem (6 minutes)

Sammy sees a black line at the bottom of the pool stretching from one end to the other. She wonders how long it is. The black line is the same length as 9 concrete slabs that make the sidewalk at the edge of the pool. One concrete slab is 5 meters long. What is the length of the black line at the bottom of the pool?

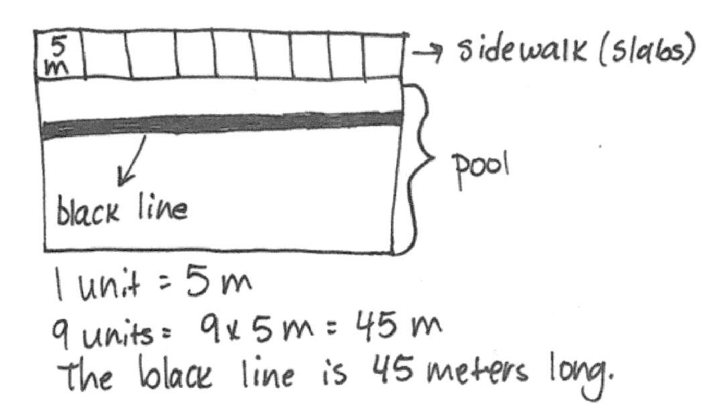

1 unit = 5 m
9 units = 9 x 5 m = 45 m
The black line is 45 meters long.

Note: This problem reviews multiplication from Modules 1 and 3. It also reviews partitioning a whole into equal parts from Topic A.

Concept Development (32 minutes)

Materials: (S) Personal white board

T: Draw a number line with endpoints 1 and 4. Label the wholes. Partition each whole into thirds. Label all of the fractions from 1 to 4.

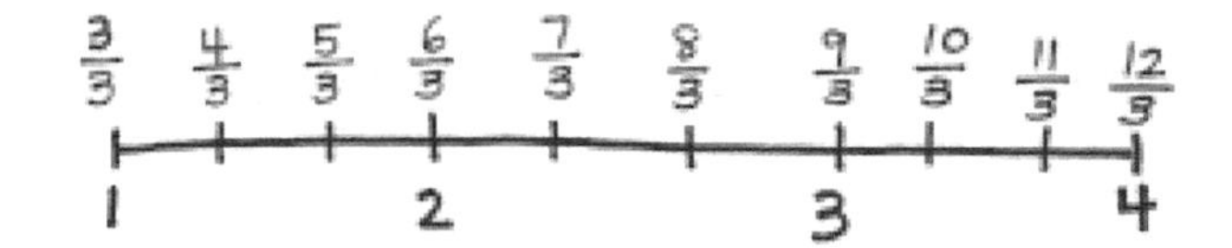

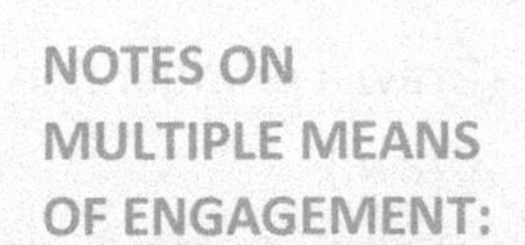

NOTES ON MULTIPLE MEANS OF ENGAGEMENT:

To help students working below grade level, locate and label fractions on the number line. Elicit answers that specify the whole and the fractional unit. Say, "Point to and count the wholes with me. How many wholes? Into what fractional unit are we partitioning the whole? Label as we count the fractions."

T: After you labeled your whole numbers, what did you think about to place your fractions?

S: Evenly spacing the marks between whole numbers to make thirds. → Writing the numbers in order: 3 thirds, 4 thirds, 5 thirds, etc. → Starting with 3 thirds because the endpoint was 1.

T: What do the fractions have in common? What do you notice?

S: All of the fractions are thirds. → All are equal to or greater than 1 whole. → The number of thirds that name whole numbers count by threes: 1 = 3 thirds, 2 = 6 thirds, 3 = 9 thirds. → $\frac{3}{3}$, $\frac{6}{3}$, $\frac{9}{3}$, and $\frac{12}{3}$ are at the same point on the number line as 1, 2, 3, and 4. Those fractions are equivalent to whole numbers.

T: Draw a number line on your board with endpoints 1 and 4.

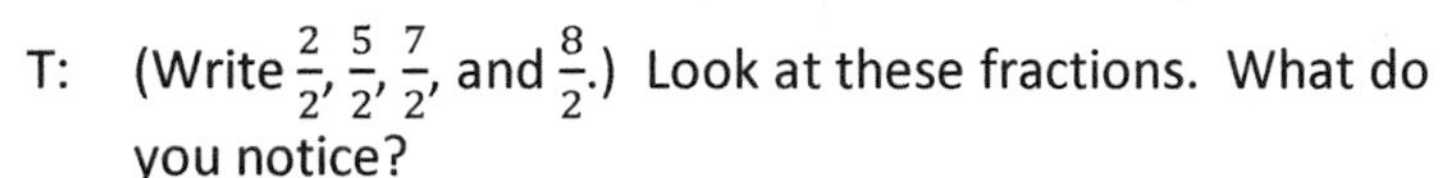

T: (Write $\frac{2}{2}$, $\frac{5}{2}$, $\frac{7}{2}$, and $\frac{8}{2}$.) Look at these fractions. What do you notice?

S: They are all halves. → They are all equal to or greater than 1. → They are in order, but some are missing.

T: Place these fractions on your number line. (After students place fractions on the number line.) Compare with your partner. Check that your number lines are the same.

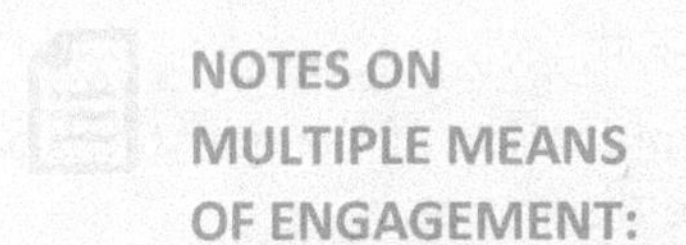

NOTES ON MULTIPLE MEANS OF ENGAGEMENT:

Ask students working above grade level this more open-ended question: "How many halves are on the number line?"

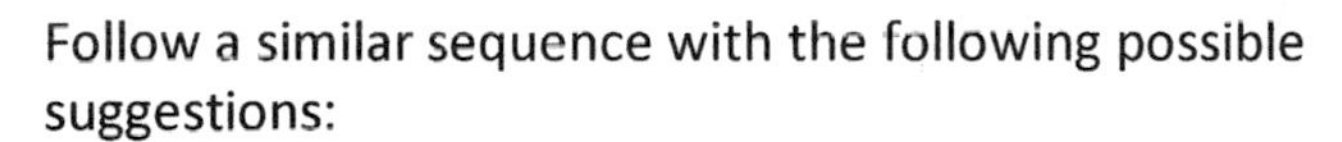

Follow a similar sequence with the following possible suggestions:

- Number line with endpoints 1 and 4, marking fractions in thirds
- Number line with endpoints 2 and 5, marking fractions in fifths
- Number line with endpoints 4 and 6, marking fractions in thirds

Close the lesson by having pairs of students generate collections of fractions to place on number lines with specified endpoints. Students might then exchange problems, challenging each other to place fractions on the number line. Students should reason aloud about how the partitioned fractional unit is chosen for each number line.

NOTES ON MULTIPLE MEANS OF ACTION AND EXPRESSION:

Support English language learners as they construct written responses. Read the prompt aloud or have students read chorally. Provide sentence starters and a word bank.

Sentence starters may include the following:

- "I think ____ has a longer pinky finger than ____ because..."

Possible words for the word bank may include the following:

less than	eighths	closer to
greater than	zero	

Problem Set (10 minutes)

Students should do their personal best to complete the Problem Set within the allotted 10 minutes. For some classes, it may be appropriate to modify the assignment by specifying which problems they work on first. Some problems do not specify a method for solving. Students should solve these problems using the RDW approach used for Application Problems.

Student Debrief (10 minutes)

Lesson Objective: Practice placing various fractions on the number line.

The Student Debrief is intended to invite reflection and active processing of the total lesson experience.

Invite students to review their solutions for the Problem Set. They should check work by comparing answers with a partner before going over answers as a class. Look for misconceptions or misunderstandings that can be addressed in the Student Debrief. Guide students in a conversation to debrief the Problem Set and process the lesson.

Any combination of the questions below may be used to lead the discussion.

- What did you think about first to help you place the fractions?
- In Problems 1–3, did you label all of the marks on each number line or just the fractions in the list? Why?
- In Problems 1–3, what was the first fraction that you placed on each number line? Why did you start with that one?
- What advice would you give an absent classmate about completing this Problem Set? What is the most important thing to remember when placing fractions on the number line?

Exit Ticket (3 minutes)

After the Student Debrief, instruct students to complete the Exit Ticket. A review of their work will help with assessing students' understanding of the concepts that were presented in today's lesson and planning more effectively for future lessons. The questions may be read aloud to the students.

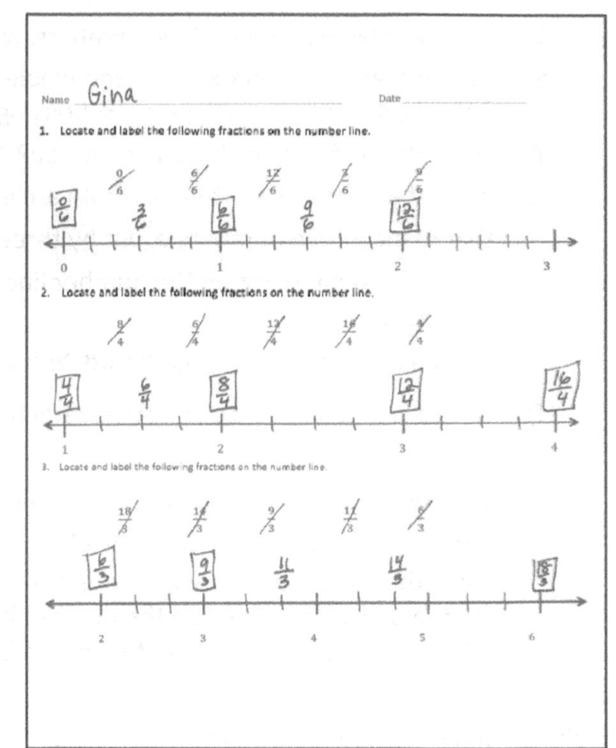

Name Gina Date

1. Locate and label the following fractions on the number line.

2. Locate and label the following fractions on the number line.

3. Locate and label the following fractions on the number line.

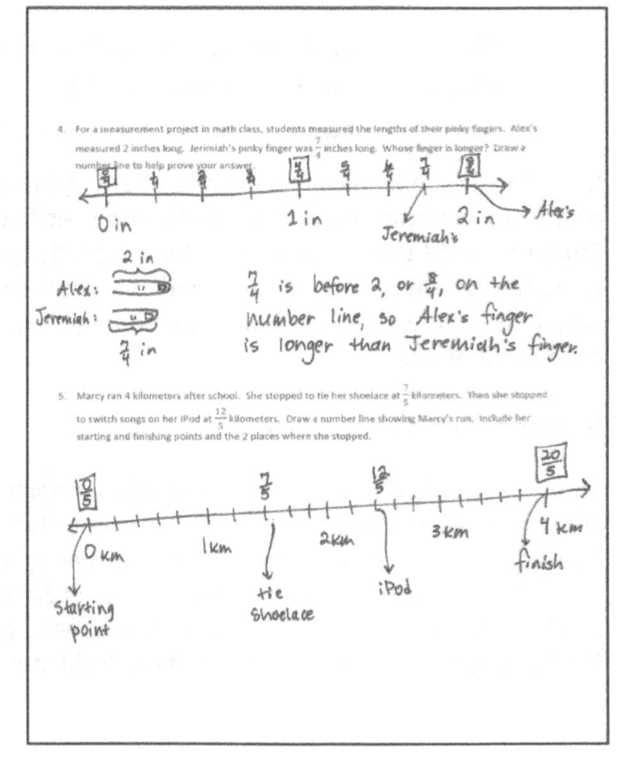

4. For a measurement project in math class, students measured the lengths of their pinky fingers. Alex's measured 2 inches long. Jerimiah's pinky finger was $\frac{7}{4}$ inches long. Whose finger is longer? Draw a number line to help prove your answer.

$\frac{7}{4}$ is before 2, or $\frac{8}{4}$, on the number line, so Alex's finger is longer than Jeremiah's finger.

5. Marcy ran 4 kilometers after school. She stopped to tie her shoelace at $\frac{7}{5}$ kilometers. Then she stopped to switch songs on her iPod at $\frac{12}{5}$ kilometers. Draw a number line showing Marcy's run. Include her starting and finishing points and the 2 places where she stopped.

A

Number Correct: _______

Division

1.	3 ÷ 3 =	
2.	4 ÷ 4 =	
3.	5 ÷ 5 =	
4.	19 ÷ 19 =	
5.	0 ÷ 1 =	
6.	0 ÷ 2 =	
7.	0 ÷ 3 =	
8.	0 ÷ 19 =	
9.	6 ÷ 3 =	
10.	9 ÷ 3 =	
11.	12 ÷ 3 =	
12.	15 ÷ 3 =	
13.	4 ÷ 2 =	
14.	6 ÷ 2 =	
15.	8 ÷ 2 =	
16.	10 ÷ 2 =	
17.	18 ÷ 3 =	
18.	12 ÷ 2 =	
19.	21 ÷ 3 =	
20.	14 ÷ 2 =	
21.	20 ÷ 10 =	
22.	20 ÷ 2 =	

23.	24 ÷ 3 =	
24.	16 ÷ 2 =	
25.	30 ÷ 10 =	
26.	30 ÷ 3 =	
27.	27 ÷ 3 =	
28.	18 ÷ 2 =	
29.	40 ÷ 10 =	
30.	40 ÷ 4 =	
31.	20 ÷ 4 =	
32.	20 ÷ 5 =	
33.	24 ÷ 4 =	
34.	30 ÷ 5 =	
35.	28 ÷ 4 =	
36.	40 ÷ 5 =	
37.	32 ÷ 4 =	
38.	45 ÷ 5 =	
39.	44 ÷ 4 =	
40.	36 ÷ 4 =	
41.	48 ÷ 6 =	
42.	63 ÷ 7 =	
43.	64 ÷ 8 =	
44.	72 ÷ 9 =	

B

Number Correct: _______

Improvement: _______

Division

1.	$2 \div 2 =$	
2.	$3 \div 3 =$	
3.	$4 \div 4 =$	
4.	$17 \div 17 =$	
5.	$0 \div 2 =$	
6.	$0 \div 3 =$	
7.	$0 \div 4 =$	
8.	$0 \div 17 =$	
9.	$4 \div 2 =$	
10.	$6 \div 2 =$	
11.	$8 \div 2 =$	
12.	$10 \div 2 =$	
13.	$6 \div 3 =$	
14.	$9 \div 3 =$	
15.	$12 \div 3 =$	
16.	$15 \div 3 =$	
17.	$12 \div 2 =$	
18.	$18 \div 3 =$	
19.	$14 \div 2 =$	
20.	$21 \div 3 =$	
21.	$20 \div 2 =$	
22.	$20 \div 10 =$	

23.	$16 \div 2 =$	
24.	$24 \div 3 =$	
25.	$30 \div 3 =$	
26.	$30 \div 10 =$	
27.	$18 \div 2 =$	
28.	$27 \div 3 =$	
29.	$40 \div 4 =$	
30.	$40 \div 10 =$	
31.	$20 \div 5 =$	
32.	$20 \div 4 =$	
33.	$30 \div 5 =$	
34.	$24 \div 4 =$	
35.	$40 \div 5 =$	
36.	$28 \div 4 =$	
37.	$45 \div 5 =$	
38.	$32 \div 4 =$	
39.	$55 \div 5 =$	
40.	$36 \div 4 =$	
41.	$54 \div 6 =$	
42.	$56 \div 7 =$	
43.	$72 \div 8 =$	
44.	$63 \div 9 =$	

Name ______________________________ Date ________________

1. Locate and label the following fractions on the number line.

$\frac{0}{6}$ $\frac{6}{6}$ $\frac{12}{6}$ $\frac{3}{6}$ $\frac{9}{6}$

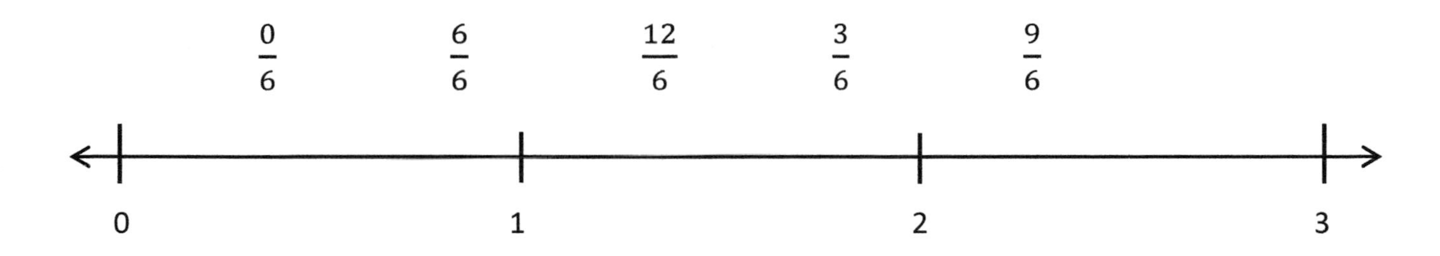

2. Locate and label the following fractions on the number line.

$\frac{8}{4}$ $\frac{6}{4}$ $\frac{12}{4}$ $\frac{16}{4}$ $\frac{4}{4}$

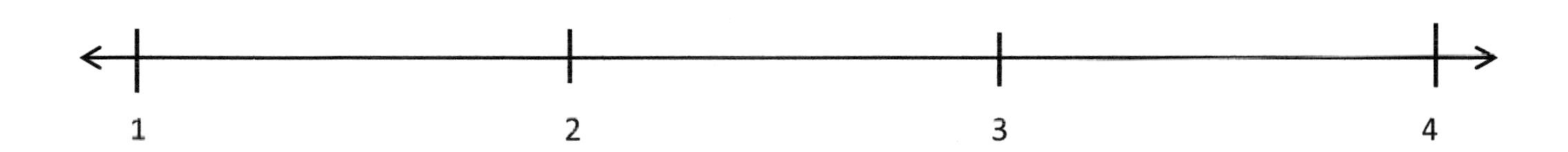

3. Locate and label the following fractions on the number line.

$\frac{18}{3}$ $\frac{14}{3}$ $\frac{9}{3}$ $\frac{11}{3}$ $\frac{6}{3}$

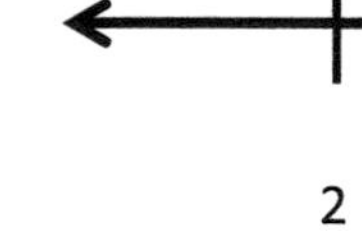

2 3 4 5 6

4. For a measurement project in math class, students measured the lengths of their pinky fingers. Alex's measured 2 inches long. Jerimiah's pinky finger was $\frac{7}{4}$ inches long. Whose finger is longer? Draw a number line to help prove your answer.

5. Marcy ran 4 kilometers after school. She stopped to tie her shoelace at $\frac{7}{5}$ kilometers. Then, she stopped to switch songs on her iPod at $\frac{12}{5}$ kilometers. Draw a number line showing Marcy's run. Include her starting and finishing points and the 2 places where she stopped.

Name ______________________________ Date ______________

1. Locate and label the following fractions on the number line.

$\frac{2}{3}$ $\frac{4}{3}$

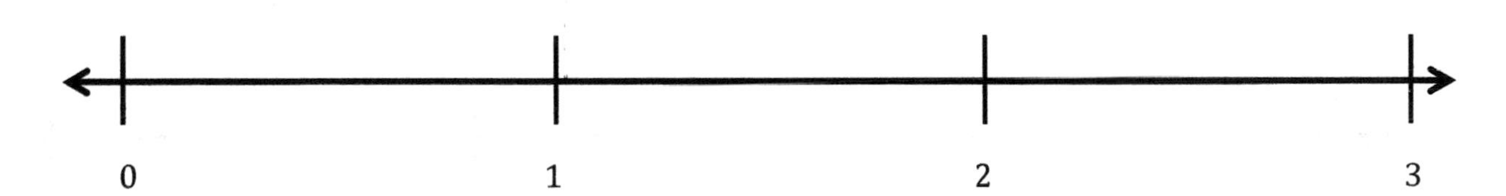

2. Katie bought 2 one-gallon bottles of juice for a party. Her guests drank $\frac{6}{4}$ gallons of juice. What fraction of a gallon of juice is left over? Draw a number line to show, and explain your answer.

Name ______________________________ Date ______________

1. Locate and label the following fractions on the number line.

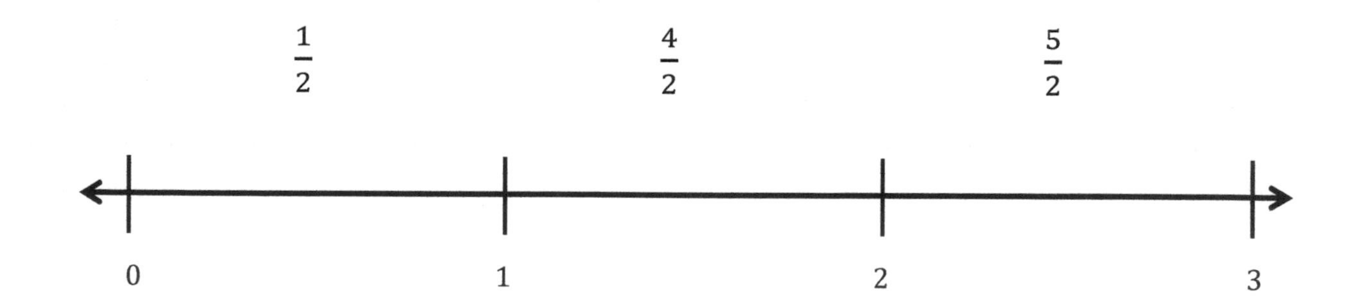

2. Locate and label the following fractions on the number line.

$\frac{11}{3}$ $\frac{6}{3}$ $\frac{8}{3}$

2 3 4

3. Locate and label the following fractions on the number line.

$\frac{20}{4}$ $\frac{13}{4}$ $\frac{23}{4}$

3 4 5 6

4. Wayne went on a 4-kilometer hike. He took a break at $\frac{4}{3}$ kilometers. He took a drink of water at $\frac{10}{3}$ kilometers. Show Wayne's hike on the number line. Include his starting and finishing place and the 2 points where he stopped.

5. Ali wants to buy a piano. The piano measures $\frac{19}{4}$ feet long. She has a space 5 feet long for the piano in her house. Does she have enough room? Draw a number line to show, and explain your answer.

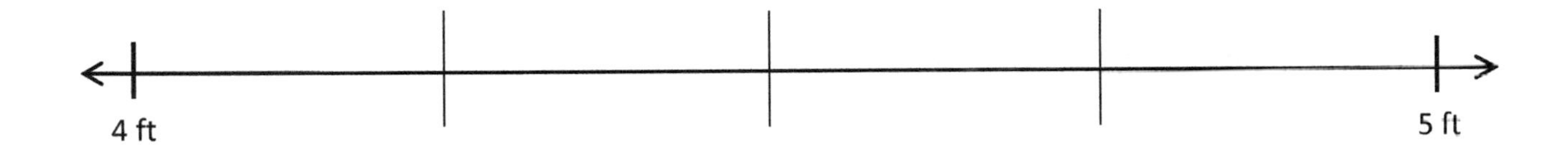

Lesson 18

Objective: Compare fractions and whole numbers on the number line by reasoning about their distance from 0.

Suggested Lesson Structure

■ Fluency Practice	(8 minutes)
■ Application Problem	(8 minutes)
■ Concept Development	(34 minutes)
■ Student Debrief	(10 minutes)
Total Time	**(60 minutes)**

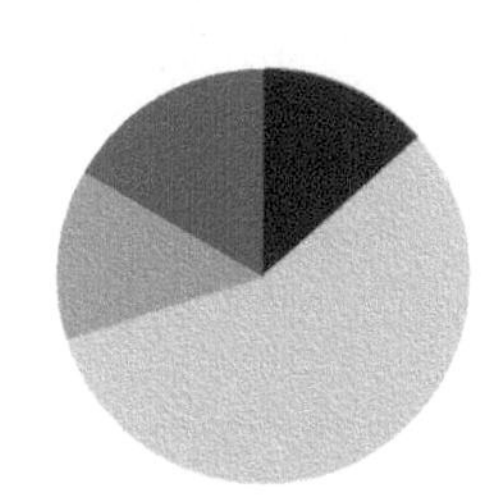

Fluency Practice (8 minutes)

- Draw Number Bonds of 1 Whole **3.NF.1** (4 minutes)
- Place Fractions on the Number Line **3.NF.2b** (4 minutes)

Draw Number Bonds of 1 Whole (4 minutes)

Materials: (S) Personal white board

Note: This activity reviews the concept of making copies of a unit fraction to build a whole.

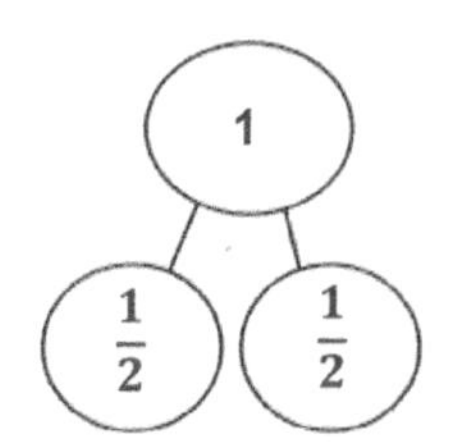

T: On your personal white board, draw a number bond to partition 1 whole into halves.

S: (Draw a number bond.)

T: How many copies of 1 half did you draw to make 1 whole?

S: 2 copies.

Continue with the following possible sequence: thirds, fourths, fifths, sixths, sevenths, and eighths. Have students draw the models side by side and compare to notice patterns at the end.

Place Fractions on the Number Line (4 minutes)

Materials: (S) Personal white board

Note: This activity reviews the concept of placing fractions on a number line from Topic D.

T: (Project a number line marked at 0, 1, 2, and 3.) Draw my number line on your board.

S: (Draw.)

T: Estimate to mark and label 1 third in the interval 0 to 1.

S: (Estimate the point between 0 and 1 and write $\frac{1}{3}$.)

T: Write 3 thirds on your number line. Label the point as a fraction.

S: (Write $\frac{3}{3}$ above the 1 on the number line.)

Continue with the following possible sequence: $\frac{6}{3}$, $\frac{9}{3}$, $\frac{4}{3}$, $\frac{7}{3}$, $\frac{2}{3}$, and $\frac{8}{3}$.

Application Problem (8 minutes)

Third-grade students are growing peppers. The student with the longest pepper wins the Green Thumb award. Jackson's pepper measured 3 inches long. Drew's measured $\frac{10}{4}$ inches long. Who won the award? Draw a number line to help prove your answer.

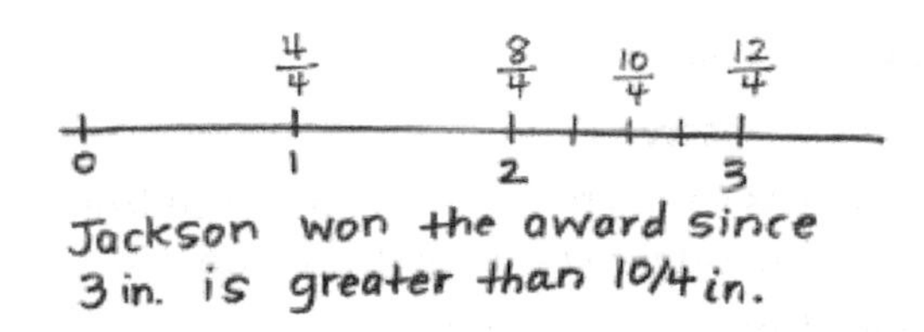

Note: This problem reviews the concept of placing fractions on a number line from Topic D. It is also used during the Concept Development to discuss a fraction's distance from 0.

Concept Development (34 minutes)

Materials: (T) Large-scale number line partitioned into thirds (description below), 4 containers, 4 beanbags (or balled-up pieces of paper), sticky notes (S) Work from Application Problem

T: Look at the number line I've created on the floor. Let's use it to measure and compare.

T: This number line shows the interval from 0 to 1. (Place sticky notes with *0* and *1* written on them in the appropriate places.) What fractional unit does the number line show?

S: Thirds.

T: Let's place containers on $\frac{1}{3}$ and $\frac{2}{3}$. (Select volunteers to place containers.)

S: (Place containers.)

T: How can we use our thirds to help us place $\frac{1}{6}$ on this number line?

S: $\frac{1}{6}$ is right in the middle of the first third. (Place a container on $\frac{1}{6}$.)

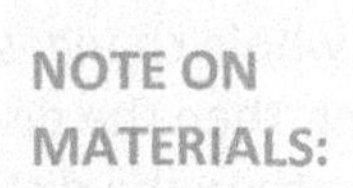

NOTE ON MATERIALS:

Before the lesson, use masking tape to make a large-scale number line from 0 to 1 on the floor or in the hallway. Partition the interval evenly into thirds. Try to make the 0 and 1 far apart.

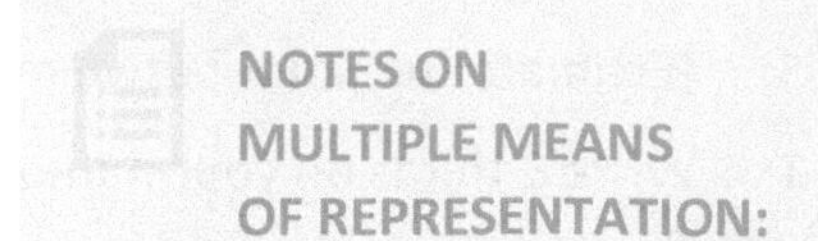

NOTES ON MULTIPLE MEANS OF REPRESENTATION:

You may want to preteach vocabulary by adding it to a math word wall before the lesson starts. Helping students connect terms such as *more than, fewer,* and *the same* to familiar symbols or words used often (greater than, less than, equal to) makes the language more accessible during the lesson.

T: Looking at the number line, where can we place our last container so that it is the greatest distance from 0?

S: On 1. → On this number line, it has to be 1 because the interval is from 0 to 1. → 1 is the farthest point from 0 on this number line. (Place a container on 1.)

T: Suppose we invite 4 volunteers to come up. Each volunteer takes a turn to stand at 0 and toss a beanbag into one of the containers. Which container will be the hardest, and which will be the easiest to toss the beanbag into? Why?

S: The container at 1 will be the hardest because it's the farthest away from 0. → The container at $\frac{1}{6}$ will be easy. It's close to 0.

T: Let's have volunteers toss. (Each volunteer tosses a beanbag into a given container. They toss in the following order: $\frac{1}{6}$, $\frac{1}{3}$, $\frac{2}{3}$, and 1 whole.)

S: (Volunteers toss while others observe.)

Guide students to discuss how each toss shows the different distance from 0 that each beanbag traveled. Emphasize the distance from 0 as an important feature of the comparison.

T: Why is a fraction's distance from 0 important for comparison?

S: (Discuss.)

T: How would the comparison change if each volunteer stood at a different place on the number line?

S: It would be hard to compare because the distances would be different. → The distance the beanbag flew wouldn't tell you how big the fraction is. → It's like measuring. When you use a ruler, you start at 0 to measure. Then you can compare the measurements. → The number line is like a giant ruler.

T: Suppose we tossed beanbags to containers at the same points from 0 to 1 on a different number line, but the distance from 0 to 1 was different. How would the comparison of the fractions change if the distance from 0 to 1 was shorter? Longer?

S: If the whole changes, the distance between fractions also changes. → So, if the number line was shorter, then the distance to toss each beanbag would also be shorter. → If the number line was longer, then the distance to toss each beanbag would also be longer. → True, but the position of each fraction within the number lines stays the same. → So, the comparisons would be the same, but the distance between 0 and each fraction would change.

Students return to their seats.

T: Think back to our Application Problem. What in the Application Problem relates to the length of the toss?

S: How big the peppers are. → The length of the peppers.

T: Talk to your partner. How did we use the distance from 0 to show the length of the peppers?

S: We saw 3 is larger than $\frac{10}{4}$. → We used the number line sort of like a ruler. We put the measurements on it. Then, we saw which one was farthest from the 0. → On the number line, you can see that the length from 0 to 3 is longer than the length from 0 to $\frac{10}{4}$.

T: Let's do the same thing we did with our big number line on the floor, pretending we measured *giant* peppers with yards instead of inches. 1 pepper measured 3 yards long, and the other measured $\frac{10}{4}$ yards. How would the comparison of the fractions change using yards rather than inches?

S: Yards are much larger than inches. → But even though the measurement units changed, $\frac{10}{4}$ yards is still less than 3 yards, just like $\frac{10}{4}$ inches is less than 3 inches.

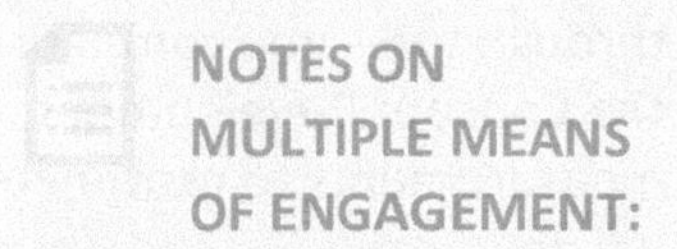

NOTES ON MULTIPLE MEANS OF ENGAGEMENT:

As students compare the giant peppers, a third pepper can be given to them to include in the comparison. If preferred, the length of the pepper can be equal to $\frac{12}{4}$.

Problem Set (10 minutes)

Students should do their personal best to complete the Problem Set within the allotted 10 minutes. For some classes, it may be appropriate to modify the assignment by specifying which problems they work on first. Some problems do not specify a method for solving. Students should solve these problems using the RDW approach used for Application Problems.

Student Debrief (10 minutes)

Lesson Objective: Compare fractions and whole numbers on the number line by reasoning about their distance from 0.

The Student Debrief is intended to invite reflection and active processing of the total lesson experience.

Invite students to review their solutions for the Problem Set. They should check work by comparing answers with a partner before going over answers as a class. Look for misconceptions or misunderstandings that can be addressed in the Student Debrief. Guide students in a conversation to debrief the Problem Set and process the lesson.

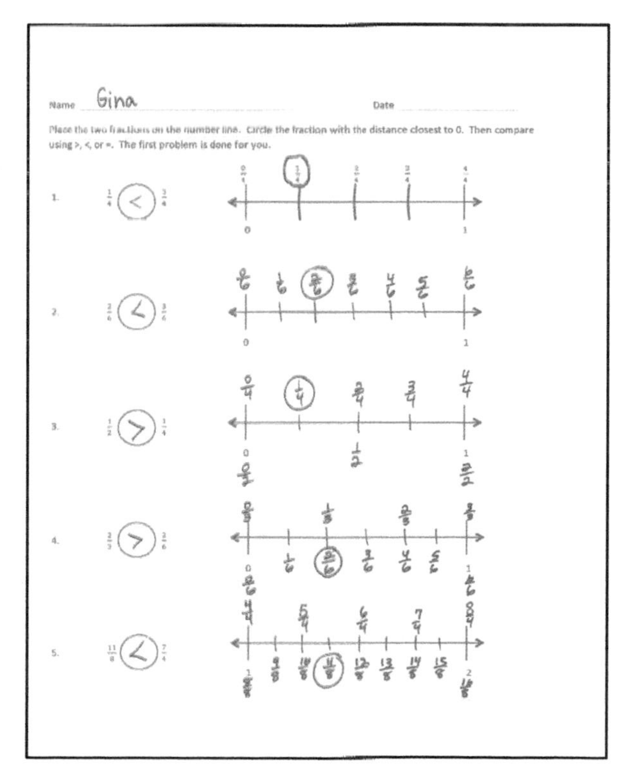

Any combination of the questions below may be used to lead the discussion.

- If necessary, review the *toss* portion of the lesson by having students draw each toss on a separate number line and then place the fractions on the same number line to compare.
- Invite students to share their work on Problems 6–8. Ensure that each student can articulate how the distance from 0 helped them figure out which fraction was greater or less.

- Extend the lesson by having students work through the same comparison given at the end of the Concept Development, this time altering the measurements to centimeters and inches.

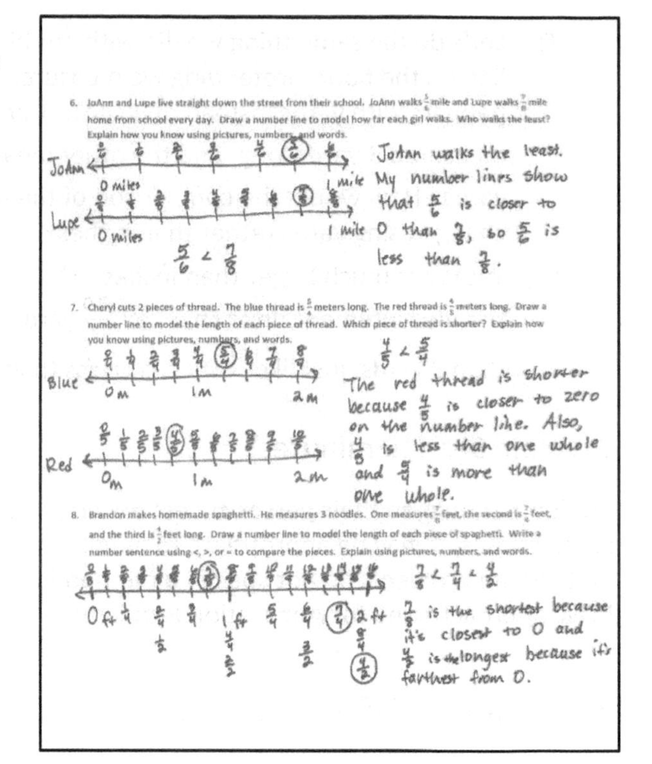

Exit Ticket (3 minutes)

After the Student Debrief, instruct students to complete the Exit Ticket. A review of their work will help with assessing students' understanding of the concepts that were presented in today's lesson and planning more effectively for future lessons. The questions may be read aloud to the students.

Name ______________________________ Date ______________

Place the two fractions on the number line. Circle the fraction with the distance closest to 0. Then, compare using >, <, or =. The first problem is done for you.

1. 
$\frac{1}{4}$ < $\frac{3}{4}$

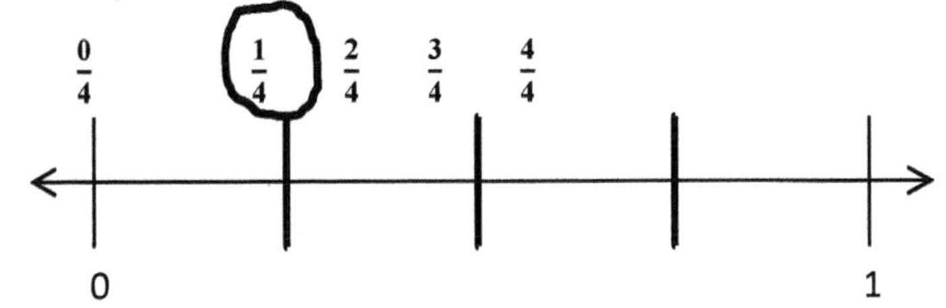

2. 
$\frac{2}{6}$ ◯ $\frac{3}{6}$

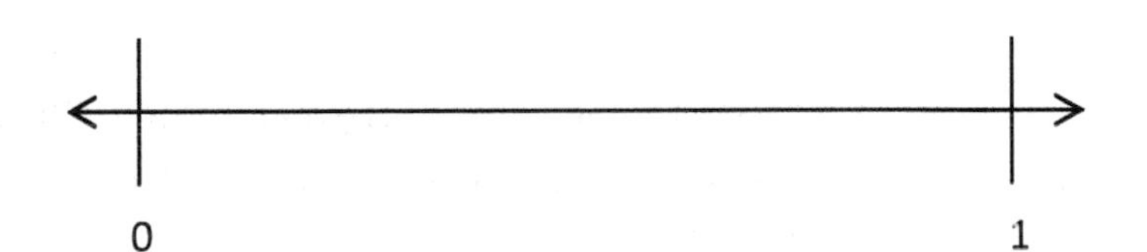

3. 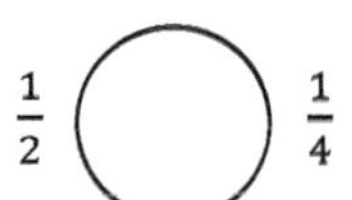
$\frac{1}{2}$ ◯ $\frac{1}{4}$

4. 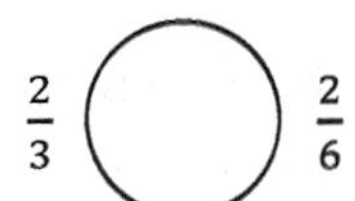
$\frac{2}{3}$ ◯ $\frac{2}{6}$

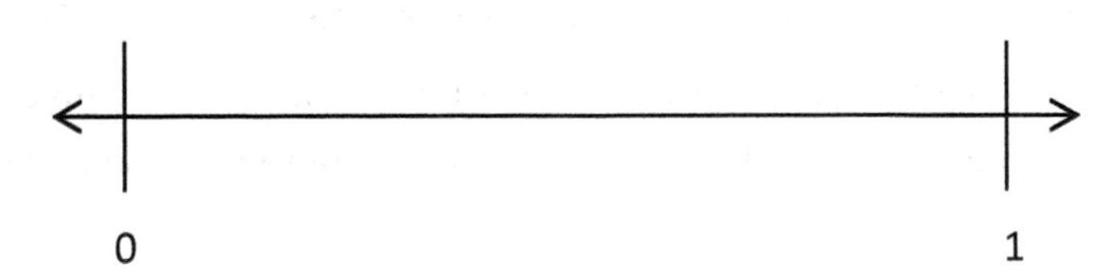

5. 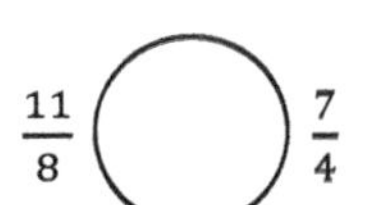
$\frac{11}{8}$ ◯ $\frac{7}{4}$

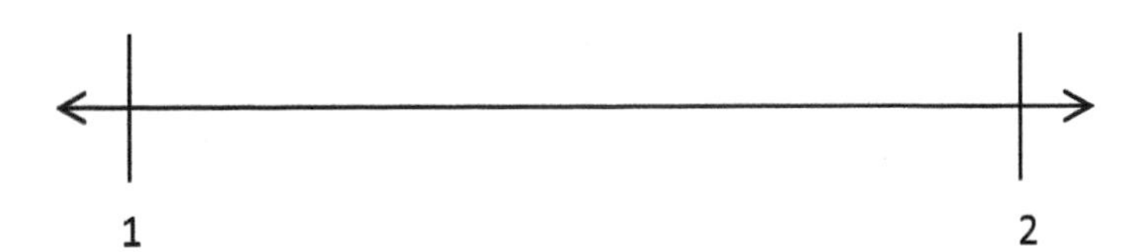

6. JoAnn and Lupe live straight down the street from their school. JoAnn walks $\frac{5}{6}$ miles and Lupe walks $\frac{7}{8}$ miles home from school every day. Draw a number line to model how far each girl walks. Who walks the least? Explain how you know using pictures, numbers, and words.

7. Cheryl cuts 2 pieces of thread. The blue thread is $\frac{5}{4}$ meters long. The red thread is $\frac{4}{5}$ meters long. Draw a number line to model the length of each piece of thread. Which piece of thread is shorter? Explain how you know using pictures, numbers, and words.

8. Brandon makes homemade spaghetti. He measures 3 noodles. One measures $\frac{7}{8}$ feet, the second is $\frac{7}{4}$ feet, and the third is $\frac{4}{2}$ feet long. Draw a number line to model the length of each piece of spaghetti. Write a number sentence using <, >, or = to compare the pieces. Explain using pictures, numbers, and words.

Name ______________________________ Date ______________

Place the two fractions on the number line. Circle the fraction with the distance closest to 0. Then, compare using >, <, or =.

1. $\frac{3}{5}$ ◯ $\frac{1}{5}$

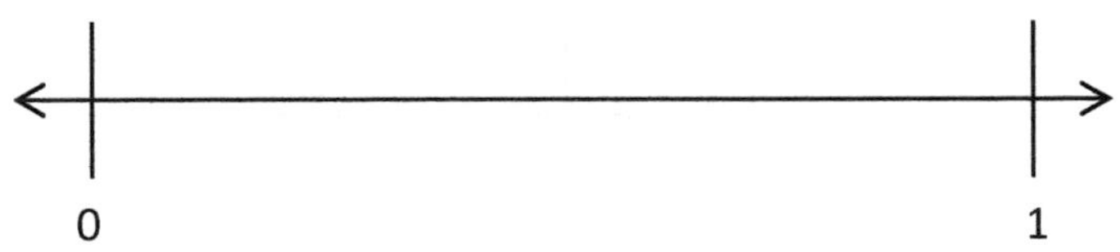

2. $\frac{1}{2}$ ◯ $\frac{3}{4}$

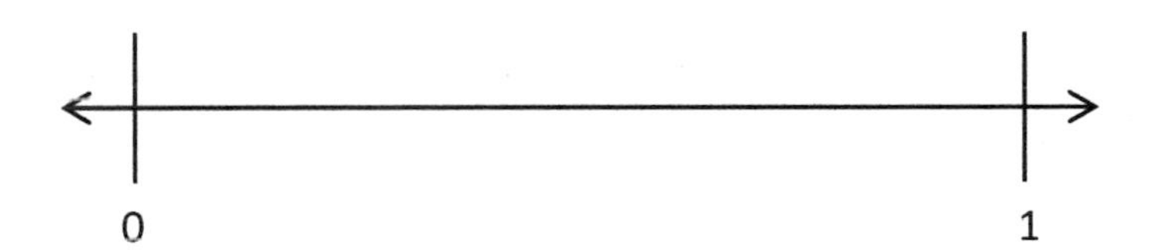

3. Mr. Brady draws a fraction on the board. Ken says it's $\frac{2}{3}$, and Dan said it's $\frac{3}{2}$. Do both of these fractions mean the same thing? If not, which fraction is larger? Draw a number line to model $\frac{2}{3}$ and $\frac{3}{2}$. Use words, pictures, and numbers to explain your comparison.

Name ______________________________ Date ________________

Place the two fractions on the number line. Circle the fraction with the distance closest to 0. Then, compare using >, <, or =.

1. $\frac{1}{3}$ ◯ $\frac{2}{3}$

0 1

2. $\frac{4}{6}$ ◯ $\frac{1}{6}$

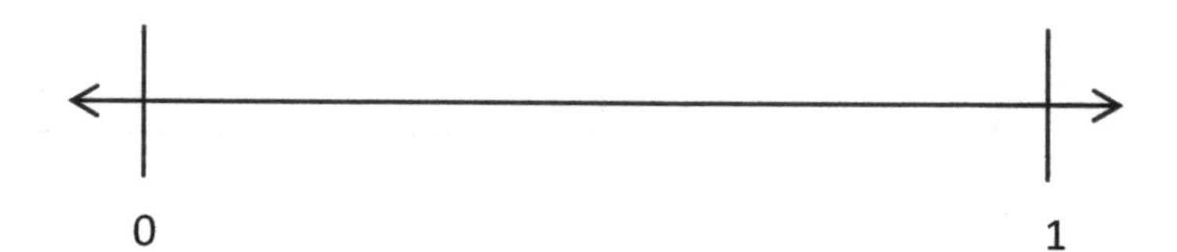

3. $\frac{1}{4}$ ◯ $\frac{1}{8}$

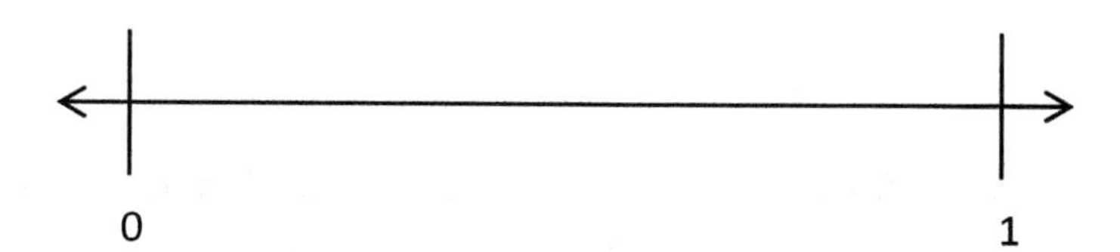

4. $\frac{4}{5}$ ◯ $\frac{4}{10}$

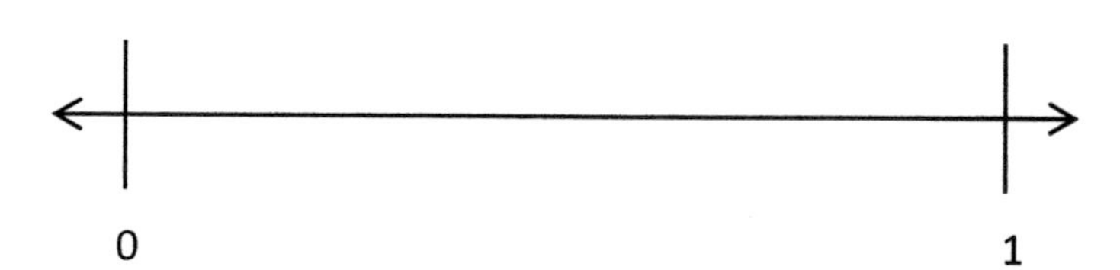

5. $\frac{8}{6}$ ◯ $\frac{5}{3}$

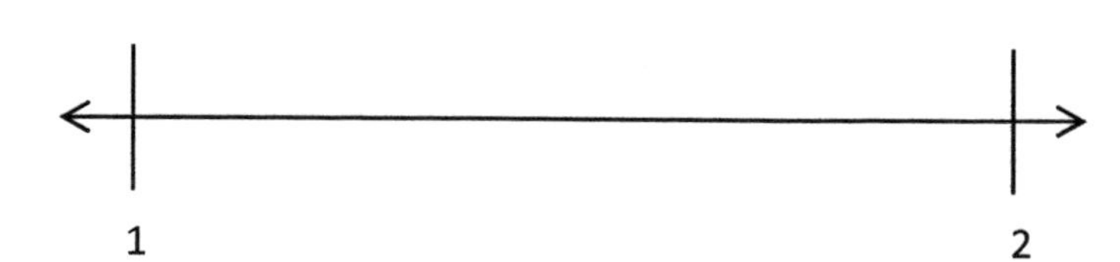

6. Liz and Jay each have a piece of string. Liz's string is $\frac{4}{6}$ yards long, and Jay's string is $\frac{5}{7}$ yards long. Whose string is longer? Draw a number line to model the length of both strings. Explain the comparison using pictures, numbers, and words.

7. In a long jump competition, Wendy jumped $\frac{9}{10}$ meters, and Judy jumped $\frac{10}{9}$ meters. Draw a number line to model the distance of each girl's long jump. Who jumped the shorter distance? Explain how you know using pictures, numbers, and words.

8. Nikki has 3 pieces of yarn. The first piece is $\frac{5}{6}$ feet long, the second piece is $\frac{5}{3}$ feet long, and the third piece is $\frac{3}{2}$ feet long. She wants to arrange them from the shortest to the longest. Draw a number line to model the length of each piece of yarn. Write a number sentence using <, >, or = to compare the pieces. Explain using pictures, numbers, and words.

Lesson 19

Objective: Understand distance and position on the number line as strategies for comparing fractions. (Optional)

Suggested Lesson Structure

Fluency Practice	(12 minutes)
Application Problem	(10 minutes)
Concept Development	(28 minutes)
Student Debrief	(10 minutes)
Total Time	**(60 minutes)**

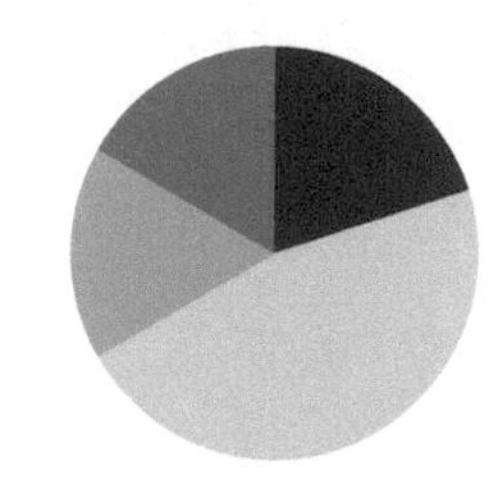

Fluency Practice (12 minutes)

- Sprint: Express Fractions as Whole Numbers **3.NF.3c** (9 minutes)
- Place Fractions on the Number Line **3.NF.2b** (3 minutes)

Sprint: Express Fractions as Whole Numbers (9 minutes)

Materials: (S) Express Fractions as Whole Numbers Sprint

Note: This Sprint reviews representing whole number fractions as whole numbers.

Place Fractions on the Number Line (3 minutes)

Materials: (S) Personal white board

Note: This activity reviews the concept of placing fractions on a number line from Topic D.

T: (Draw a number line marked at 0, 1, 2, and 3.) Draw my number line on your personal white board.

S: (Draw.)

T: Estimate to mark and label 1 third on the interval 0 to 1.

S: (Estimate the point between 0 and 1, and write $\frac{1}{3}$.)

T: Write 3 thirds on your number line. Label the point as a fraction.

S: (Write $\frac{3}{3}$ above the 1 on the number line.)

Continue with the following possible sequence, drawing a new number line for the different fractional units: $\frac{6}{3}$, $\frac{9}{3}$, $\frac{4}{3}$, $\frac{7}{3}$, $\frac{2}{3}$, $\frac{8}{3}$, $\frac{1}{2}$, $\frac{2}{2}$, $\frac{4}{2}$, $\frac{3}{2}$, $\frac{5}{2}$, and $\frac{6}{2}$.

Application Problem (10 minutes)

Thomas has 2 sheets of paper. He wants to punch 4 equally spaced holes along the edge of each sheet.

Draw Thomas's 2 sheets of paper next to each other so the ends meet. Label a number line from 0 at the start of his first paper to 2 at the end of his second paper. Show Thomas where to hole-punch his papers and label the fractions. What fraction is labeled at the eighth hole?

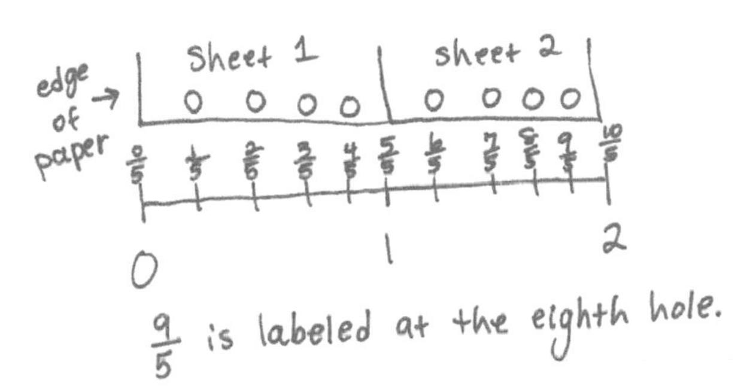

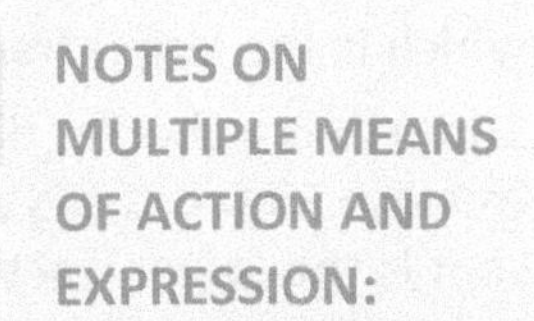

NOTES ON MULTIPLE MEANS OF ACTION AND EXPRESSION:

Students working below grade level may benefit from acting out the Application Problem, lining up 2 sheets of paper to make a concrete example.

Note that this problem is different from the problem with the ribbon in Lesson 16. The first hole is not marking 0. 0 is the edge of the paper. Students working below grade level may not pick up on this. For all students, it is important when measuring to be clear about the location of 0.

Note: This problem reviews the concept of placing fractions on a number line from Topic D. Also, this Application Problem is used during the Concept Development to discuss the difference between the position of a fraction on a number line and the fraction's distance from 0.

Concept Development (28 minutes)

Materials: (S) Personal white board

T: Draw 2 same-sized rectangles on your board, and partition both into 4 equal parts. Shade your top rectangle to show 1 fourth, and shade the bottom to show 3 copies of 1 fourth.

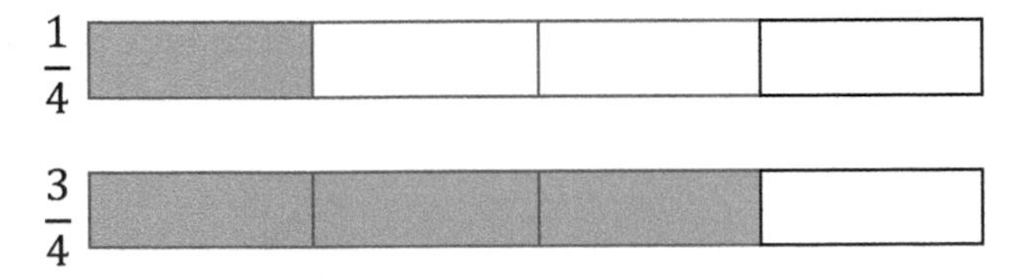

T: Compare the models. Which shaded fraction is larger? Tell your partner how you know.

S: I know 3 fourths is larger because 3 parts is greater than just 1 part of the same size.

T: Use your rectangles to measure and draw a number line from 0 to 1. Partition it into fourths. Label the wholes and fractions on your number line.

S: (Draw and label the number line.)

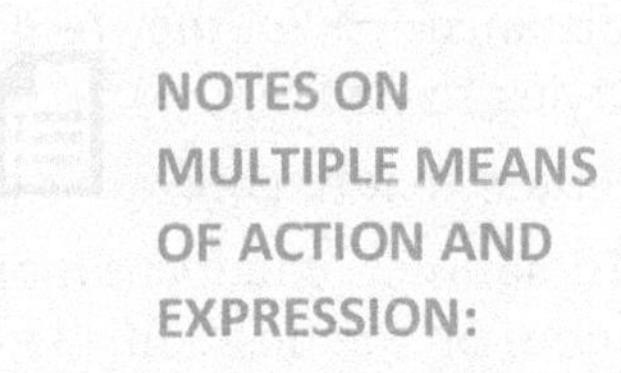

NOTES ON MULTIPLE MEANS OF ACTION AND EXPRESSION:

For English language learners, model the directions or use gestures to clarify English language (e.g., extend both arms to demonstrate *long*).

Give English language learners a little more time to discuss with a partner their math thinking in English.

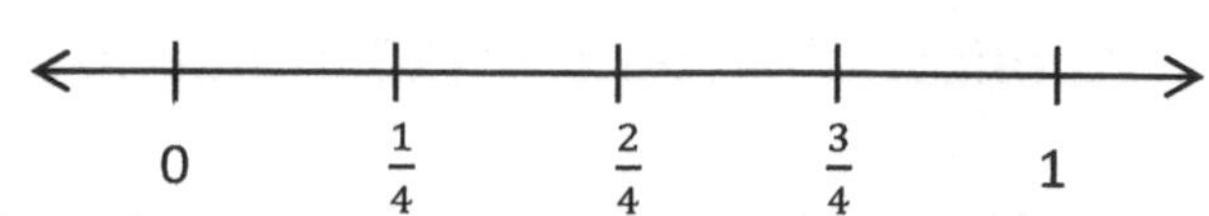

T: Talk with your partner to compare 1 fourth to 3 fourths using the number line. How do you know which is the larger fraction?

S: 1 fourth is a shorter distance from 0, so it is the smaller fraction. 3 fourths is a greater distance away from 0, so it is the larger fraction.

T: Many of you are comparing the fractions by seeing their distance from 0. You're right; 1 unit is a shorter distance from 0 than 3 units. If we know where 0 is on the number line, how can it help us find the smaller or larger fraction?

S: The smaller fraction will always be to the left of the larger fraction.

T: How do you know?

S: Because the farther you go to the right on the number line, the farther the distance from 0. → That means the fraction to the left is always smaller. It's closer to 0.

T: Think back to our Application Problem. What were we trying to find? The length of the page from the edge to each hole? Or were we simply finding the location of each hole?

S: The location of each hole.

T: Remember the pepper problem from yesterday? What were we comparing? The length of the peppers or the location of the peppers?

S: We were looking for the length of each pepper.

T: Talk to a partner: What is the same and what is different about the way we solved these problems?

S: In both, we placed fractions on the number line. → To do that, we actually had to find the distance of each from 0, too. → Yes, but in Thomas's, we were more worried about the position of each fraction, so he'd put the holes in the right places. → And in the pepper problem, the distance from 0 to the fraction told us the length of each pepper, and then we compared that.

T: How do distance and position relate to each other when we compare fractions on the number line?

S: You use the distance from 0 to find the fraction's placement. → Or you use the placement to find the distance. → So, they're both part of comparing. The part you focus on just depends on what you're trying to find out.

T: Relate that to your work on the pepper and hole-punch problems.

S: Sometimes, you focus more on the distance, like in the pepper problem, and sometimes you focus more on the position, like in Thomas's problem. It depends on what the problem is asking.

T: Try and use both ways of thinking about comparing as you work through the problems on today's Problem Set.

Problem Set (10 minutes)

Students should do their personal best to complete the Problem Set within the allotted 10 minutes. For some classes, it may be appropriate to modify the assignment by specifying which problems they work on first. Some problems do not specify a method for solving. Students should solve these problems using the RDW approach used for Application Problems.

Student Debrief (10 minutes)

Lesson Objective: Understand distance and position on the number line as strategies for comparing fractions. (Optional)

The Student Debrief is intended to invite reflection and active processing of the total lesson experience.

Invite students to review their solutions for the Problem Set. They should check work by comparing answers with a partner before going over answers as a class. Look for misconceptions or misunderstandings that can be addressed in the Student Debrief. Guide students in a conversation to debrief the Problem Set and process the lesson.

Any combination of the questions below may be used to lead the discussion.

- Invite students to share their work on Problems 3–5. Students should have slightly different explanations for Problems 4 and 5. Invite a variety of responses so that both explanations are heard.
- Extend the lesson by having students work together (or guide them) to create word problems with real world contexts that emphasize different types of comparisons:
  - Create word problems with a context that emphasizes placement of the fraction on a number line (such as the hole-punch problem).
  - Create word problems with a context that emphasizes the distance of the fraction from 0 (such as the pepper problem).
  - Have students solve the problems together and discuss how the context of the problem affects the way in which the solution is delivered.

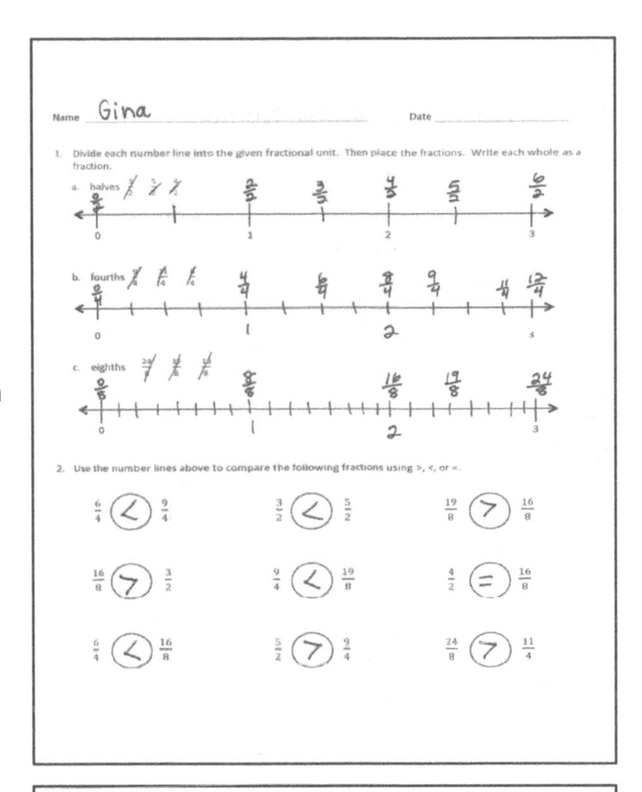

Name Gina Date

1. Divide each number line into the given fractional unit. Then place the fractions. Write each whole as a fraction.

a. halves

b. fourths

c. eighths

2. Use the number lines above to compare the following fractions using >, <, or =.

$\frac{6}{4} < \frac{9}{4}$	$\frac{3}{2} < \frac{5}{2}$	$\frac{19}{8} > \frac{16}{8}$
$\frac{16}{8} > \frac{3}{2}$	$\frac{9}{4} < \frac{19}{8}$	$\frac{4}{2} = \frac{16}{8}$
$\frac{6}{4} < \frac{16}{8}$	$\frac{5}{2} > \frac{9}{4}$	$\frac{24}{8} > \frac{11}{4}$

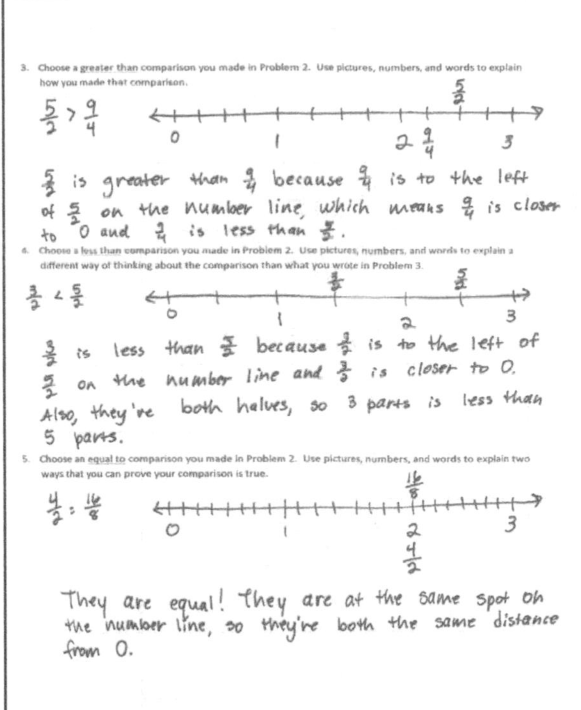

3. Choose a greater than comparison you made in Problem 2. Use pictures, numbers, and words to explain how you made that comparison.

$\frac{5}{2} > \frac{9}{4}$

$\frac{5}{2}$ is greater than $\frac{9}{4}$ because $\frac{9}{4}$ is to the left of $\frac{5}{2}$ on the number line, which means $\frac{9}{4}$ is closer to 0 and $\frac{9}{4}$ is less than $\frac{5}{2}$.

4. Choose a less than comparison you made in Problem 2. Use pictures, numbers, and words to explain a different way of thinking about the comparison than what you wrote in Problem 3.

$\frac{3}{2} < \frac{5}{2}$

$\frac{3}{2}$ is less than $\frac{5}{2}$ because $\frac{3}{2}$ is to the left of $\frac{5}{2}$ on the number line and $\frac{3}{2}$ is closer to 0. Also, they're both halves, so 3 parts is less than 5 parts.

5. Choose an equal to comparison you made in Problem 2. Use pictures, numbers, and words to explain two ways that you can prove your comparison is true.

$\frac{4}{2} = \frac{16}{8}$

They are equal! They are at the same spot on the number line, so they're both the same distance from 0.

Exit Ticket (3 minutes)

After the Student Debrief, instruct students to complete the Exit Ticket. A review of their work will help with assessing students' understanding of the concepts that were presented in today's lesson and planning more effectively for future lessons. The questions may be read aloud to the students.

A

Number Correct: _______

Express Fractions as Whole Numbers

1.	$^2/_1$ =		23.	$^6/_3$ =	
2.	$^2/_2$ =		24.	$^3/_3$ =	
3.	$^4/_2$ =		25.	$^3/_1$ =	
4.	$^6/_2$ =		26.	$^9/_3$ =	
5.	$^{10}/_2$ =		27.	$^{16}/_4$ =	
6.	$^8/_2$ =		28.	$^{20}/_4$ =	
7.	$^5/_1$ =		29.	$^{12}/_3$ =	
8.	$^5/_5$ =		30.	$^{15}/_3$ =	
9.	$^{10}/_5$ =		31.	$^{70}/_{10}$ =	
10.	$^{15}/_5$ =		32.	$^{12}/_2$ =	
11.	$^{25}/_5$ =		33.	$^{14}/_2$ =	
12.	$^{20}/_5$ =		34.	$^{90}/_{10}$ =	
13.	$^{10}/_{10}$ =		35.	$^{30}/_5$ =	
14.	$^{50}/_{10}$ =		36.	$^{35}/_5$ =	
15.	$^{30}/_{10}$ =		37.	$^{60}/_{10}$ =	
16.	$^{10}/_1$ =		38.	$^{18}/_2$ =	
17.	$^{20}/_{10}$ =		39.	$^{40}/_5$ =	
18.	$^{40}/_{10}$ =		40.	$^{80}/_{10}$ =	
19.	$^8/_4$ =		41.	$^{16}/_2$ =	
20.	$^4/_4$ =		42.	$^{45}/_5$ =	
21.	$^4/_1$ =		43.	$^{27}/_3$ =	
22.	$^{12}/_4$ =		44.	$^{32}/_4$ =	

B

Number Correct: _______

Improvement: _______

Express Fractions as Whole Numbers

1.	$5/1$ =			23.	$8/4$ =	
2.	$5/5$ =			24.	$4/4$ =	
3.	$10/5$ =			25.	$4/1$ =	
4.	$15/5$ =			26.	$12/4$ =	
5.	$25/5$ =			27.	$12/3$ =	
6.	$20/5$ =			28.	$15/3$ =	
7.	$2/1$ =			29.	$16/4$ =	
8.	$2/2$ =			30.	$20/4$ =	
9.	$4/2$ =			31.	$90/10$ =	
10.	$6/2$ =			32.	$30/5$ =	
11.	$10/2$ =			33.	$35/5$ =	
12.	$8/2$ =			34.	$70/10$ =	
13.	$10/1$ =			35.	$12/2$ =	
14.	$10/10$ =			36.	$14/2$ =	
15.	$50/10$ =			37.	$80/10$ =	
16.	$30/10$ =			38.	$45/5$ =	
17.	$20/10$ =			39.	$16/2$ =	
18.	$40/10$ =			40.	$60/10$ =	
19.	$6/3$ =			41.	$18/2$ =	
20.	$3/3$ =			42.	$40/5$ =	
21.	$3/1$ =			43.	$36/4$ =	
22.	$9/3$ =			44.	$24/3$ =	

Name ______________________________ Date ________________

1. Divide each number line into the given fractional unit. Then, place the fractions. Write each whole as a fraction.

   a. halves $\frac{3}{2}$ $\frac{5}{2}$ $\frac{4}{2}$

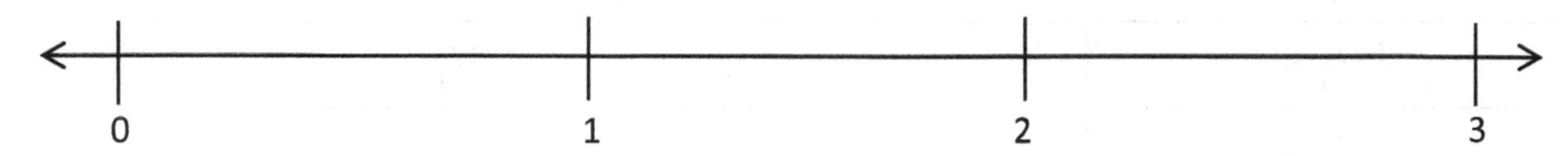

   b. fourths $\frac{9}{4}$ $\frac{11}{4}$ $\frac{6}{4}$

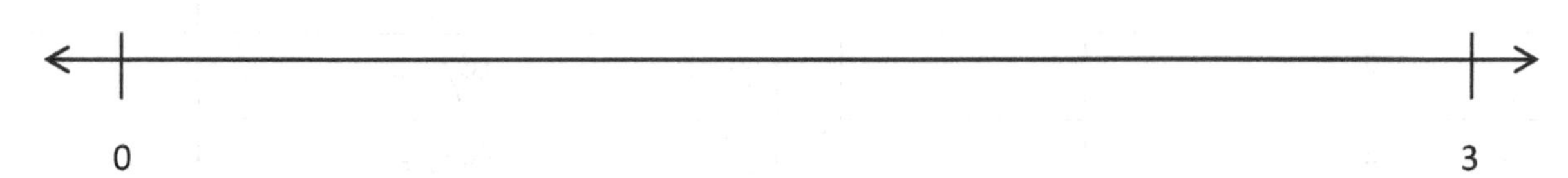

   c. eighths $\frac{24}{8}$ $\frac{19}{8}$ $\frac{16}{8}$

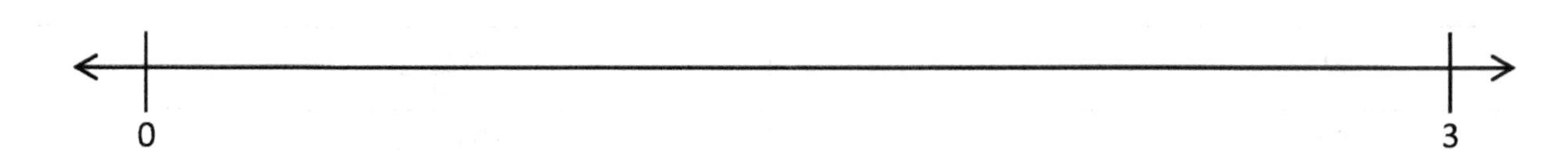

2. Use the number lines above to compare the following fractions using >, <, or =.

$\frac{6}{4}$ ◯ $\frac{9}{4}$ $\frac{3}{2}$ ◯ $\frac{5}{2}$ $\frac{19}{8}$ ◯ $\frac{16}{8}$

$\frac{16}{8}$ ◯ $\frac{3}{2}$ $\frac{9}{4}$ ◯ $\frac{19}{8}$ $\frac{4}{2}$ ◯ $\frac{16}{8}$

$\frac{6}{4}$ ◯ $\frac{16}{8}$ $\frac{5}{2}$ ◯ $\frac{9}{4}$ $\frac{24}{8}$ ◯ $\frac{11}{4}$

3. Choose a *greater than* comparison you made in Problem 2. Use pictures, numbers, and words to explain how you made that comparison.

4. Choose a *less than* comparison you made in Problem 2. Use pictures, numbers, and words to explain a different way of thinking about the comparison than what you wrote in Problem 3.

5. Choose an *equal to* comparison you made in Problem 2. Use pictures, numbers, and words to explain two ways that you can prove your comparison is true.

Name ____________________________ Date ____________

1. Divide the number line into the given fractional unit. Then, place the fractions. Write each whole as a fraction.

   fourths $\frac{2}{4}$ $\frac{10}{4}$ $\frac{7}{4}$

2. Use the number line above to compare the following fractions using >, <, or =.

   $\frac{3}{4}$ ◯ $\frac{5}{4}$ $\frac{7}{4}$ ◯ $\frac{4}{4}$ 3 ◯ $\frac{6}{4}$

3. Use the number line from Problem 1. Which is larger: 2 wholes or $\frac{9}{4}$? Use words, pictures, and numbers to explain your answer.

Name ______________________________ Date ______________

1. Divide each number line into the given fractional unit. Then, place the fractions. Write each whole as a fraction.

a. thirds $\frac{6}{3}$ $\frac{5}{3}$ $\frac{8}{3}$

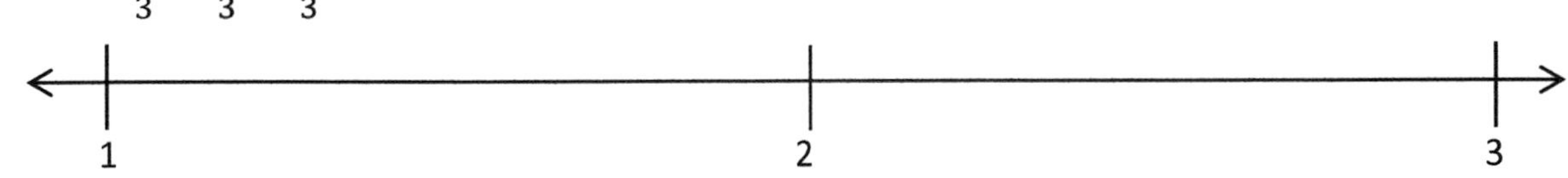

b. sixths $\frac{10}{6}$ $\frac{18}{6}$ $\frac{15}{6}$

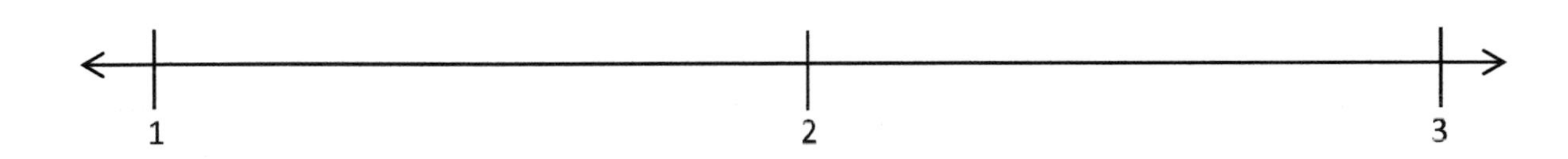

c. fifths $\frac{14}{5}$ $\frac{7}{5}$ $\frac{11}{5}$

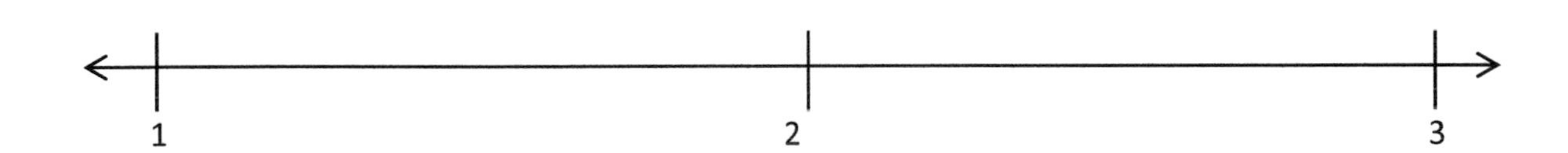

2. Use the number lines above to compare the following fractions using >, <, or =.

$\frac{17}{6}$ ◯ $\frac{15}{6}$ $\frac{7}{3}$ ◯ $\frac{9}{3}$ $\frac{11}{5}$ ◯ $\frac{8}{5}$

$\frac{4}{3}$ ◯ $\frac{8}{6}$ $\frac{13}{6}$ ◯ $\frac{8}{3}$ $\frac{11}{6}$ ◯ $\frac{5}{3}$

$\frac{10}{6}$ ◯ $\frac{3}{3}$ $\frac{6}{3}$ ◯ $\frac{12}{6}$ $\frac{15}{5}$ ◯ $\frac{5}{3}$

3. Use fractions from the number lines in Problem 1. Complete the sentence. Use words, pictures, or numbers to explain how you made that comparison.

 ____________ is *greater than* ____________.

4. Use fractions from the number lines in Problem 1. Complete the sentence. Use words, pictures, or numbers to explain how you made that comparison.

 ____________ is *less than* ____________.

5. Use fractions from the number lines in Problem 1. Complete the sentence. Use words, pictures, or numbers to explain how you made that comparison.

 ____________ is *equal to* ____________.

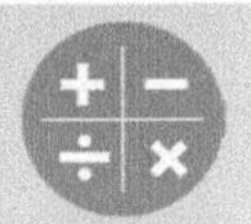

Topic E

Equivalent Fractions

3.NF.3a–c

Focus Standard:	3.NF.3	Explain equivalence of fractions in special cases, and compare fractions by reasoning about their size. a. Understand two fractions as equivalent (equal) if they are the same size, or the same point on a number line. b. Recognize and generate simple equivalent fractions, e.g., 1/2 = 2/4, 4/6 = 2/3. Explain why the fractions are equivalent, e.g., by using a visual fraction model. c. Express whole numbers as fractions, and recognize fractions that are equivalent to whole numbers. *Examples: Express 3 in the form 3 = 3/1; recognize that 6/1 = 6; locate 4/4 and 1 at the same point of a number line diagram.*
Instructional Days:	8	
Coherence -Links from:	G2–M8	Time, Shapes, and Fractions as Equal Parts of Shapes
-Links to:	G4–M5	Fraction Equivalence, Ordering, and Operations

In Topic D, students practiced placing and comparing fractions on a number line. In Topic E, they identify equivalent fractions using fraction strips, number bonds, and the number line as models. Students compare fractions on the number line to recognize that equivalent fractions refer to the same whole. They say $\frac{1}{2} = \frac{2}{4}$, assuming they are comparing like units (e.g. $\frac{1}{2}$ gallon = $\frac{2}{4}$ gallon rather than $\frac{1}{2}$ cup = $\frac{2}{4}$ gallon). Likewise, when they model $\frac{1}{2} = \frac{2}{4}$ on the number line, both fractions are in reference to the same length unit, the same whole. Equivalent fractions are different ways to represent the same number, the same point on the number line. Initially, students find equivalence in fractions less than 1 whole (e.g., 1 half = 2 fourths). They then express whole numbers as fractions, using number bonds and number lines, to show how many copies of a unit are needed to make the whole (e.g., 4 copies of 1 fourth equals 1 whole). They reason about why whole numbers can be written as fractions with a denominator of 1. Finally, students explain equivalence through manipulating units.

A Teaching Sequence Toward Mastery of Equivalent Fractions	
Objective 1:	**Recognize and show that equivalent fractions have the same size, though not necessarily the same shape. (Lesson 20)**
Objective 2:	**Recognize and show that equivalent fractions refer to the same point on the number line. (Lesson 21)**
Objective 3:	**Generate simple equivalent fractions by using visual fraction models and the number line. (Lessons 22–23)**
Objective 4:	**Express whole numbers as fractions and recognize equivalence with different units. (Lesson 24)**
Objective 5:	**Express whole number fractions on the number line when the unit interval is 1. (Lesson 25)**
Objective 6:	**Decompose whole number fractions greater than 1 using whole number equivalence with various models. (Lesson 26)**
Objective 7:	**Explain equivalence by manipulating units and reasoning about their size. (Lesson 27)**

Lesson 20

Objective: Recognize and show that equivalent fractions have the same size, though not necessarily the same shape.

Suggested Lesson Structure

■ Fluency Practice	(9 minutes)
■ Application Problem	(8 minutes)
■ Concept Development	(33 minutes)
■ Student Debrief	(10 minutes)
Total Time	**(60 minutes)**

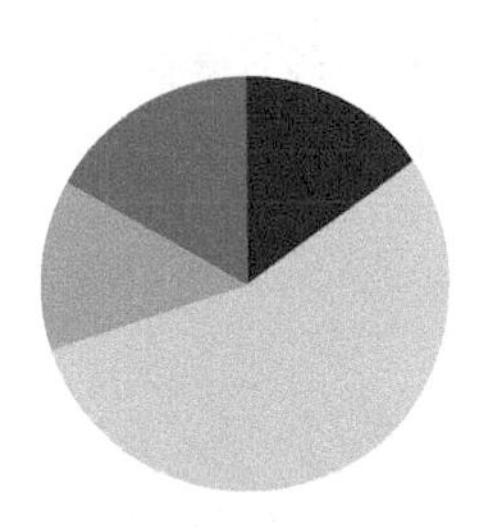

Fluency Practice (9 minutes)

- Multiply by 7 **3.OA.4** (9 minutes)

Multiply by 7 (9 minutes)

Materials: (S) Multiply by 7 (1–5) Pattern Sheet

Note: This Pattern Sheet supports fluency with multiplication using units of 7.

T: Skip-count by sevens. (Write multiples horizontally as students count.)

S: 7, 14, 21, 28, 35, 42, 49, 56, 63, 70.

T: (Write 5 × 7 = ____.) Let's skip-count by sevens to find the answer. (Count with fingers to 5 as students count.)

S: 7, 14, 21, 28, 35.

T: (Circle 35 and write 5 × 7 = 35 above it. Write 3 × 7 = ____.) Let's skip-count up by sevens again. (As students count, show fingers to count with them.)

S: 7, 14, 21.

T: Let's see how we can skip-count down to find the answer, too. Start at 35. (Count down with fingers as students say numbers.)

S: 35, 28, 21.

T: (Write 9 × 7 = ____.) Let's skip-count up by sevens. (Count with fingers to 9 as students count.)

S: 7, 14, 21, 28, 35, 42, 49, 56, 63.

T: Let's see how we can skip-count down to find the answer, too. Start at 70. (Count down with fingers as student say numbers.)

S: 70, 63.

Continue with the following possible sequence: 6×7, 8×7, and 4×7.

T: (Distribute the Multiply by 7 Pattern Sheet.) Let's practice multiplying by 7. Be sure to work left to right across the page.

Directions for administration of Multiply-By Pattern Sheet are as follows:

1. Distribute the Pattern Sheet.
2. Allow a maximum of two minutes for students to complete as many problems as possible.
3. Direct students to work left to right across the page.
4. Encourage skip-counting strategies to solve unknown facts.

Application Problem (8 minutes)

Max ate $\frac{2}{3}$ of his pizza for lunch. He wanted to eat a small snack in the afternoon, so he cut the leftover pizza in half and ate 1 slice. How much of the pizza was left? Draw a picture to help you think about the pizza.

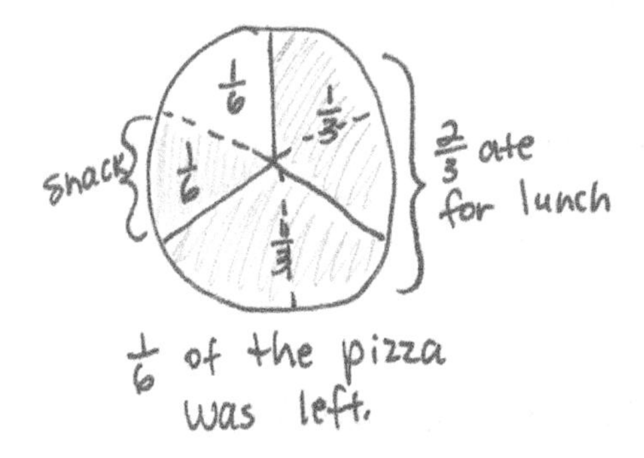

Note: This problem reviews partitioning a whole into equal parts from Topic A. Invite students to share their models and discuss why $\frac{1}{2}$ is not a reasonable answer, even though Max cut the leftover pizza in half.

NOTES ON MULTIPLE MEANS OF REPRESENTATION:

Empower English language learners to solve the Application Problem by connecting its context to their prior knowledge. Discuss their experiences at lunch, eating pizza, having leftovers, and eating a snack.

Concept Development (33 minutes)

Materials: (T) Linking cubes in 2 colors (S) Thirds (Template), red crayon, scissors, glue stick, and blank paper

Use linking cubes to create Model 1, as shown to the right.

Model 1

T: The whole is all of the cubes. Whisper to your partner the fraction of cubes that are blue.

S: (Whisper $\frac{1}{4}$.)

Use linking cubes to create Model 2, as shown to the right.

Model 2

T: Again, the whole is all of the cubes. Whisper to your partner the fraction of cubes that are blue.

S: (Whisper $\frac{1}{4}$.)

T: Discuss with your partner whether the fraction of cubes that are blue in these models is equal, even though the models are not the same shape.

S: They don't look the same, so they are different. → I disagree. They are equal because they are both $\frac{1}{4}$ blue. → They are equal because the units are still the same size, and the wholes have the same number of units. They are in a different shape.

T: I hear you noticing that the units make a different shape in the second model. It's square rather than rectangular. Good observation. Take another minute to notice what is similar about our models.

S: They both use the same linking cubes as units. → They both have the same amount of blues and reds. → Both wholes have the same number of units, and the units are the same size.

T: The size of the units and the size of the whole didn't change. That means $\frac{1}{4}$ and $\frac{1}{4}$ are equal, or what we call equivalent fractions, even though the shapes of our wholes are different.

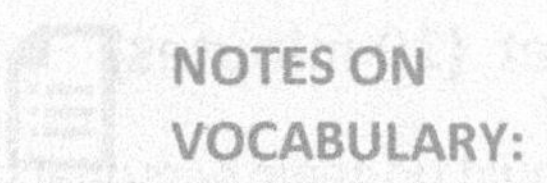

The concept of equivalent fractions was first introduced in Lesson 16 in reference to fractions that are at the same point on the number line. In this lesson, the students' understanding of equivalent fractions expands to include pictorial models, where the equivalent fractions name the same size. Guide students to recognize the differences and similarities between these methods for finding equivalent fractions.

If necessary, do other examples to demonstrate the point made with Model 2.

Use linking cubes to create Model 3, as shown to the right.

Model 3

T: Why isn't the fraction represented by the blue cubes equal to the other fractions we made with cubes?

S: This fraction shows $\frac{2}{4}$ of the cubes are blue.

T: When we are finding equivalent fractions, the shapes of the wholes can be different. However, equivalent fractions must describe parts of the whole that are the same size.

Equivalent Shapes Collage Activity

Thirds Template

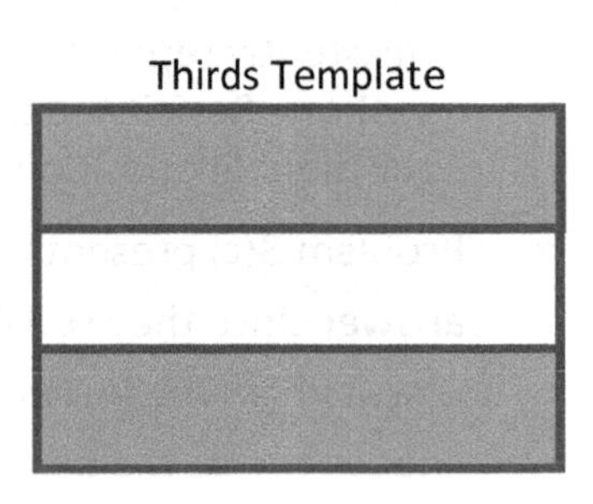

Students use the thirds template, and follow the directions below to create various representations of 2 thirds.

Directions for this activity are as follows:

1. Color the white 1 third red.
2. Cut out the rectangle. Cut it into 2–4 smaller shapes.
3. Reassemble all of the pieces into a new shape with no overlaps.
4. Glue the new shape onto a blank paper.

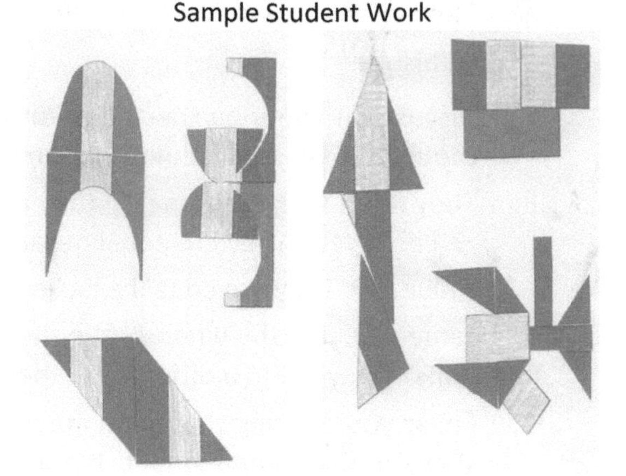
Sample Student Work

Invite students to look at their classmates' work and discuss the equivalence represented by these shapes. Each of the 6 shapes pictured to the right is an example of possible student work. These shapes are equivalent because they all show $\frac{2}{3}$ grey, although clearly in different shapes.

Problem Set (10 minutes)

Students should do their personal best to complete the Problem Set within the allotted 10 minutes. For some classes, it may be appropriate to modify the assignment by specifying which problems they work on first. Some problems do not specify a method for solving. Students should solve these problems using the RDW approach used for Application Problems.

Student Debrief (10 minutes)

Lesson Objective: Recognize and show that equivalent fractions have the same size, though not necessarily the same shape.

The Student Debrief is intended to invite reflection and active processing of the total lesson experience.

Invite students to review their solutions for the Problem Set. They should check work by comparing answers with a partner before going over answers as a class. Look for misconceptions or misunderstandings that can be addressed in the Student Debrief. Guide students in a conversation to debrief the Problem Set and process the lesson.

Any combination of the questions below may be used to lead the discussion.

MP.6

- Invite students to share their models for Problems 2(a) and 2(b). Although answers will vary, students should consistently represent equivalent fractions for each question. Revisit the different work from the Equivalent Shapes Collage Activity.
- Problem 3(c) presents seeing triangles as halves of squares. Some students might put $\frac{4}{8}$ as the answer since they see 8 units. You may want to pose the question, "Are all 8 parts equal units?" Discuss how the answer can be $\frac{4}{12}$ if students choose to use the base unit of triangles or $\frac{2}{6}$ if they choose to use the base unit of squares. Guide them to see that the two fractions are equivalent.

NOTES ON MULTIPLE MEANS OF ACTION AND EXPRESSION:

For students working below grade level, break the task of labeling fractions on the Problem Set into steps with sentence frames:

- There are ____ equal parts.
- ____ parts are shaded.
- The fraction shaded is _____.

The open-ended questions on the Problem Set are just right for students working above grade level who enjoy independence. Communicate high expectations for explaining their reasoning clearly with evidence.

- Problem 4 also presents an interesting discussion topic because of the use of containers that are different shapes with the same capacity. Without reading carefully, students are likely to make a mistake in their answer. This may provide an opportunity to further explore the difference between different-sized wholes and different-looking wholes.
- Earlier, you learned that equivalent fractions are at the same point on the number line. How did your understanding of equivalent fractions change today?

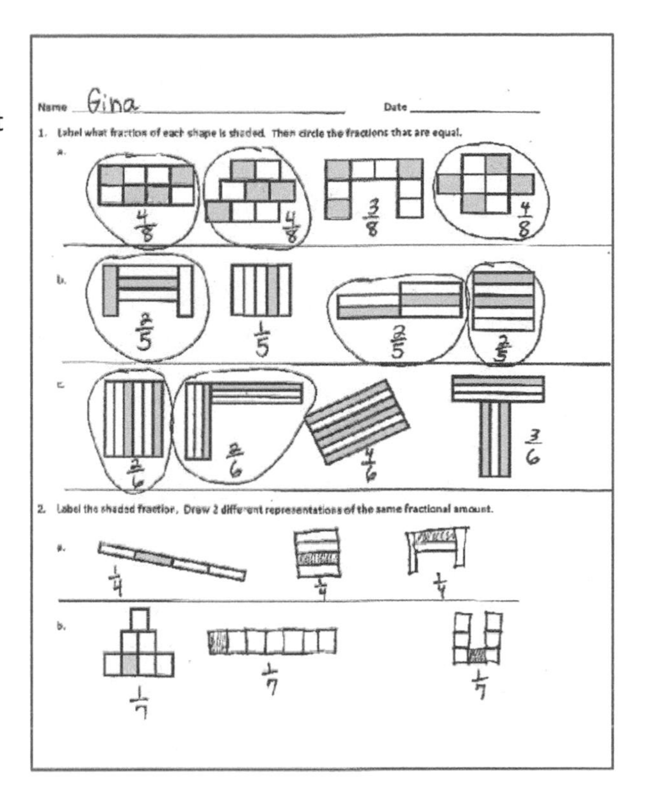

Name Gina Date

1. Label what fraction of each shape is shaded. Then circle the fractions that are equal.

a. $\frac{4}{8}$ $\frac{4}{8}$ $\frac{3}{8}$ $\frac{4}{8}$

b. $\frac{2}{5}$ $\frac{1}{5}$ $\frac{2}{5}$ $\frac{2}{5}$

c. $\frac{2}{6}$ $\frac{2}{6}$ $\frac{4}{6}$ $\frac{3}{6}$

2. Label the shaded fraction. Draw 2 different representations of the same fractional amount.

a. $\frac{1}{4}$ $\frac{1}{4}$ $\frac{1}{4}$

b. $\frac{1}{7}$ $\frac{1}{7}$ $\frac{1}{7}$

Exit Ticket (3 minutes)

After the Student Debrief, instruct students to complete the Exit Ticket. A review of their work will help with assessing students' understanding of the concepts that were presented in today's lesson and planning more effectively for future lessons. The questions may be read aloud to the students.

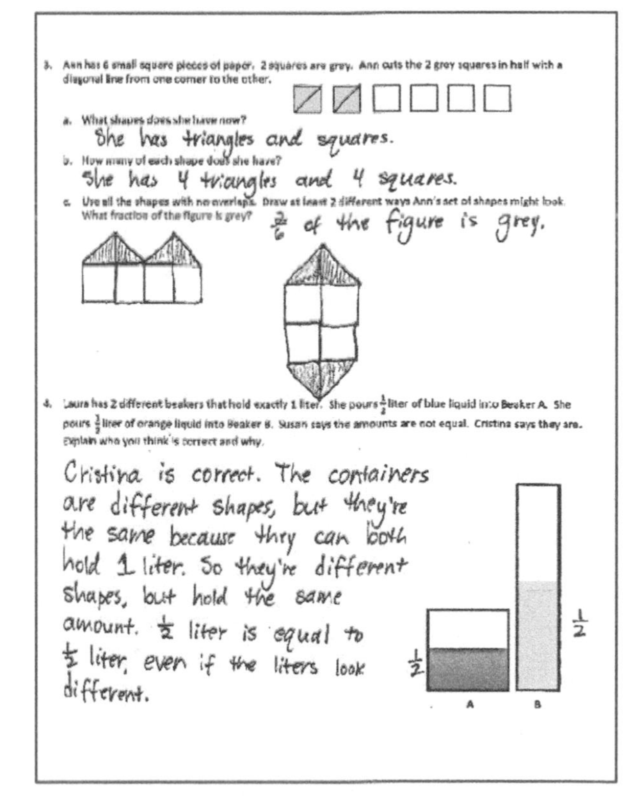

3. Ann has 6 small square pieces of paper. 2 squares are grey. Ann cuts the 2 grey squares in half with a diagonal line from one corner to the other.

a. What shapes does she have now?
She has triangles and squares.

b. How many of each shape does she have?
She has 4 triangles and 4 squares.

c. Use all the shapes with no overlaps. Draw at least 2 different ways Ann's set of shapes might look. What fraction of the figure is grey?
$\frac{2}{6}$ of the figure is grey.

4. Laura has 2 different beakers that hold exactly 1 liter. She pours $\frac{1}{2}$ liter of blue liquid into Beaker A. She pours $\frac{1}{2}$ liter of orange liquid into Beaker B. Susan says the amounts are not equal. Cristina says they are. Explain who you think is correct and why.

Cristina is correct. The containers are different shapes, but they're the same because they can both hold 1 liter. So they're different shapes, but hold the same amount. $\frac{1}{2}$ liter is equal to $\frac{1}{2}$ liter, even if the liters look different.

Multiply.

7 x 1 = ______	7 x 2 = ______	7 x 3 = ______	7 x 4 = ______
7 x 5 = ______	7 x 1 = ______	7 x 2 = ______	7 x 1 = ______
7 x 3 = ______	7 x 1 = ______	7 x 4 = ______	7 x 1 = ______
7 x 5 = ______	7 x 1 = ______	7 x 2 = ______	7 x 3 = ______
7 x 2 = ______	7 x 4 = ______	7 x 2 = ______	7 x 5 = ______
7 x 2 = ______	7 x 1 = ______	7 x 2 = ______	7 x 3 = ______
7 x 1 = ______	7 x 3 = ______	7 x 2 = ______	7 x 3 = ______
7 x 4 = ______	7 x 3 = ______	7 x 5 = ______	7 x 3 = ______
7 x 4 = ______	7 x 1 = ______	7 x 4 = ______	7 x 2 = ______
7 x 4 = ______	7 x 3 = ______	7 x 4 = ______	7 x 5 = ______
7 x 4 = ______	7 x 5 = ______	7 x 1 = ______	7 x 5 = ______
7 x 2 = ______	7 x 5 = ______	7 x 3 = ______	7 x 5 = ______
7 x 4 = ______	7 x 2 = ______	7 x 4 = ______	7 x 3 = ______
7 x 5 = ______	7 x 3 = ______	7 x 2 = ______	7 x 4 = ______
7 x 3 = ______	7 x 5 = ______	7 x 2 = ______	7 x 4 = ______

multiply by 7 (1–5)

Name ______________________________ Date ______________

1. Label what fraction of each shape is shaded. Then, circle the fractions that are equal.

a.

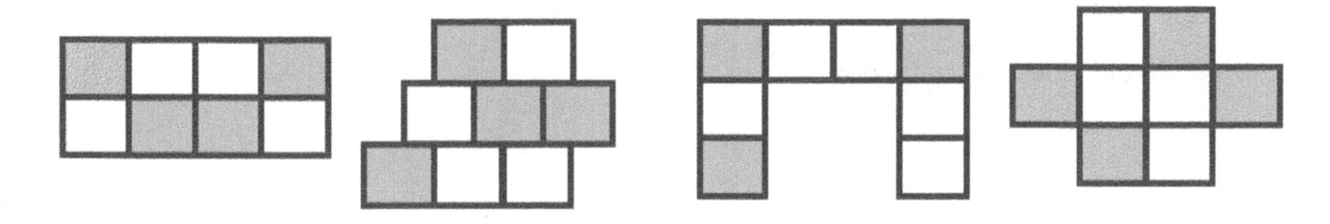

b.

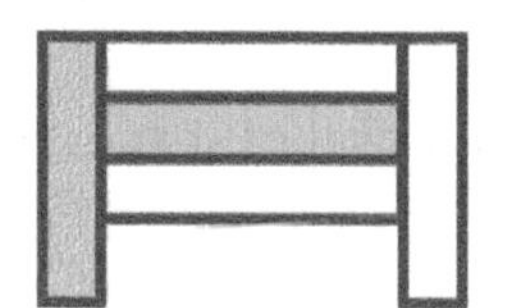

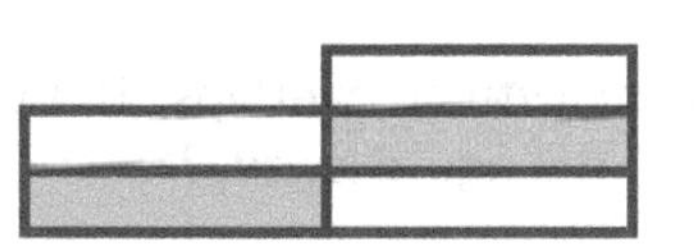
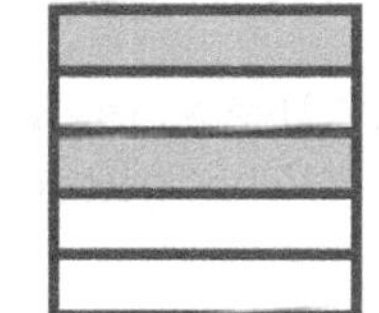

c.

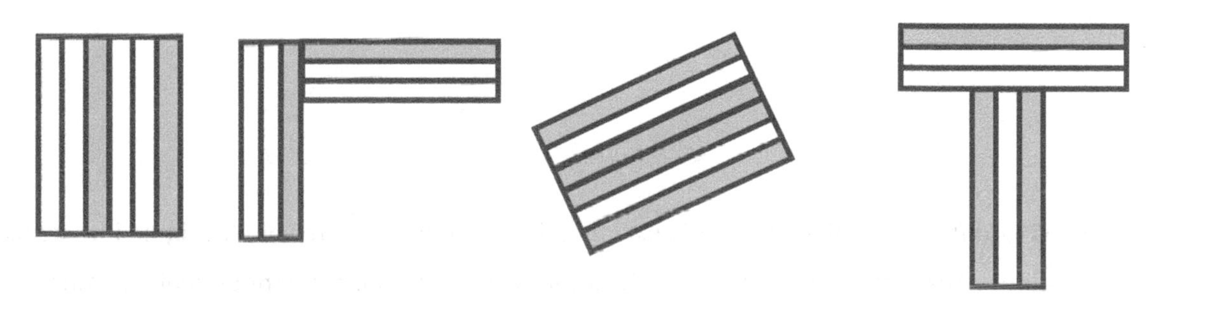

2. Label the shaded fraction. Draw 2 different representations of the same fractional amount.

a.

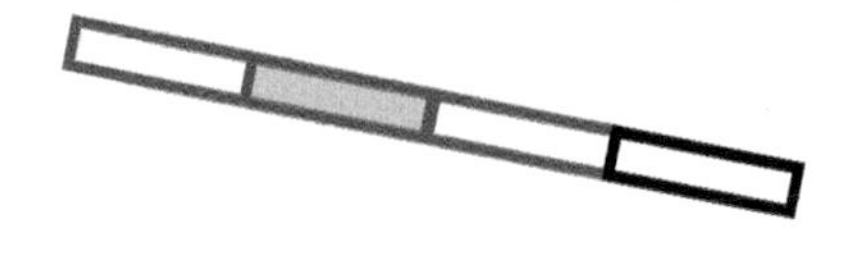

b.

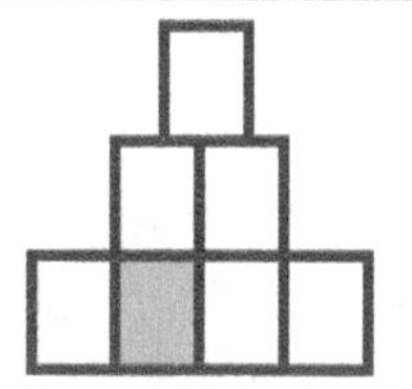

3. Ann has 6 small square pieces of paper. 2 squares are grey. Ann cuts the 2 grey squares in half with a diagonal line from one corner to the other.

   a. What shapes does she have now?

   b. How many of each shape does she have?

   c. Use all the shapes with no overlaps. Draw at least 2 different ways Ann's set of shapes might look. What fraction of the figure is grey?

4. Laura has 2 different beakers that hold exactly 1 liter. She pours $\frac{1}{2}$ liter of blue liquid into Beaker A. She pours $\frac{1}{2}$ liter of orange liquid into Beaker B. Susan says the amounts are not equal. Cristina says they are. Explain who you think is correct and why.

A B

Name ______________________________ Date ______________

1. Label what fraction of the figure is shaded. Then, circle the fractions that are equal.

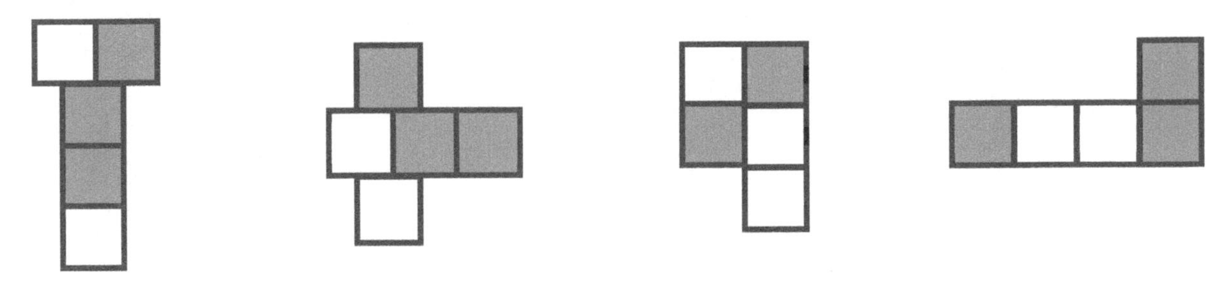

2. Label the shaded fraction. Draw 2 different representations of the same fractional amount.

a.

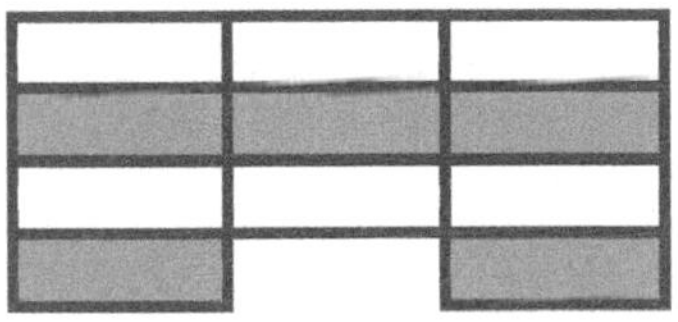

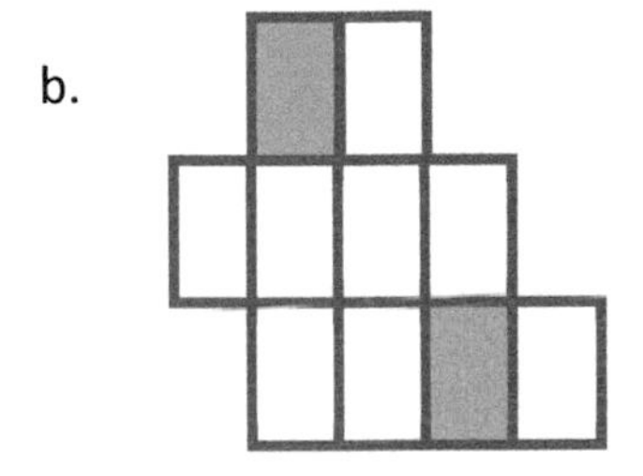

b.

Name ______________________________ Date ______________

1. Label the shaded fraction. Draw 2 different representations of the same fractional amount.

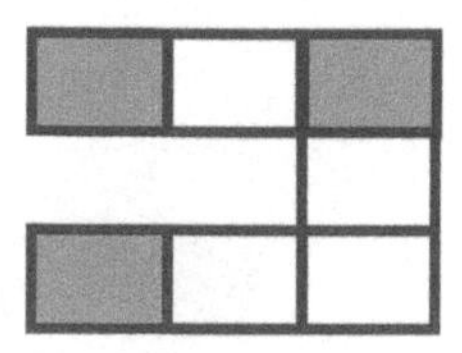

2. These two shapes both show $\frac{4}{5}$.

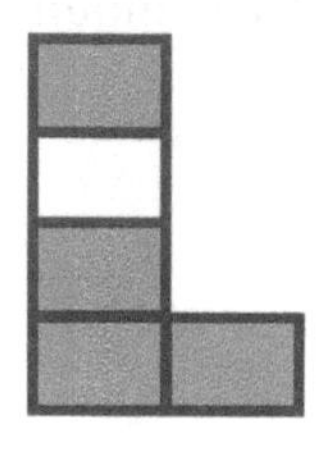

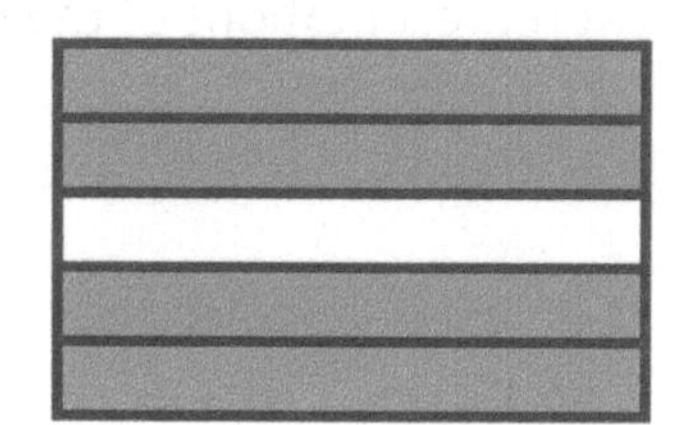

a. Are the shapes equivalent? Why or why not?

b. Draw two different representations of $\frac{4}{5}$ that are equivalent.

3. Diana ran a quarter mile straight down the street. Becky ran a quarter mile on a track. Who ran more? Explain your thinking.

Diana ______________

Becky 

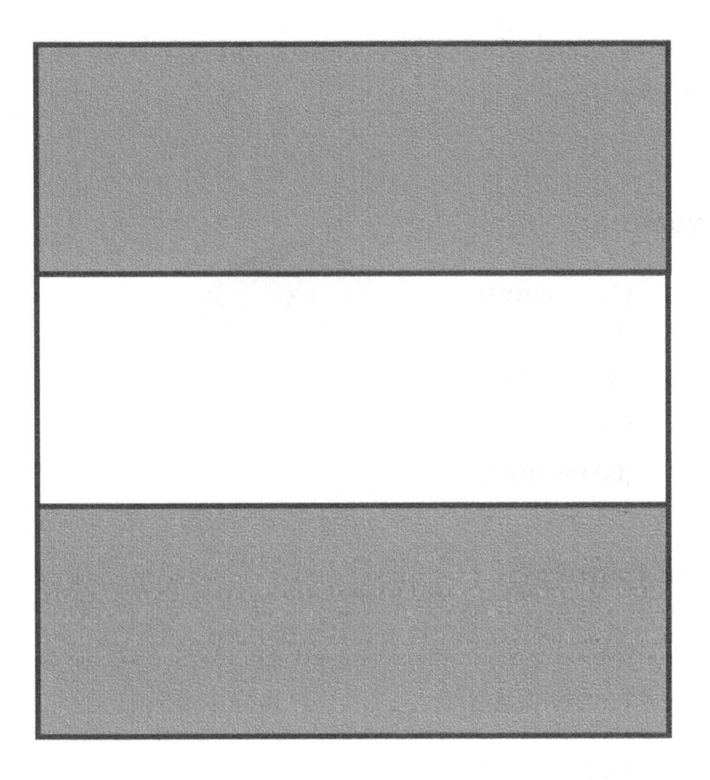

thirds

Lesson 21

Objective: Recognize and show that equivalent fractions refer to the same point on the number line.

Suggested Lesson Structure

■ Fluency Practice	(12 minutes)
■ Application Problem	(8 minutes)
■ Concept Development	(30 minutes)
■ Student Debrief	(10 minutes)
Total Time	**(60 minutes)**

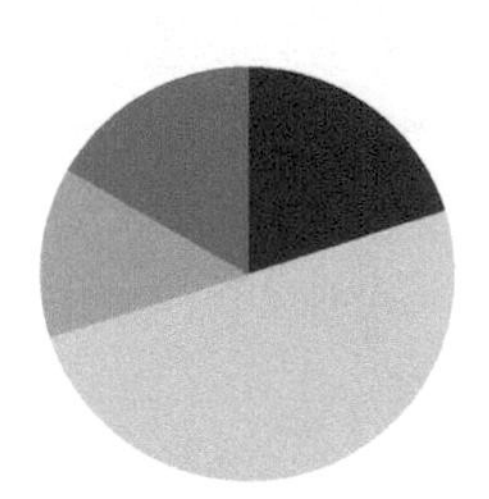

Fluency Practice (12 minutes)

- Whole Number Division **3.OA.7** (8 minutes)
- 1 Whole Expressed as Unit Fractions **3.NF.1** (4 minutes)

Whole Number Division (8 minutes)

Materials: (S) Blank paper

Note: This activity supports fluency with division. Steps 1 and 2 are timed for two minutes. Step 3 is timed for 1 minute of testing for each partner. Step 4 is timed for two minutes.

1. Students self-select a number and write a set of multiples up to that number's multiple of 10 vertically down the left-hand side of the page (e.g., 6, 12, 18, 24, 30, 36, 42, 48, 54, 60).
2. Select a multiple, and divide it by the original number (e.g., 24 ÷ 6 = 4).
3. Change papers and test a partner by selecting multiples out of order (e.g., "What is 24 ÷ 6?" "What is 54 ÷ 6?" "What is 12 ÷ 6?").
4. Redo Steps 1 and 2 to see improvement.

Let students know that the same activity will be done the next day, so they have a chance to practice and improve further, possibly advancing to the next number, which might further challenge them.

1 Whole Expressed as Unit Fractions (4 minutes)

Materials: (S) Personal white board

Note: This problem reviews the concept of using a number bond to decompose 1 whole into unit fractions from Topic A.

T: Draw a number bond that partitions a whole into 3 equal parts.

S: (Draw a number bond.)

T: What is the unit fraction?

S: 1 third.

Continue with the following possible sequence: halves, fourths, fifths, sixths, and eighths.

Application Problem (8 minutes)

Dorothea is training to run a 2-mile race. She marks off her starting point and the finish line. To track her progress, she places a mark at 1 mile. She then places a mark halfway between her starting position and 1 mile, and another mark halfway between 1 mile and the finish line.

a. Draw and label a number line to show the points Dorothea marks along her run.

b. What fractional unit does Dorothea make as she marks the points on her run?

c. What fraction of her run has she completed when she reaches the third marker?

a. $\frac{0}{4}$ $\frac{1}{4}$ $\frac{2}{4}$ $\frac{3}{4}$ $\frac{4}{4}$
0 miles start, 1 mile, 2 miles finish

b. Dorothea marks fourths.

c. She has finished $\frac{3}{4}$ of her run when she gets to the third marker.

Note: This problem reviews the importance of specifying the whole from Topic C. Invite students to discuss why the fractional units are fourths instead of halves.

Concept Development (30 minutes)

Materials: (S) $4\frac{1}{4}$-inch × 1-inch fraction strips (5 per student), math journal, crayons, glue, personal white board

T: We're going to make different fractional units with our fraction strips. Fold your first strip into halves.

S: (Fold fraction strip.)

T: Label each part with a unit fraction. Then, use a crayon to shade in 1 half.

S: (Label and shade.)

T: Glue your fraction strip at the top of a new page in your math journal.

S: (Glue fraction strip.)

T: Fold another fraction strip to make fourths. Label each part with a unit fraction. Then, glue your fraction strip directly below the first one in your math journal. Make sure that the ends are lined up.

S: (Fold, label, place, and glue fraction strip.)

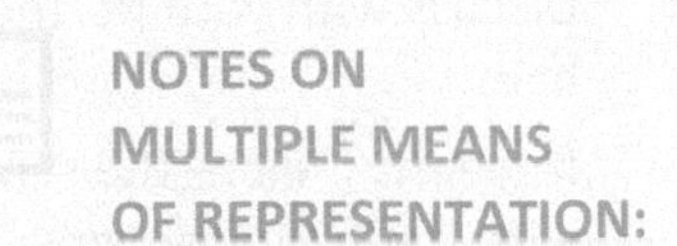

The vocabulary word equivalent has the advantage of cognates in many languages. Build English language learners' understanding of equivalent fractions through discussion, word webs, and questioning.

Ask the following:

- How are these equivalent fractions related?
- What particular property do they have in common?
- When might it be useful to interchange equivalent fractions?

T: Now, shade the number of fourths that are equivalent to the shaded half. Whisper to your partner how many units you shaded.

S: (Shade and whisper *2*.)

Guide students through the same sequence for a fraction strip folded into eighths.

T: Write the shaded fraction to the right of each fraction strip in your journal.

S: (Write $\frac{1}{2}$, $\frac{2}{4}$, and $\frac{4}{8}$.)

T: The fractional units are different. Discuss with a partner whether the fractions are equal or equivalent.

S: Since the fractional units are different, then they are not equal. → They have a different number of shaded parts, so I'm not sure. → The same amount of the fraction strip is shaded for each one. That must mean they're equal.

MP.7

T: I hear some uncertainty. Besides our fraction strips, what's another tool we can use to test their equivalence?

S: We can place them on a number line.

T: Let's do that. Place your personal white board under the fraction strip folded into halves. Use the fraction strip to measure a number line from 0 to 1. Label 0 halves, rename the whole, and then label $\frac{1}{2}$.

S: (Measure, draw, and label a number line.)

T: Move your board down so that your number line is under your fourths fraction strip. On the same number line, label the fourths. See if any fractions are located at the same point on the number line.

S: Hey, $\frac{1}{2}$ and $\frac{2}{4}$ are at the same point! → So are $\frac{2}{2}$, $\frac{4}{4}$, and 1. → Zeros too, but we already knew that!

T: Discuss with your partner what it means when two fractions are at the same point on the number line.

S: It means they're the same. → It proves what we saw with the fraction strips. They had the same amount shaded before, and now they're in the same place on the number line. → The fractions must be equivalent because they are at the same point.

T: I can use the equal sign to show that the fractions are equivalent when I write them. (Write $\frac{1}{2} = \frac{2}{4}$.) The equal sign is like a balance. It means *is the same as*. We might read this as $\frac{1}{2}$ is the same as $\frac{2}{4}$ because they have the same value. We just proved that with our number line! As long as the total values on both sides of the equal sign are the same, we can use it to show equivalence. (Write $\frac{2}{2} = \frac{4}{4} = 1$.) Turn and tell your partner: Is this statement true?

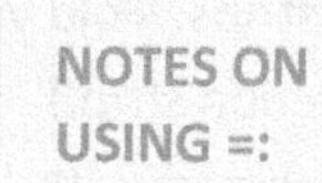

NOTES ON USING =:

It is worth spending a moment to ensure that students are clear on the meaning of the equal sign in this lesson because it is an important symbol throughout Topic E. Students become accustomed to associating its use with an operation and an answer, not fully understanding its application in a context such as $\frac{1}{2} = \frac{2}{4}$.

S: The equal sign works when there are two things, not three. → But the value of $\frac{2}{2}$ is 1, and $\frac{4}{4}$ is 1, and 1 is 1, so I think it's true. → Remember, we can also say *equals* as *is the same as*? $\frac{2}{2}$ is the same as $\frac{4}{4}$. Those are the same as 1. They are written differently, but they have the same value.

Instruct students to follow the same process to label eighths independently.

T: Fold your last 2 fraction strips. One should be thirds, and the other should be sixths. Label the parts with unit fractions, and glue these strips below the others in your math journal in order from greatest to least. Shade 1 third. Then, shade the number of sixths equal to 1 third.

S: (Fold, label, glue, and shade fraction strips.)

T: Now, work with your partner to measure and draw a new number line using your thirds and sixths. Then, using your other strips, find and label all of the fractions that are equivalent to thirds and sixths.

NOTES ON MULTIPLE MEANS OF ENGAGEMENT:

Slip the Problem Set into a clear plastic sheet protector. Using a dry erase marker, students working below grade level can highlight the intervals, shade unit fractions as they count, and circle equivalent fractions.

Present an open-ended alternative for students working above grade level who may enjoy finding unlimited equivalent fractions for a given point on the number line. Ask (for example), "How many equivalent fractions can you model for 3 halves?"

Note: If math journals are not used in the classroom, have students store these fraction strips in a safe place. They are used again in Lesson 22.

Problem Set (10 minutes)

Students should do their personal best to complete the Problem Set within the allotted 10 minutes. For some classes, it may be appropriate to modify the assignment by specifying which problems they work on first. Some problems do not specify a method for solving. Students should solve these problems using the RDW approach used for Application Problems.

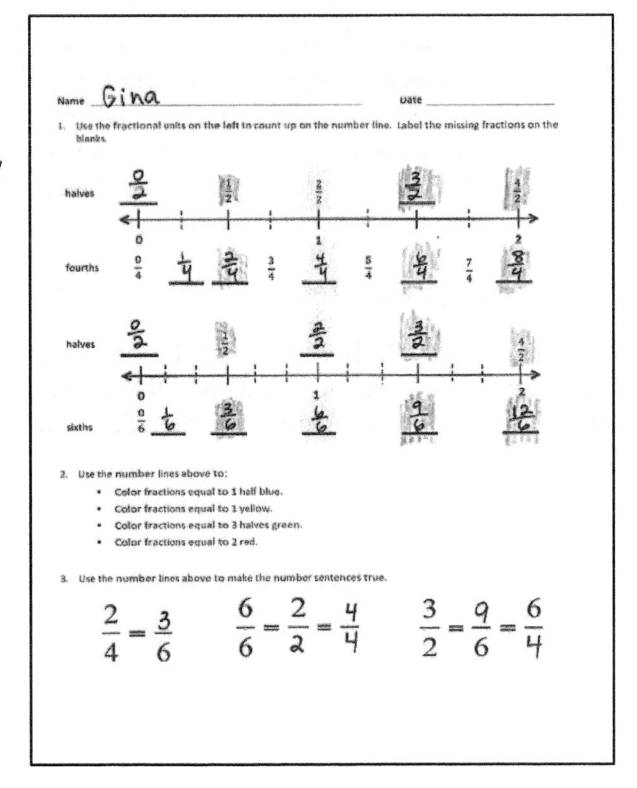

Name Gina Date

1. Use the fractional units on the left to count up on the number line. Label the missing fractions on the blanks.

halves: $\frac{0}{2}$, $\frac{1}{2}$, $\frac{2}{2}$, $\frac{3}{2}$, $\frac{4}{2}$

0, 1, 2

fourths: $\frac{0}{4}$, $\frac{1}{4}$, $\frac{2}{4}$, $\frac{3}{4}$, $\frac{4}{4}$, $\frac{5}{4}$, $\frac{6}{4}$, $\frac{7}{4}$, $\frac{8}{4}$

halves: $\frac{0}{2}$, $\frac{1}{2}$, $\frac{2}{2}$, $\frac{3}{2}$, $\frac{4}{2}$

0, 1, 2

sixths: $\frac{0}{6}$, $\frac{1}{6}$, $\frac{3}{6}$, $\frac{6}{6}$, $\frac{9}{6}$, $\frac{12}{6}$

2. Use the number lines above to:
   - Color fractions equal to 1 half blue.
   - Color fractions equal to 1 yellow.
   - Color fractions equal to 3 halves green.
   - Color fractions equal to 2 red.

3. Use the number lines above to make the number sentences true.

$\frac{2}{4} = \frac{3}{6}$ $\frac{6}{6} = \frac{2}{2} = \frac{4}{4}$ $\frac{3}{2} = \frac{9}{6} = \frac{6}{4}$

Student Debrief (10 minutes)

Lesson Objective: Recognize and show that equivalent fractions refer to the same point on the number line.

The Student Debrief is intended to invite reflection and active processing of the total lesson experience.

Invite students to review their solutions for the Problem Set. They should check work by comparing answers with a partner before going over answers as a class. Look for misconceptions or misunderstandings that can be addressed in the Student Debrief. Guide students in a conversation to debrief the Problem Set and process the lesson.

Any combination of the questions below may be used to lead the discussion.

- After students have checked their work for Problems 4 and 5, ask them to use the fraction strips in their math journals to see if they can name another equivalent fraction. ($\frac{3}{6}$ is the only possibility.) Ask students to talk about how they know the fractions are equivalent and possibly plot them on the same number line to emphasize the lesson objective.
- Guide students to articulate that equivalent fractions refer to the same point on the number line. They are different ways to show the same number! Ensure students are clear on what the word equivalent means and are comfortable using it.
- In anticipation of Lesson 22, ask students to look at Problem 4. Ask them to study the fractions equivalent to 1 whole. Have students notice that the number of shaded parts is the same as the total number of parts (numerator and denominator are the same). Have them use the pattern to name other fractions equivalent to 1 whole. Generate excitement by encouraging them to use extremely large numbers, as well as those that are more familiar.

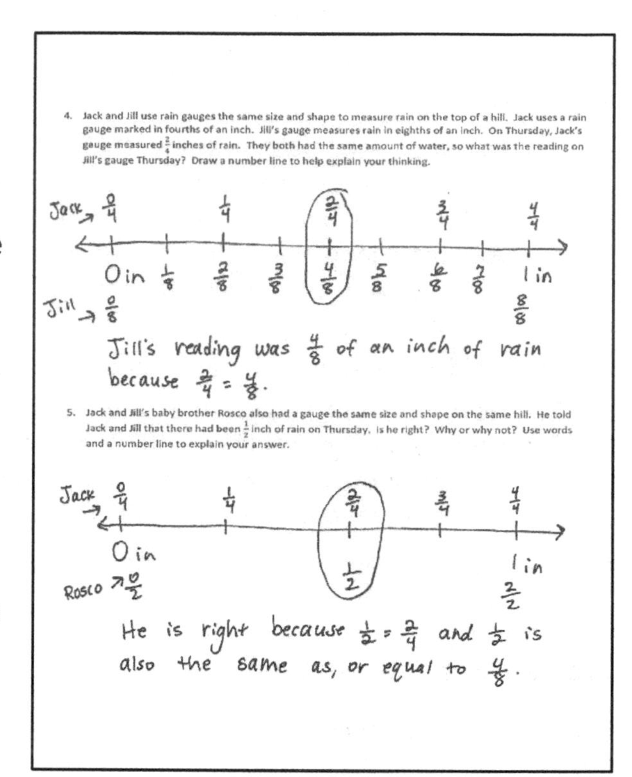
4. Jack and Jill use rain gauges the same size and shape to measure rain on the top of a hill. Jack uses a rain gauge marked in fourths of an inch. Jill's gauge measures rain in eighths of an inch. On Thursday, Jack's gauge measured $\frac{2}{4}$ inches of rain. They both had the same amount of water, so what was the reading on Jill's gauge Thursday? Draw a number line to help explain your thinking.

5. Jack and Jill's baby brother Rosco also had a gauge the same size and shape on the same hill. He told Jack and Jill that there had been $\frac{1}{2}$ inch of rain on Thursday. Is he right? Why or why not? Use words and a number line to explain your answer.

Exit Ticket (3 minutes)

After the Student Debrief, instruct students to complete the Exit Ticket. A review of their work will help with assessing students' understanding of the concepts that were presented in today's lesson and planning more effectively for future lessons. The questions may be read aloud to the students.

Name ______________________________ Date ______________

1. Use the fractional units on the left to count up on the number line. Label the missing fractions on the blanks.

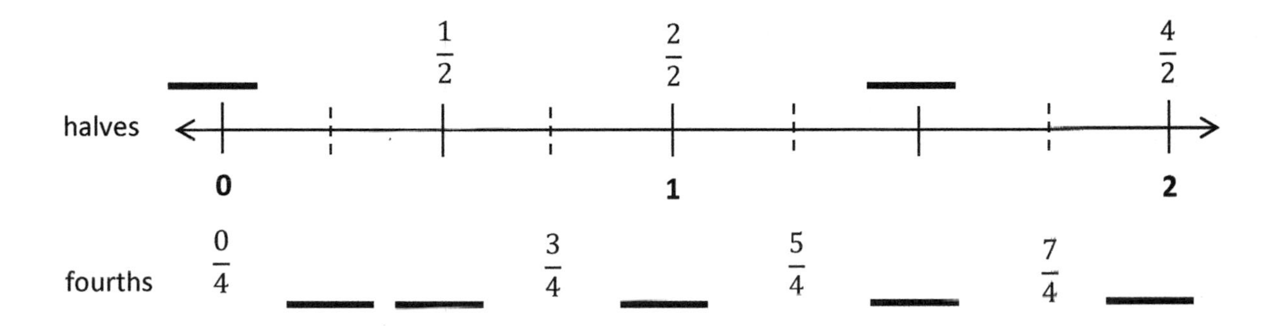

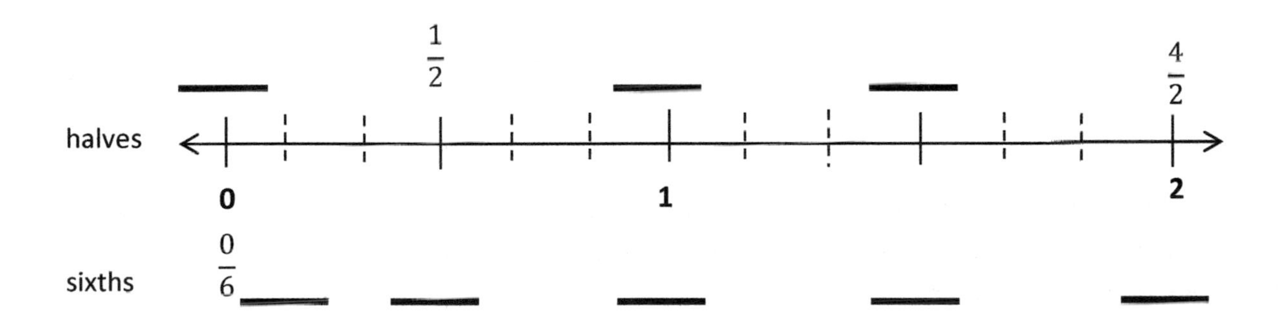

2. Use the number lines above to:
   - Color fractions equal to 1 half blue.
   - Color fractions equal to 1 yellow.
   - Color fractions equal to 3 halves green.
   - Color fractions equal to 2 red.

3. Use the number lines above to make the number sentences true.

$$\frac{2}{4} = \frac{\_}{6} \qquad \frac{6}{6} = \frac{2}{\_} = \frac{\_}{\_} \qquad \frac{3}{2} = \frac{\_}{6} = \frac{6}{\_}$$

4. Jack and Jill use rain gauges the same size and shape to measure rain on the top of a hill. Jack uses a rain gauge marked in fourths of an inch. Jill's gauge measures rain in eighths of an inch. On Thursday, Jack's gauge measured $\frac{2}{4}$ inches of rain. They both had the same amount of water, so what was the reading on Jill's gauge Thursday? Draw a number line to help explain your thinking.

5. Jack and Jill's baby brother Rosco also had a gauge the same size and shape on the same hill. He told Jack and Jill that there had been $\frac{1}{2}$ inch of rain on Thursday. Is he right? Why or why not? Use words and a number line to explain your answer.

Name ______________________________ Date ________________

Claire went home after school and told her mother that 1 whole is the same as $\frac{2}{2}$ and $\frac{6}{6}$. Her mother asked why, but Claire couldn't explain. Use a number line and words to help Claire show and explain why $1 = \frac{2}{2} = \frac{6}{6}$.

Name ______________________________ Date ______________

1. Use the fractional units on the left to count up on the number line. Label the missing fractions on the blanks.

fourths

eighths

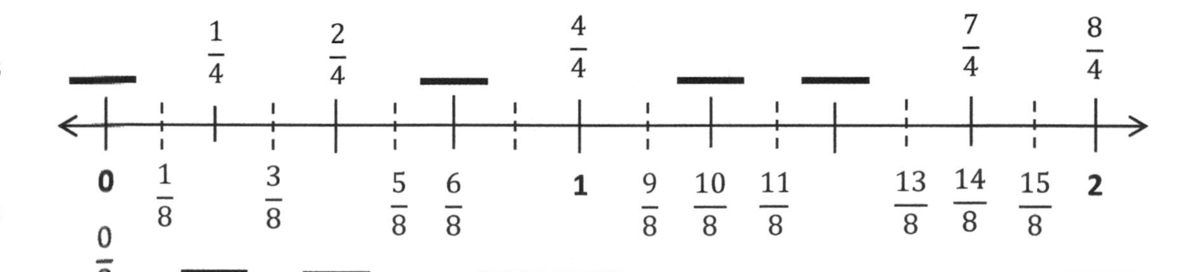

thirds

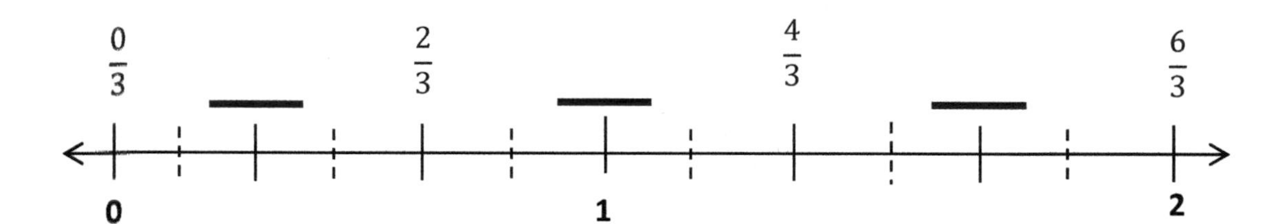

sixths

2. Use the number lines above to:
   - Color fractions equal to 1 purple.
   - Color fractions equal to 2 fourths yellow.
   - Color fractions equal to 2 blue.
   - Color fractions equal to 5 thirds green.
   - Write a pair of fractions that are equivalent.

______________ = ______________

3. Use the number lines on the previous page to make the number sentences true.

$\frac{1}{4} = \frac{\quad}{8}$ $\frac{6}{4} = \frac{12}{\quad}$ $\frac{2}{3} = \frac{\quad}{6}$

$\frac{6}{3} = \frac{12}{\quad}$ $\frac{3}{3} = \frac{\quad}{6}$ $2 = \frac{8}{4} = \frac{\quad}{8}$

4. Mr. Fairfax ordered 3 large pizzas for a class party. Group A ate $\frac{6}{6}$ of the first pizza, and Group B ate $\frac{8}{6}$ of the remaining pizza. During the party, the class discussed which group ate more pizza.

   a. Did Group A or B eat more pizza? Use words and pictures to explain your answer to the class.

   b. Later, Group C ate all remaining slices of pizza. What fraction of the pizza did group C eat? Use words and pictures to explain your answer.

Lesson 22

Objective: Generate simple equivalent fractions by using visual fraction models and the number line.

Suggested Lesson Structure

Fluency Practice	(12 minutes)
Application Problem	(8 minutes)
Concept Development	(30 minutes)
Student Debrief	(10 minutes)
Total Time	**(60 minutes)**

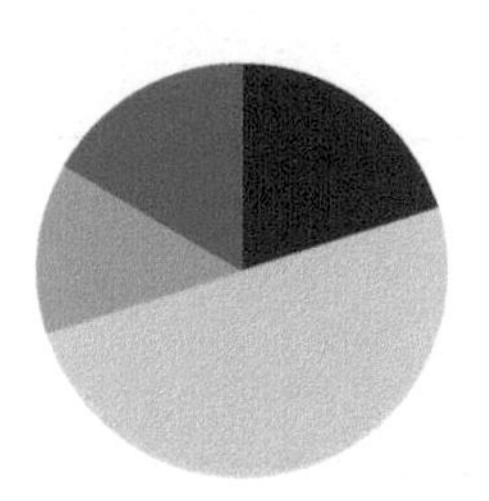

Fluency Practice (12 minutes)

- Whole Number Division **3.OA.7** (8 minutes)
- Counting by Fractions Equal to Whole Numbers on the Number Line **3.NF.3a** (4 minutes)

Whole Number Division (8 minutes)

Materials: (S) Blank paper

Note: This activity supports fluency with division. Steps 1 and 2 are timed for two minutes each. Step 3 is timed for one minute of testing for each partner. Step 4 is timed for two minutes.

1. Students self-select a number and write a set of multiples up to that number's multiple of 10 vertically down the left-hand side of the page (e.g., 6, 12, 18, 24, 30, 36, 42, 48, 54, 60).
2. Select a multiple, and divide it by the original number (e.g., 24 ÷ 6 = 4).
3. Change papers and test a partner by selecting multiples out of order (e.g., "What is 24 ÷ 6?" "What is 54 ÷ 6?" "What is 12 ÷ 6?").
4. Redo Steps 1 and 2 to see improvement.

Counting by Fractions Equal to Whole Numbers on the Number Line (4 minutes)

Materials: (S) Personal white board

Note: This activity reviews the concept of naming equivalent fractions on the number line from Lesson 21.

T: (Project a number line partitioned into 12 thirds.) Count by thirds. (Write fractions as students count.)

S: 1 third, 2 thirds, 3 thirds, 4 thirds, 5 thirds, 6 thirds, 7 thirds, 8 thirds, 9 thirds, 10 thirds, 11 thirds, 12 thirds.

T: On your personal white board, write the fractions equal to whole numbers in order from least to greatest. Continue beyond those shown on our number line if you finish early.

S: (Write $\frac{3}{3}$, $\frac{6}{3}$, $\frac{9}{3}$, and $\frac{12}{3}$.)

Continue with the following possible sequence: halves and fourths.

Application Problem (8 minutes)

Mr. Ramos wants to put a wire on the wall. He puts 9 nails equally spaced along the wire. Draw a number line representing the wire. Label it from 0 at the start of the wire to 1 at the end. Mark each fraction where Mr. Ramos puts each nail.

a. Build a number bond with unit fractions to 1 whole.

b. Write the fraction of the nail that is equivalent to $\frac{1}{2}$ of the wire.

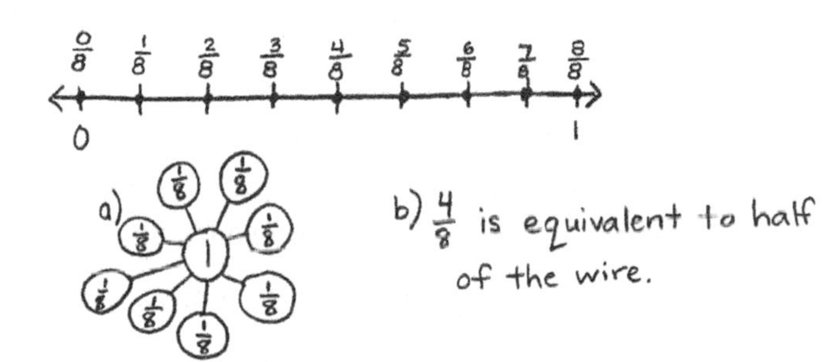

Note: This problem reviews placing fractions on a number line, decomposing 1 whole into unit fractions, and naming equivalent fractions. The first nail is located at $\frac{0}{8}$, which represents no length of wire. This results in 9 nails rather than 8 nails, even though the number line is partitioned into eighths. Watch for and discuss misconceptions that arise.

Concept Development (30 minutes)

Materials: (S) Math journal or fraction strips made in Lesson 21, new $4\frac{1}{4}$-inch × 1-inch fraction strips (3 per student), crayons, personal white board, glue

> **NOTES ON MULTIPLE MEANS OF ENGAGEMENT:**
>
> Students working below grade level may alternatively use two fraction strips—one partitioned into sixths, the other partitioned into fourths—to compare 3 sixths and 2 fourths. Or, have students draw number lines on personal white boards so that they may erase partitioned sixths before partitioning fourths.

T: Take out your math journal, and turn to the page where you glued your fraction strips yesterday. Name the fraction that is equivalent to 1 third.

S: $\frac{2}{6}$.

T: Now, name the fractions that are equivalent to 1 half.

S: $\frac{2}{4}$, $\frac{4}{8}$, and $\frac{3}{6}$.

T: During our Debrief yesterday, I challenged you to find another fraction equivalent to 1 half, even though it wasn't shaded. You came up with $\frac{3}{6}$.

MP.3

T: Now, I want you to work with a partner to look at your fraction strips again. See if you can find other equivalent fractions, shaded or unshaded. Draw and label them on your personal white board. For example, using my fraction strips, I can see that $\frac{2}{2}$ and $\frac{4}{4}$ are equivalent. Fourths are just halves cut in half again. Be ready to explain how you know, just like I did.

S: (Possible answers other than those already discussed:

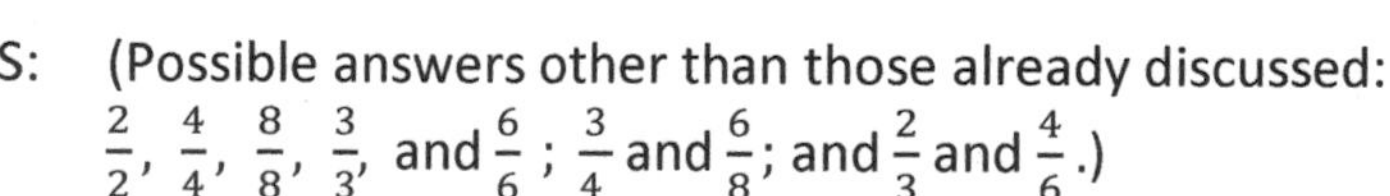

$\frac{2}{2}$, $\frac{4}{4}$, $\frac{8}{8}$, $\frac{3}{3}$, and $\frac{6}{6}$; $\frac{3}{4}$ and $\frac{6}{8}$; and $\frac{2}{3}$ and $\frac{4}{6}$.)

T: (Have students share their work.) Let's look at $\frac{2}{3}$ and $\frac{4}{6}$. Talk with your partner. Do you notice a relationship between the numbers in these fractions?

NOTES ON MULTIPLE MEANS OF ENGAGEMENT:

Challenge students working above grade level to collect the data presented (e.g., sets of equivalencies) and organize it in a table or graph. Guide them to analyze the organized data and draw conclusions. Ask (for example), "Which fraction has more equivalent fractions? Why?"

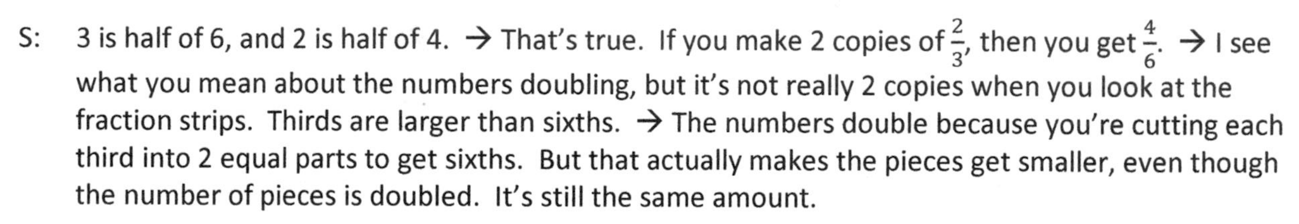

S: 3 is half of 6, and 2 is half of 4. → That's true. If you make 2 copies of $\frac{2}{3}$, then you get $\frac{4}{6}$. → I see what you mean about the numbers doubling, but it's not really 2 copies when you look at the fraction strips. Thirds are larger than sixths. → The numbers double because you're cutting each third into 2 equal parts to get sixths. But that actually makes the pieces get smaller, even though the number of pieces is doubled. It's still the same amount.

T: Now, look at $\frac{3}{4}$ and $\frac{6}{8}$. Does the same pattern you just noticed apply to these fractions?

S: (Discuss.)

It may be a good idea to have students repeat the process with whole number fractions if they are unsure.

T: I'm hearing you say that the numbers in these equivalent fractions doubled. Look again at these equivalent fractions: $\frac{2}{3}$ and $\frac{4}{6}$. What fraction would we get if we doubled the 4 and 6 in $\frac{4}{6}$?

S: $\frac{8}{12}$.

T: (Pass out 3 fraction strips to each student.) Fold your strips into thirds, sixths, and twelfths. Label the unit fractions. Then, shade $\frac{2}{3}$, $\frac{4}{6}$, and $\frac{8}{12}$ to compare. Is $\frac{8}{12}$ equivalent to $\frac{2}{3}$ and $\frac{4}{6}$?

S: They are equivalent!

T: What did we do to the equal parts each time to make the number of shaded parts and total number of parts double?

S: We cut them in 2. Thirds get cut in 2 to make sixths, and sixths get cut in 2 to make twelfths.

T: Did the whole change?

S: Nope, it just has more equal parts.

T: What happens to the shaded area?

S: It stays the same size.

T: So, the fractions are...?

S: Equivalent!

Have students glue the equivalent fractions into their math journals and label them.

Show the pictorial models to the right.

T: Let's look at a different model. These 3 wholes are the same. Name the shaded fraction as I point to the model.

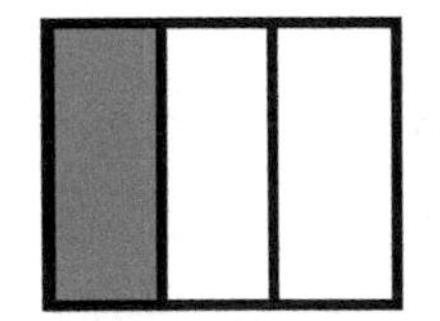

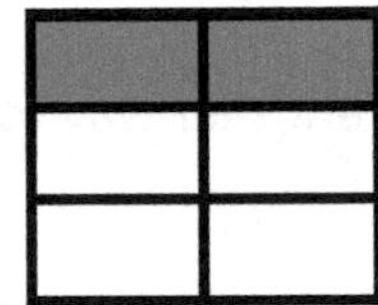

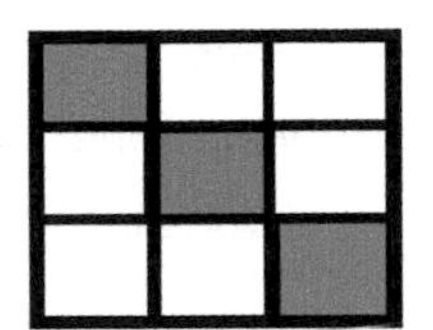

While pointing to each model, label with student responses: $\frac{1}{3}$, $\frac{2}{6}$, and $\frac{3}{9}$.

T: Are these fractions equivalent? Work with your partner to use the number line to prove your answer. Be ready to share your thinking.

After students work, have pairs share at tables, or select partners to present different methods to the class. Provide other examples using pictorial models and the number line as necessary.

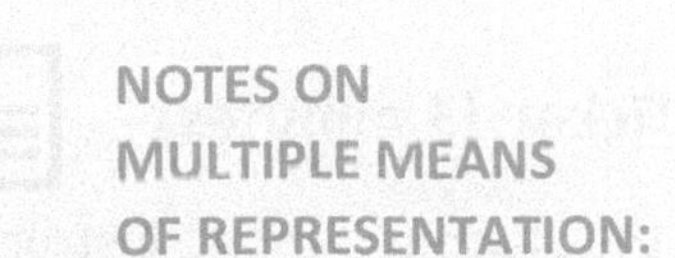

NOTES ON MULTIPLE MEANS OF REPRESENTATION:

Celebrate and encourage English language learners' use of math language in English. In the Problem Set, encourage learners to whisper the unit fraction, whisper count the shaded units, and whisper the shaded fraction as they write.

Problem Set (10 minutes)

Students should do their personal best to complete the Problem Set within the allotted 10 minutes. For some classes, it may be appropriate to modify the assignment by specifying which problems they work on first. Some problems do not specify a method for solving. Students should solve these problems using the RDW approach used for Application Problems.

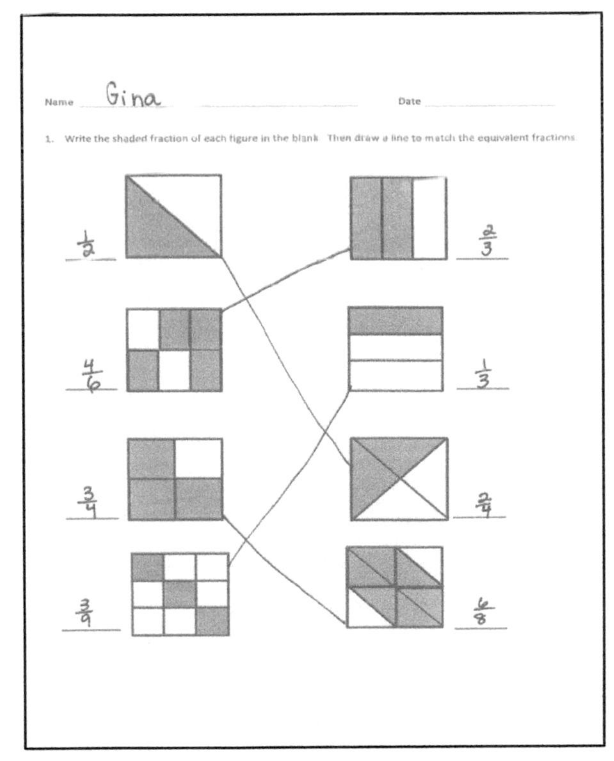

Student Debrief (10 minutes)

Lesson Objective: Generate simple equivalent fractions by using visual fraction models and the number line.

The Student Debrief is intended to invite reflection and active processing of the total lesson experience.

Invite students to review their solutions for the Problem Set. They should check work by comparing answers with a partner before going over answers as a class. Look for misconceptions or misunderstandings that can be addressed in the Student Debrief. Guide students in a conversation to debrief the Problem Set and process the lesson.

Any combination of the questions below may be used to lead the discussion.

- What did you notice about the models in Problem 1?
- In Problem 1, which shapes were most difficult to match? Why?
- What might be another way to draw a fraction equivalent to $\frac{3}{4}$?
- Look at Problem 2. What pattern do you notice between the 3 sets of models?
- How does the pattern you noticed in Problem 2 relate to other parts of today's lesson?

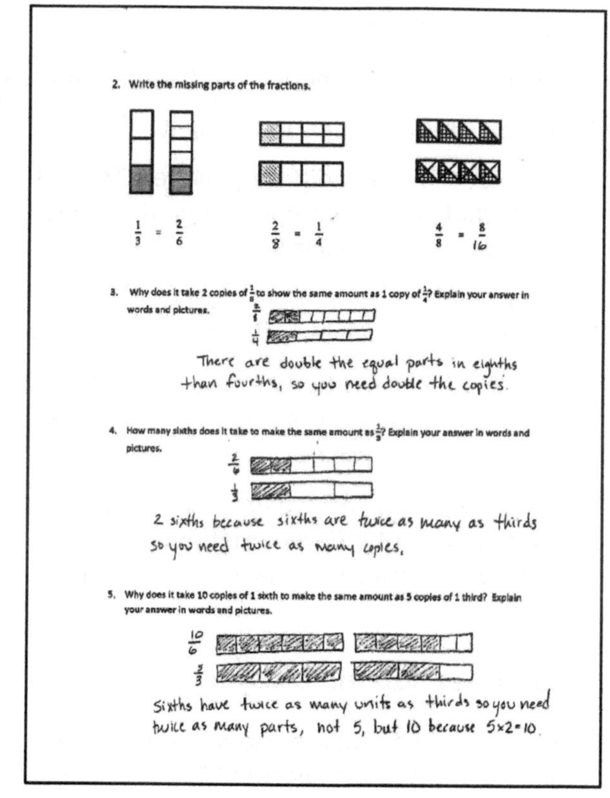

2. Write the missing parts of the fractions.

$\frac{1}{3} = \frac{2}{6}$ $\frac{2}{8} = \frac{1}{4}$ $\frac{4}{8} = \frac{8}{16}$

3. Why does it take 2 copies of $\frac{1}{8}$ to show the same amount as 1 copy of $\frac{1}{4}$? Explain your answer in words and pictures.

$\frac{2}{8}$ $\frac{1}{4}$

There are double the equal parts in eighths than fourths, so you need double the copies.

4. How many sixths does it take to make the same amount as $\frac{1}{3}$? Explain your answer in words and pictures.

$\frac{2}{6}$ $\frac{1}{3}$

2 sixths because sixths are twice as many as thirds so you need twice as many copies.

5. Why does it take 10 copies of 1 sixth to make the same amount as 5 copies of 1 third? Explain your answer in words and pictures.

$\frac{10}{6}$ $\frac{5}{3}$

Sixths have twice as many units as thirds so you need twice as many parts, not 5, but 10 because $5 \times 2 = 10$.

Exit Ticket (3 minutes)

After the Student Debrief, instruct students to complete the Exit Ticket. A review of their work will help with assessing students' understanding of the concepts that were presented in today's lesson and planning more effectively for future lessons. The questions may be read aloud to the students.

Name ______________________________ Date ________________

1. Write the shaded fraction of each figure on the blank. Then, draw a line to match the equivalent fractions.

_____ 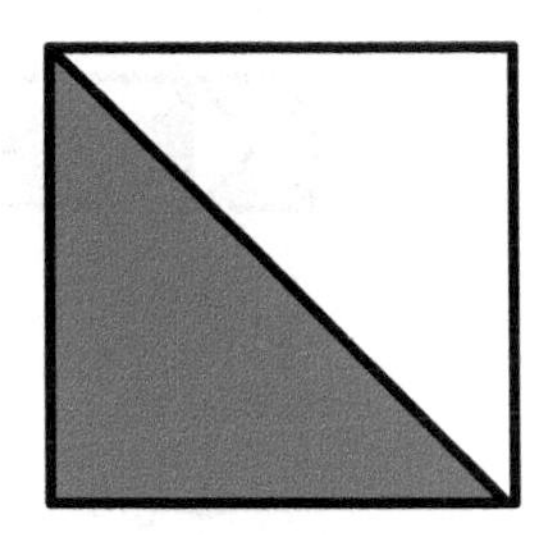

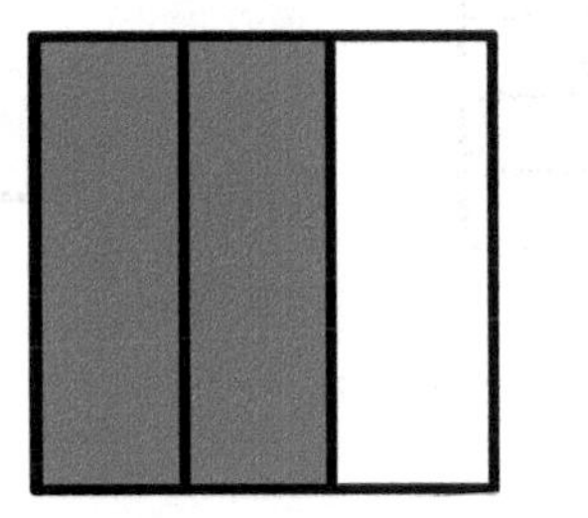 _____

_____ 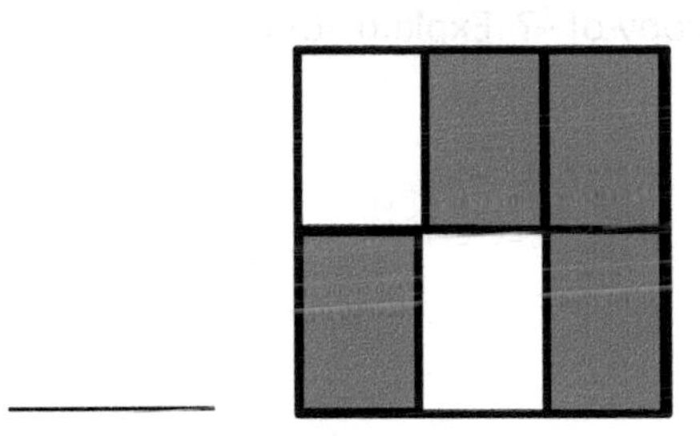

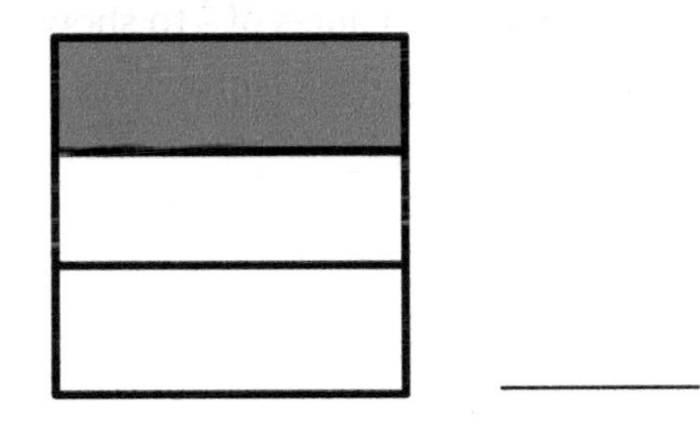 _____

_____ 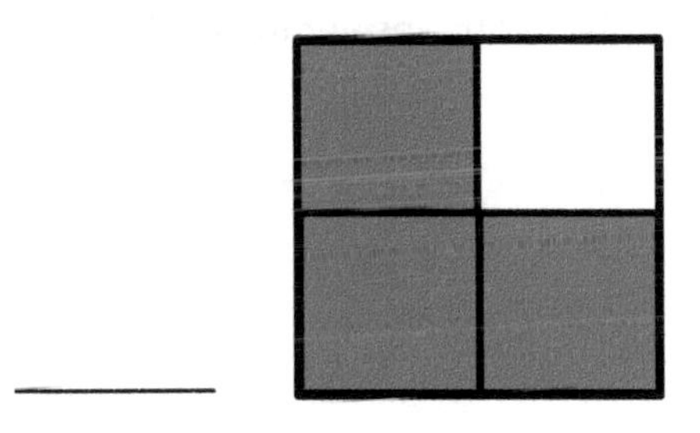

 _____

_____ 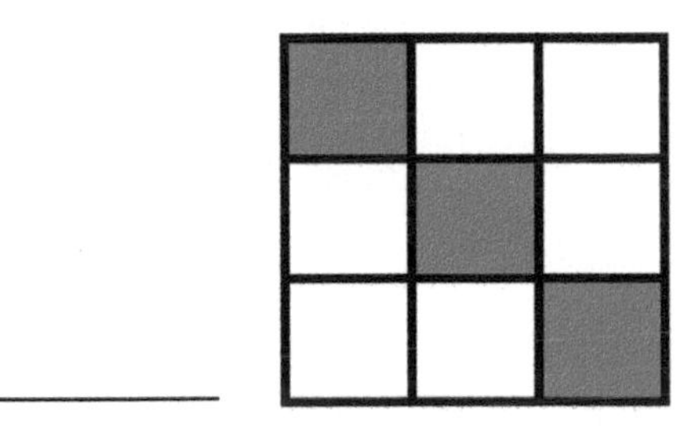

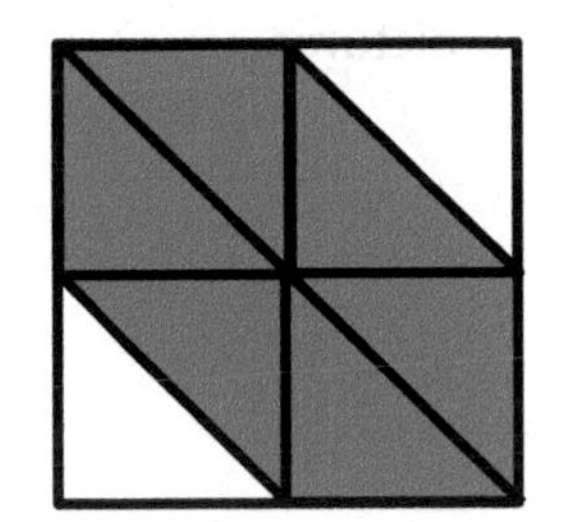 _____

2. Write the missing parts of the fractions.

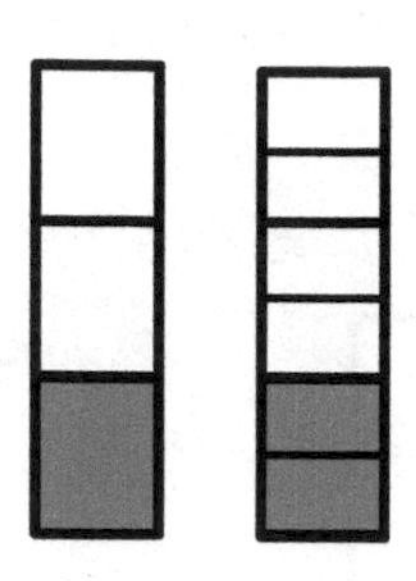
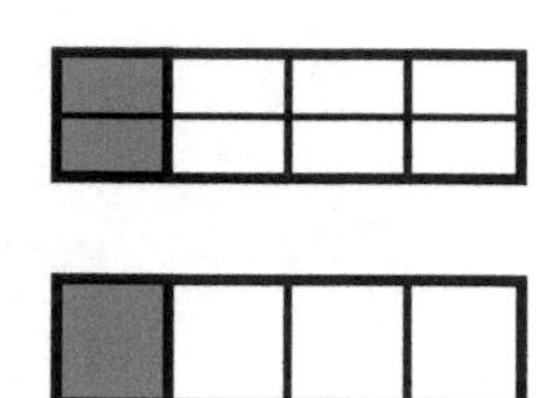
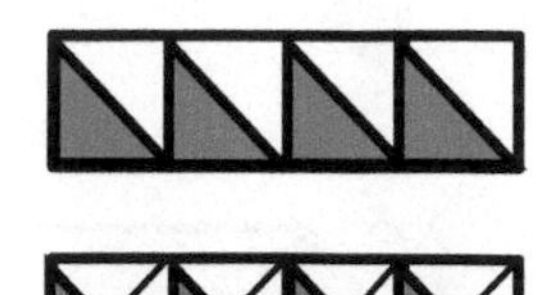

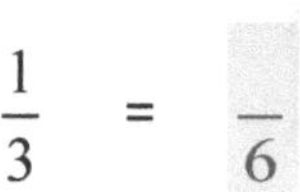

$\frac{1}{3} = \frac{\ }{6}$ $\frac{2}{\ } = \frac{1}{4}$ $\frac{4}{8} = \frac{8}{\ }$

3. Why does it take 2 copies of $\frac{1}{8}$ to show the same amount as 1 copy of $\frac{1}{4}$? Explain your answer in words and pictures.

4. How many sixths does it take to make the same amount as $\frac{1}{3}$? Explain your answer in words and pictures.

5. Why does it take 10 copies of 1 sixth to make the same amount as 5 copies of 1 third? Explain your answer in words and pictures.

Name ____ Date ____________________

1. Draw and label two models that show equivalent fractions.

2. Draw a number line that proves your thinking about Problem 1.

Name ______________________ Date ____________

1. Write the shaded fraction of each figure on the blank. Then, draw a line to match the equivalent fractions.

____ ____

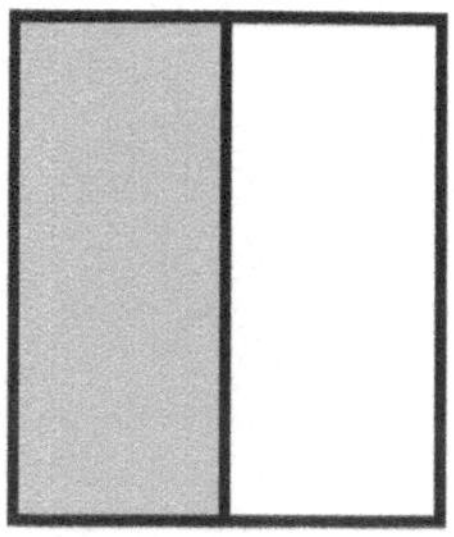

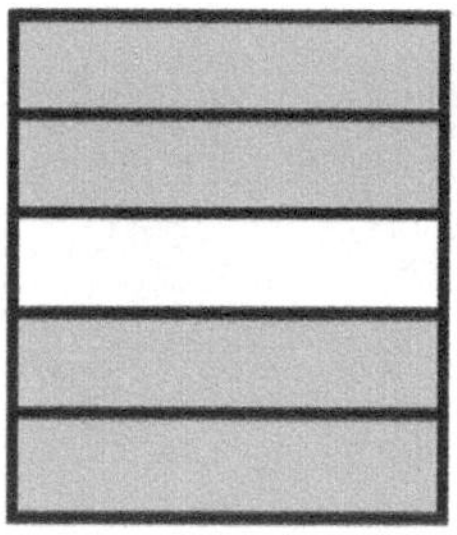

____ ____

____ ____

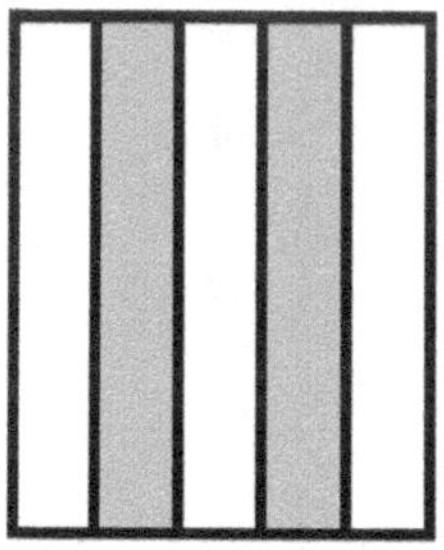

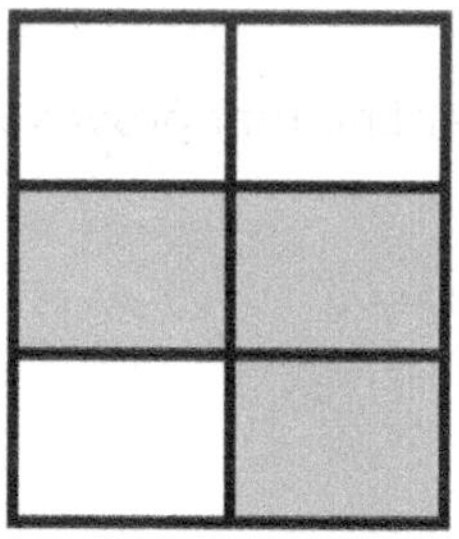

____ ____

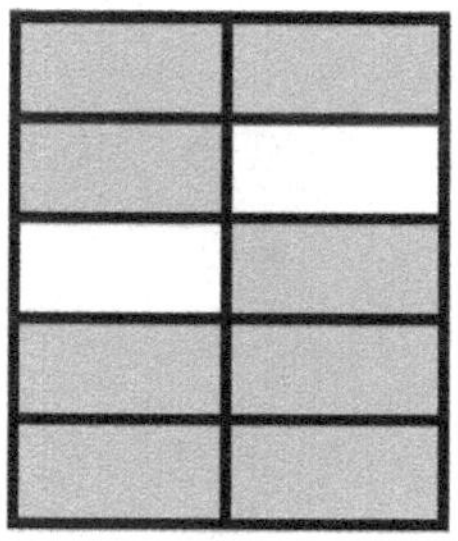

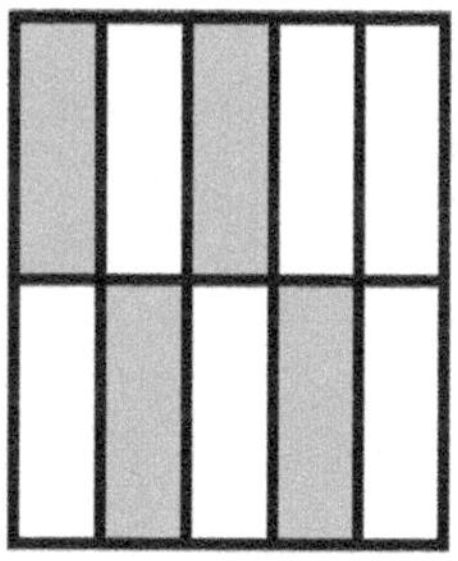

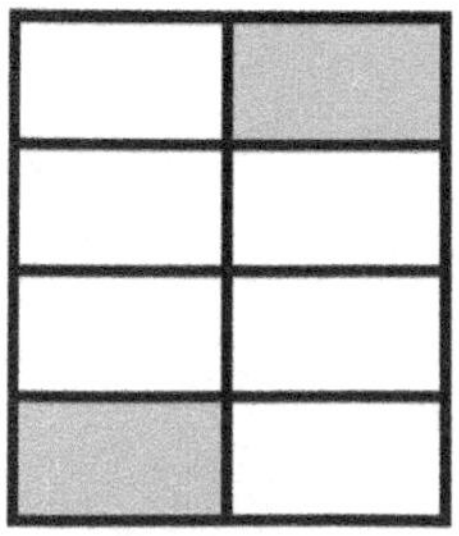

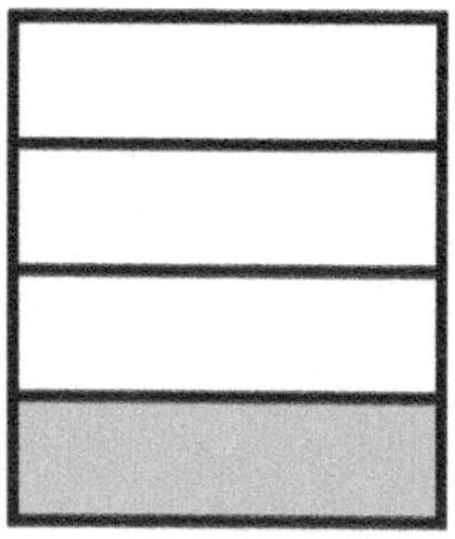

2. Complete the fractions to make true statements.

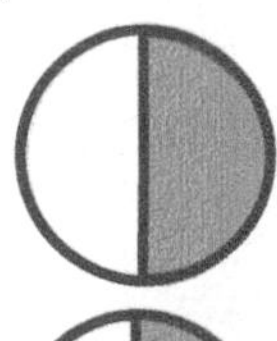
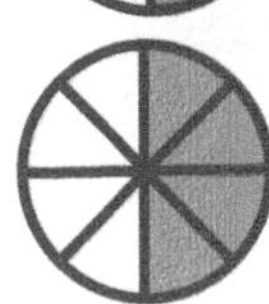
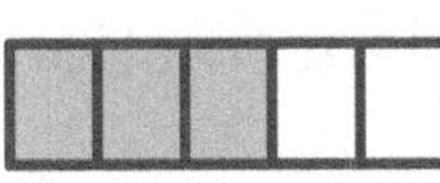
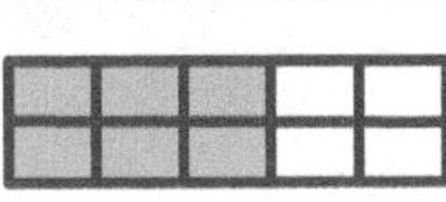

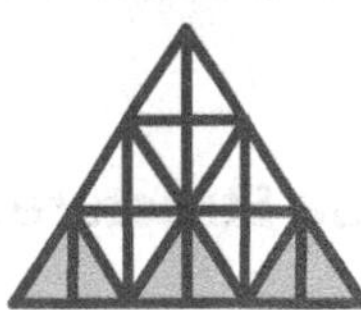

$\frac{1}{2} = \frac{4}{}$ $\frac{3}{5} = \frac{}{10}$ $\frac{3}{9} = \frac{6}{}$

3. Why does it take 3 copies of $\frac{1}{6}$ to show the same amount as 1 copy of $\frac{1}{2}$? Explain your answer in words and pictures.

4. How many ninths does it take to make the same amount as $\frac{1}{3}$? Explain your answer in words and pictures.

5. A pie was cut into 8 equal slices. If Ruben ate $\frac{3}{4}$ of the pie, how many slices did he eat? Explain your answer using a number line and words.

Lesson 23

Objective: Generate simple equivalent fractions by using visual fraction models and the number line.

Suggested Lesson Structure

■ Fluency Practice	(12 minutes)
■ Application Problem	(8 minutes)
■ Concept Development	(30 minutes)
■ Student Debrief	(10 minutes)
Total Time	**(60 minutes)**

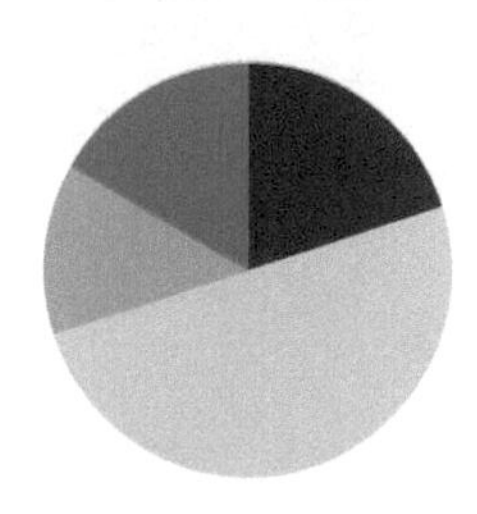

Fluency Practice (12 minutes)

- Sprint: Add by Six **2.NBT.5** (8 minutes)
- Find the Equivalent Fraction **3.NF.3d** (4 minutes)

Sprint: Add by Six (8 minutes)

Materials: (S) Add by Six Sprint

Note: This Sprint supports fluency with addition by 6.

Find the Equivalent Fraction (4 minutes)

Materials: (T) Prepared fraction images (S) Personal white board

Note: This activity reviews finding equivalent fractions from Lesson 20.

T: (Project a square partitioned into 2 parts with 1 part shaded in.) Say the shaded fraction.

S: 1 half.

T: (Write $\frac{1}{2}$ underneath the square.) Copy my picture and fraction on your personal white board.

S: (Copy the image and fraction on the board.)

T: (Project an identical square to the right of the first square.) On your board, draw a second identical square.

S: (Draw a second identical square.)

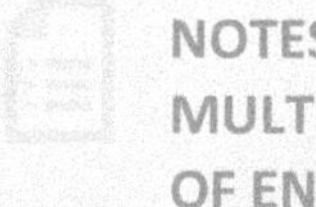

NOTES ON MULTIPLE MEANS OF ENGAGEMENT:

Students working below grade level may enjoy counting fractions more than once—first, with the addition of models (e.g., shading fourths) and then without, gradually increasing speed with each repetition.

T: (Below the squares, write $\frac{1}{2} = \frac{}{4}$.) On your board, partition your second square to make fourths, and fill in the number sentence.

S: (Draw a horizontal line to show 2 parts of 4 shaded, and write $\frac{1}{2} = \frac{2}{4}$.)

Continue with the following possible sequence: $\frac{1}{2} = \frac{}{6}$, $\frac{2}{8} = \frac{}{4}$, and $\frac{5}{10} = \frac{}{20}$.

Application Problem (8 minutes)

Shannon stood at the end of a 100-meter long soccer field and kicked the ball to her teammate. She kicked it 20 meters. The commentator said she kicked it a quarter of the way down the field. Is that true? If not, what fraction should the commentator have said? Prove your answer by using a number line.

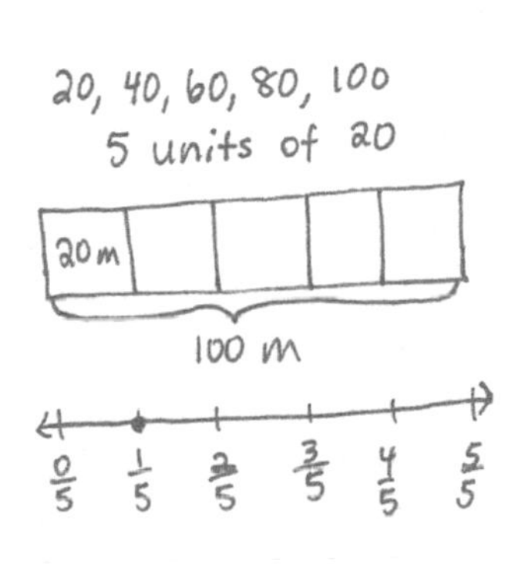

She did not kick it a quarter ($\frac{1}{4}$) of the way. She kicked it $\frac{1}{5}$. The commentator should've said $\frac{1}{5}$ of the way.

Note: This problem reviews partitioning a whole into equal parts from Topic A.

Concept Development (30 minutes)

Materials: (S) Index card (1 per pair, described below), sentence strip (1 per pair), chart paper (1 per group), markers, glue, math journal

Students work in pairs. Each pair receives one sentence strip and an index card. The index card designates endpoints on a number line and a unit with which to partition (examples on the right).

Divide the class so each group is composed of pairs (each group contains more than one pair). Create the following index cards, and distribute one card to each pair per group.

Group A: Interval 3–5, thirds and sixths

Group B: Interval 1–3, sixths and twelfths

Group C: Interval 3–5, halves and fourths

Group D: Interval 1–3, fourths and eighths

Group E: Interval 4–6, sixths and twelfths

Group F: Interval 6–8, halves and fourths

Example Index Cards for Group A

Group A
Interval: 3–5
Unit: thirds

Group A
Interval: 3–5
Unit: sixths

Note: Differentiate the activity by strategically assigning *just right* intervals and units to pairs of students.

T: With your partner, use your sentence strip to make a number line with your given interval. Then, estimate to partition into your given unit by folding your sentence strip. Label the endpoints and fractions. Rename the wholes.

S: (Work in pairs.)

T: (Give one piece of chart paper to a member of each letter group.) Now, stand up and find your other letter group members. Once you've found them, glue your number lines in a column so that the ends match up on your chart paper. Compare number lines to find equivalent fractions. Record all possible equivalent fractions in your math journals.

S: (Find letter group members, and glue fraction strips onto chart paper. Letter group members discuss and record equivalent fractions.)

T: (Hang each chart paper around the room.) Now, we're going to do a *museum walk.* As a letter group, you will visit the other groups' chart papers. One person in each group will be the recorder. You can switch recorders each time you visit a new chart paper. Your job will be to find and list all of the equivalent fractions you see at each chart paper.

S: (Go to another letter group's chart paper and begin.)

T: (Rotate groups briskly so that, at the beginning, students don't finish finding all fractions at 1 station. As letter groups rotate and chart papers fill up, challenge groups to check others' work to ensure no fractions are missing.)

T: (After rotation is complete.) Go back to your own chart paper with your letter group. Take your math journals, and check your friends' work. Did they name the same equivalent fractions you found?

NOTES ON MULTIPLE MEANS OF ENGAGEMENT:

Challenge students working above grade level to write more than two equivalent fractions on the Problem Set. As they begin to generate equivalencies mentally and rapidly, guide students to articulate the pattern and its rule.

Problem Set (10 minutes)

Students should do their personal best to complete the Problem Set within the allotted 10 minutes. For some classes, it may be appropriate to modify the assignment by specifying which problems they work on first. Some problems do not specify a method for solving. Students should solve these problems using the RDW approach used for Application Problems.

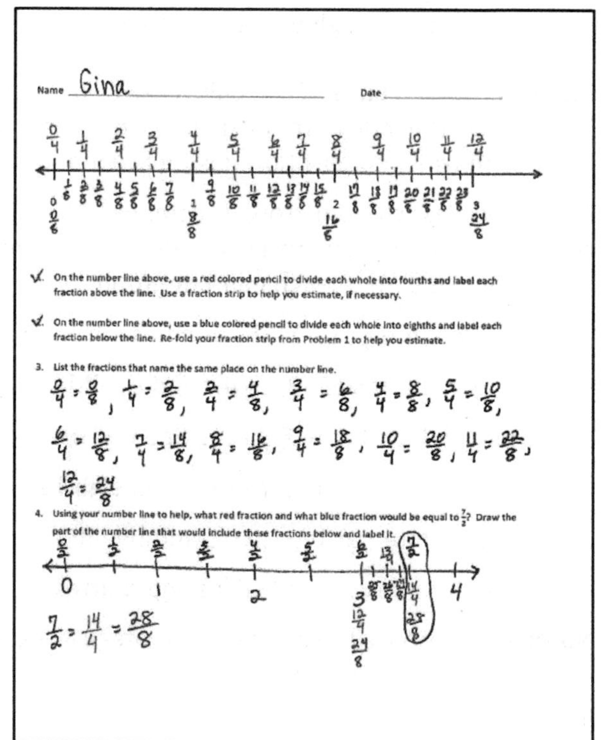

Student Debrief (10 minutes)

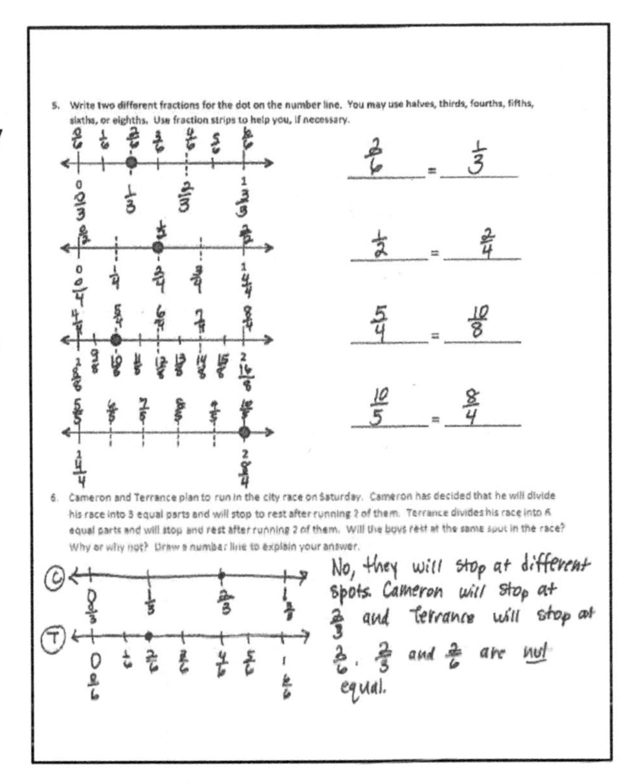
5. Write two different fractions for the dot on the number line. You may use halves, thirds, fourths, fifths, sixths, or eighths. Use fraction strips to help you, if necessary.

$\frac{2}{6} = \frac{1}{3}$

$\frac{1}{2} = \frac{2}{4}$

$\frac{5}{4} = \frac{10}{8}$

$\frac{10}{5} = \frac{8}{4}$

6. Cameron and Terrance plan to run in the city race on Saturday. Cameron has decided that he will divide his race into 3 equal parts and will stop to rest after running 2 of them. Terrance divides his race into 6 equal parts and will stop and rest after running 2 of them. Will the boys rest at the same spot in the race? Why or why not? Draw a number line to explain your answer.

No, they will stop at different spots. Cameron will stop at $\frac{2}{3}$ and Terrance will stop at $\frac{2}{6}$. $\frac{2}{3}$ and $\frac{2}{6}$ are not equal.

Lesson Objective: Generate simple equivalent fractions by using visual fraction models and the number line.

The Student Debrief is intended to invite reflection and active processing of the total lesson experience.

Invite students to review their solutions for the Problem Set. They should check work by comparing answers with a partner before going over answers as a class. Look for misconceptions or misunderstandings that can be addressed in the Debrief. Guide students in a conversation to debrief the Problem Set and process the lesson.

Any combination of the questions below may be used to lead the discussion.

- Could you have compared the number line you made in today's lesson to a number line from a different group? What would the result be?
- How did your work change when the interval on your number line was no longer from 0 to 1?
- Could we sequentially connect the number lines you made in today's lesson even though they are partitioned into different units? What would happen then?
- Compare all of the answers for Problem 5. (Use this comparison to advance the idea that the world of fractions is endless. There are many different fractions that label a single point.)

NOTES ON MULTIPLE MEANS OF REPRESENTATION:

To assist comprehension, develop multiple ways of asking the same question. For example, the question, "Could we sequentially ...?" can be changed to, "What if we put all the number lines together in numerical order?" or "What do you think of a number line whose intervals are partitioned into different fractional units?"

Exit Ticket (3 minutes)

After the Student Debrief, instruct students to complete the Exit Ticket. A review of their work will help with assessing students' understanding of the concepts that were presented in today's lesson and planning more effectively for future lessons. The questions may be read aloud to the students.

A

Number Correct: _______

Add by Six

1.	0 + 6 =	
2.	1 + 6 =	
3.	2 + 6 =	
4.	3 + 6 =	
5.	4 + 6 =	
6.	6 + 4 =	
7.	6 + 3 =	
8.	6 + 2 =	
9.	6 + 1 =	
10.	6 + 0 =	
11.	15 + 6 =	
12.	25 + 6 =	
13.	35 + 6 =	
14.	45 + 6 =	
15.	55 + 6 =	
16.	85 + 6 =	
17.	6 + 6 =	
18.	16 + 6 =	
19.	26 + 6 =	
20.	36 + 6 =	
21.	46 + 6 =	
22.	76 + 6 =	

23.	7 + 6 =	
24.	17 + 6 =	
25.	27 + 6 =	
26.	37 + 6 =	
27.	47 + 6 =	
28.	77 + 6 =	
29.	8 + 6 =	
30.	18 + 6 =	
31.	28 + 6 =	
32.	38 + 6 =	
33.	48 + 6 =	
34.	78 + 6 =	
35.	9 + 6 =	
36.	19 + 6 =	
37.	29 + 6 =	
38.	39 + 6 =	
39.	89 + 6 =	
40.	6 + 75 =	
41.	6 + 56 =	
42.	6 + 77 =	
43.	6 + 88 =	
44.	6 + 99 =	

B

Number Correct: _______

Improvement: _______

Add by Six

1.	6 + 0 =		23.	7 + 6 =	
2.	6 + 1 =		24.	17 + 6 =	
3.	6 + 2 =		25.	27 + 6 =	
4.	6 + 3 =		26.	37 + 6 =	
5.	6 + 4 =		27.	47 + 6 =	
6.	4 + 6 =		28.	67 + 6 =	
7.	3 + 6 =		29.	8 + 6 =	
8.	2 + 6 =		30.	18 + 6 =	
9.	1 + 6 =		31.	28 + 6 =	
10.	0 + 6 =		32.	38 + 6 =	
11.	5 + 6 =		33.	48 + 6 =	
12.	15 + 6 =		34.	88 + 6 =	
13.	25 + 6 =		35.	9 + 6 =	
14.	35 + 6 =		36.	19 + 6 =	
15.	45 + 6 =		37.	29 + 6 =	
16.	75 + 6 =		38.	39 + 6 =	
17.	6 + 6 =		39.	79 + 6 =	
18.	16 + 6 =		40.	6 + 55 =	
19.	26 + 6 =		41.	6 + 76 =	
20.	36 + 6 =		42.	6 + 57 =	
21.	46 + 6 =		43.	6 + 98 =	
22.	86 + 6 =		44.	6 + 89 =	

Name ______________________________ Date ________________

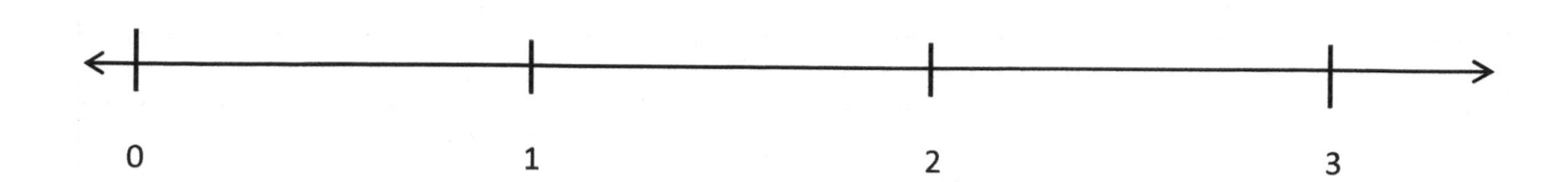

1. On the number line above, use a red colored pencil to divide each whole into fourths, and label each fraction above the line. Use a fraction strip to help you estimate, if necessary.

2. On the number line above, use a blue colored pencil to divide each whole into eighths, and label each fraction below the line. Refold your fraction strip from Problem 1 to help you estimate.

3. List the fractions that name the same place on the number line.

4. Using your number line to help, what red fraction and what blue fraction would be equal to $\frac{7}{2}$? Draw the part of the number line below that would include these fractions, and label it.

5. Write two different fractions for the dot on the number line. You may use halves, thirds, fourths, fifths, sixths, or eighths. Use fraction strips to help you, if necessary.

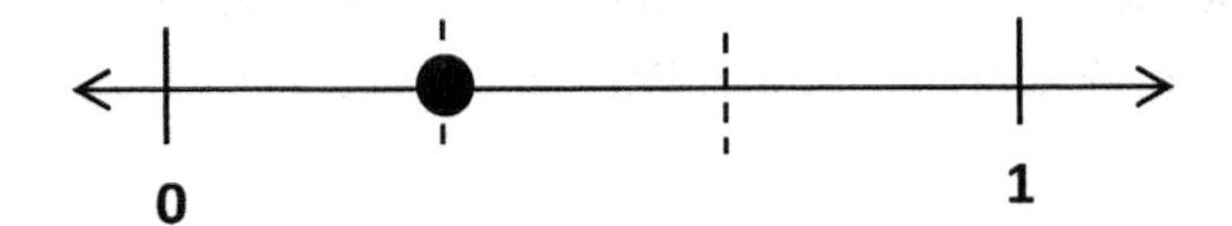

______ = ______

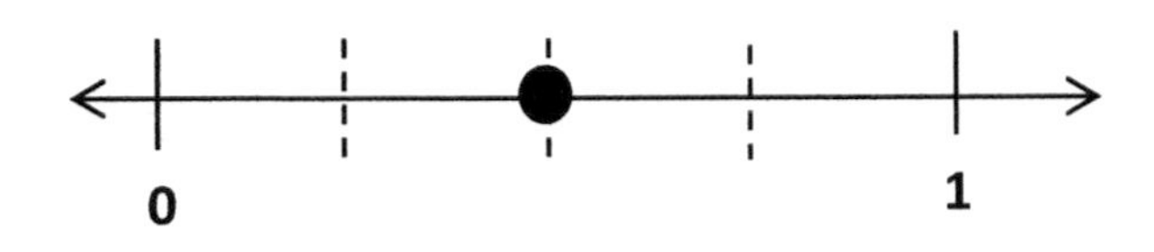

______ = ______

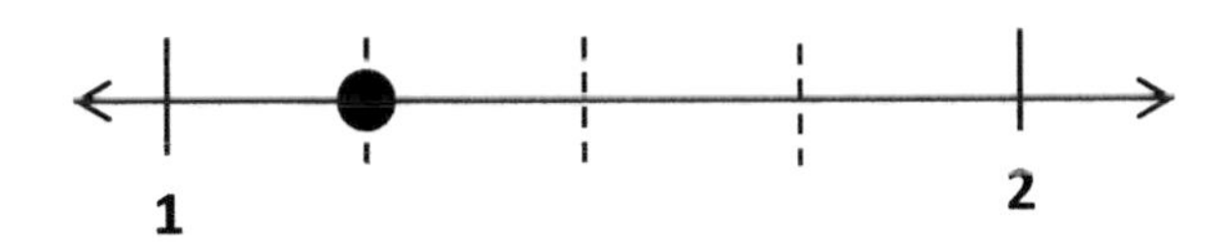

______ = ______

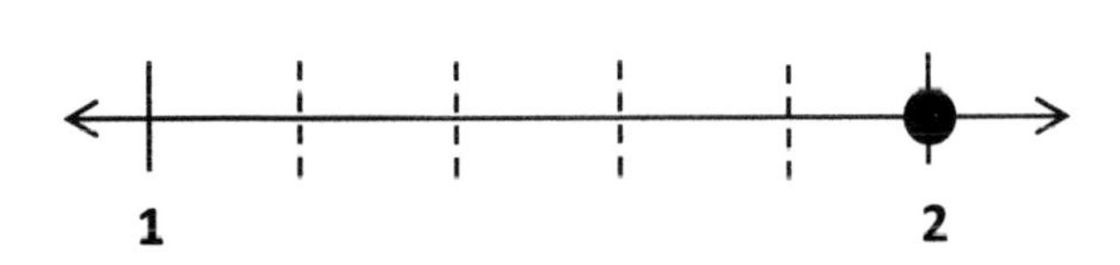

______ = ______

6. Cameron and Terrance plan to run in the city race on Saturday. Cameron has decided that he will divide his race into 3 equal parts and will stop to rest after running 2 of them. Terrance divides his race into 6 equal parts and will stop and rest after running 2 of them. Will the boys rest at the same spot in the race? Why or why not? Draw a number line to explain your answer.

Name ___________________________________ Date ________________

Henry and Maddie were in a pie-eating contest. The pies were cut either into thirds or sixths. Henry picked up a pie cut into sixths and ate $\frac{4}{6}$ of it in 1 minute. Maddie picked up a pie cut into thirds. What fraction of her pie does Maddie have to eat in 1 minute to tie with Henry? Draw a number line, and use words to explain your answer.

Name ______________________________ Date ______________

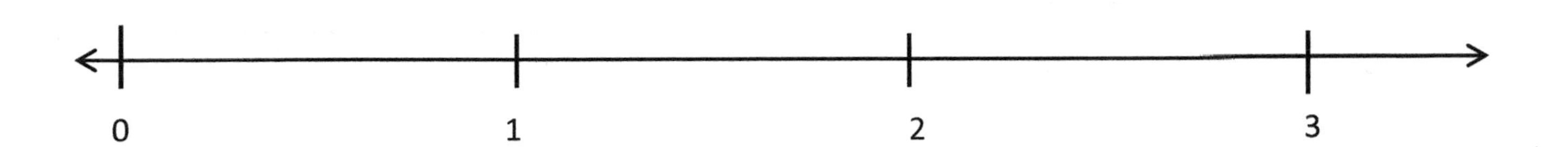

1. On the number line above, use a colored pencil to divide each whole into thirds and label each fraction above the line.

2. On the number line above, use a different colored pencil to divide each whole into sixths and label each fraction below the line.

3. Write the fractions that name the same place on the number line.

4. Using your number line to help, name the fraction equivalent to $\frac{20}{6}$. Name the fraction equivalent to $\frac{12}{3}$. Draw the part of the number line that would include these fractions below, and label it.

$\frac{20}{6} = \frac{\ }{3}$ $\frac{12}{3} = \frac{\ }{6}$

5. Write two different fraction names for the dot on the number line. You may use halves, thirds, fourths, fifths, sixths, eighths, or tenths.

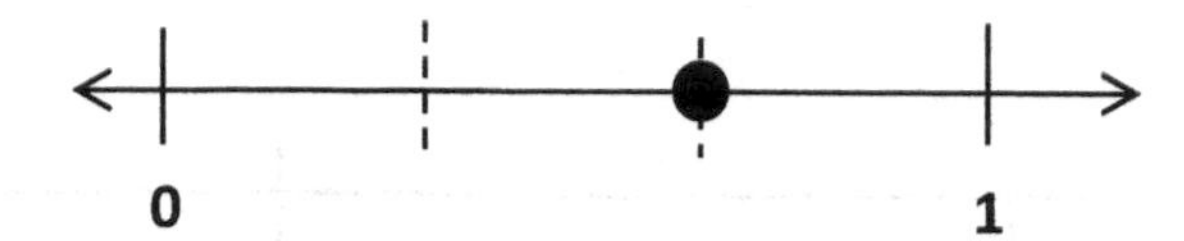

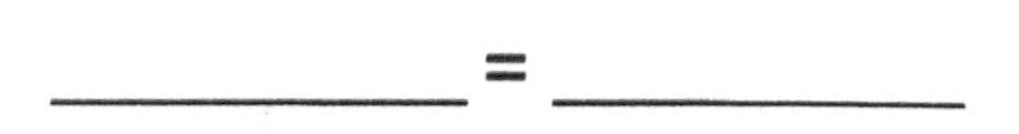

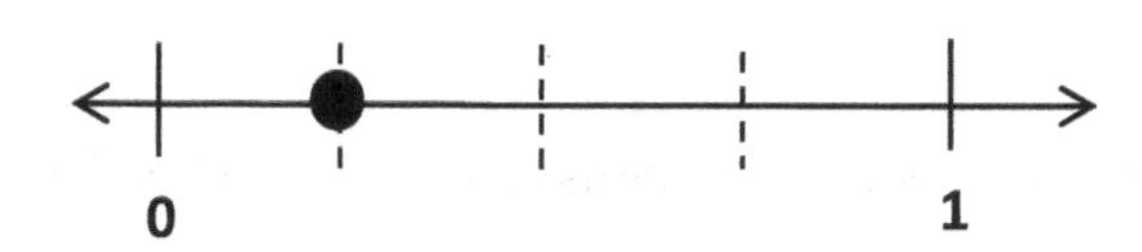

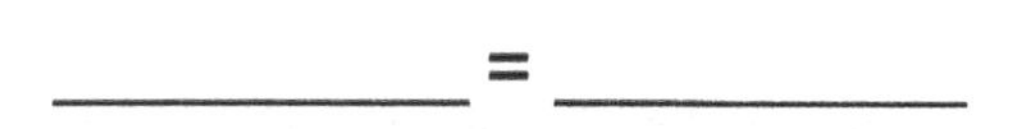

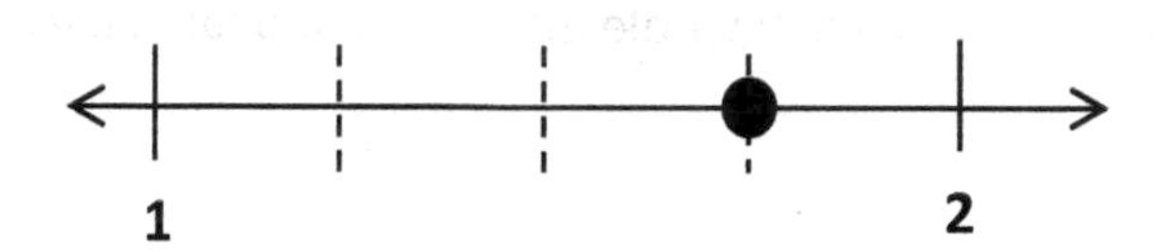

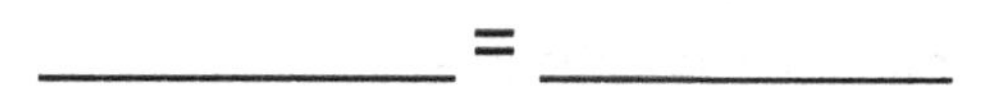

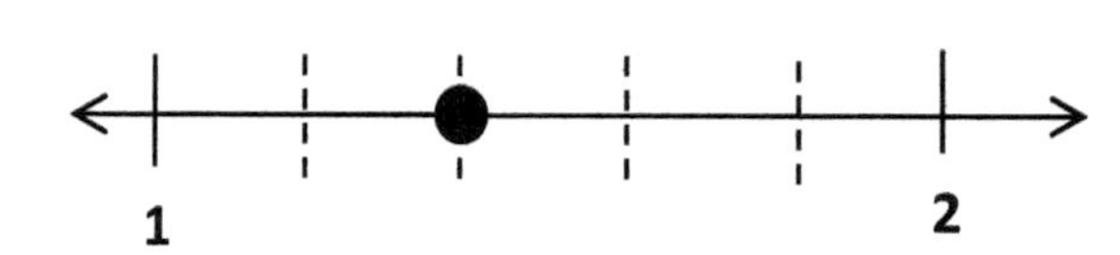

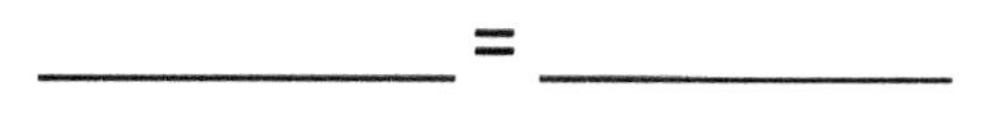

6. Danielle and Mandy each ordered a large pizza for dinner. Danielle's pizza was cut into sixths, and Mandy's pizza was cut into twelfths. Danielle ate 2 sixths of her pizza. If Mandy wants to eat the same amount of pizza as Danielle, how many slices of pizza will she have to eat? Write the answer as a fraction. Draw a number line to explain your answer.

Lesson 24

Objective: Express whole numbers as fractions and recognize equivalence with different units.

Suggested Lesson Structure

Fluency Practice	(12 minutes)
Application Problem	(5 minutes)
Concept Development	(33 minutes)
Student Debrief	(10 minutes)
Total Time	**(60 minutes)**

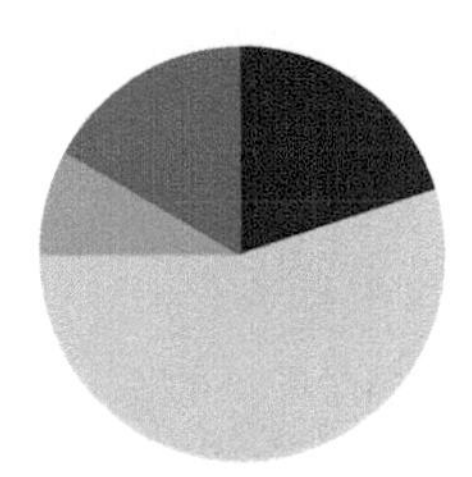

Fluency Practice (12 minutes)

- Sprint: Add by 7 **2.NBT.5** (8 minutes)
- Write Equal Fractions **3.NF.3d** (4 minutes)

Sprint: Add by Seven (8 minutes)

Materials: (S) Add by Seven Sprint

Note: This Sprint supports fluency with addition by 7.

Write Equal Fractions (4 minutes)

Materials: (S) Personal white board

Note: This activity reviews the skill of finding equivalent fractions on the number line from Topic E.

T: (Project number line with endpoints 0 and 1 partitioned into 2 equal parts by a dotted line.) Say the unit fraction represented by the dotted line.

S: 1 half.

T: (Write $\frac{1}{2}$ below the dotted line. To the right of the number line, write $\frac{1}{2} = \frac{}{4}$.) On your personal white board, write the number sentence, and fill in the blank.

S: (Write $\frac{1}{2} = \frac{2}{4}$.)

T: (Write $\frac{2}{4}$ below $\frac{1}{2}$ on the number line.)

Continue with the following possible sequence, drawing a new number line for each example: $\frac{1}{3} = \frac{2}{}$ and $\frac{1}{4} = \frac{}{8}$.

Application Problem (5 minutes)

The zipper on Robert's jacket is 1 foot long. It breaks on the first day of winter. He can only zip it $\frac{8}{12}$ of the way before it gets stuck. Draw and label a number line to show how far Robert can zip his jacket.

a. Divide and label the number line in thirds. What fraction of the way can he zip his jacket in thirds?

b. What fraction of Robert's jacket is not zipped? Write your answer in twelfths and thirds.

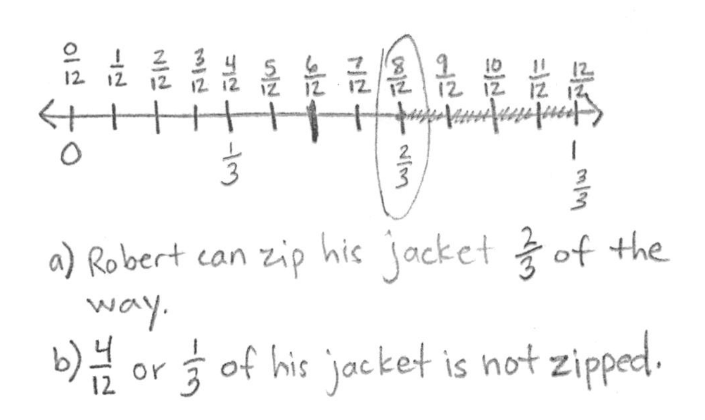

NOTES ON MULTIPLE MEANS OF ENGAGEMENT:

Partitioning the interval into two different fractional units is a stimulating challenge for students working above grade level.

Students working below grade level can draw two separate number lines or use fraction strips to solve.

Note: This problem reviews the skill of finding equivalent fractions on the number line from Topic E. Invite students to share their strategies for partitioning the number line into thirds and twelfths.

Concept Development (33 minutes)

Materials: (S) Fraction pieces (Template), scissors, envelope, personal white board, sentence strip, crayons

1 whole
halves
fourths
thirds
sixths

Each student starts with the fraction pieces, an envelope, and scissors.

T: Cut out all of the rectangles on the fraction pieces, and initial each rectangle so you know which ones are yours.

S: (Cut and initial.)

T: Place the rectangle that says *1 whole* on your personal white board. Take another rectangle. How many halves make 1 whole? Show by folding and labeling each unit fraction.

S: (Fold the second rectangle in half, and label $\frac{1}{2}$ on each of the 2 parts.)

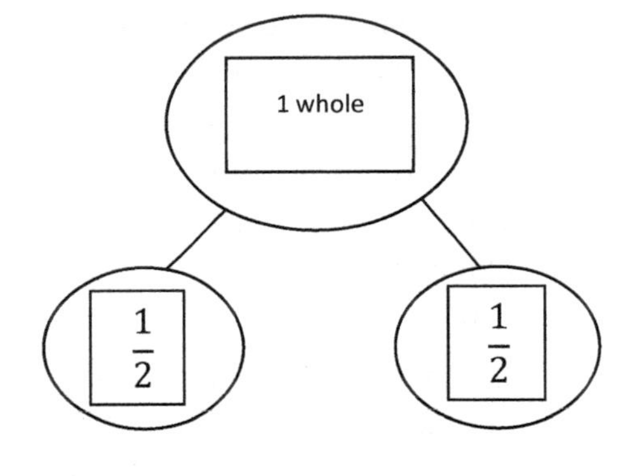

T: Now, cut on the fold. Draw circles around your whole and your parts to make a number bond.

S: (Draw a number bond using the shapes to represent wholes and parts.)

T: In your whole, write an equality that shows how many halves are equal to 1 whole. Remember, the equal sign is like a balance. Both sides have the same value.

S: (Write 1 whole = $\frac{2}{2}$ in the *1 whole* rectangle.)

T: Put your halves inside your envelope.

Follow the same sequence for each rectangle so that students cut all pieces indicated. Have students update the equality on their *1 whole* rectangle each time they cut a new piece. At the end, it should read: 1 whole = $\frac{2}{2} = \frac{3}{3} = \frac{4}{4} = \frac{6}{6}$. Discuss the equality with students to ensure that they understand the meaning of the equal sign and the role it plays in this number sentence.

Project or show Image 1, shown to the right.

Image 1

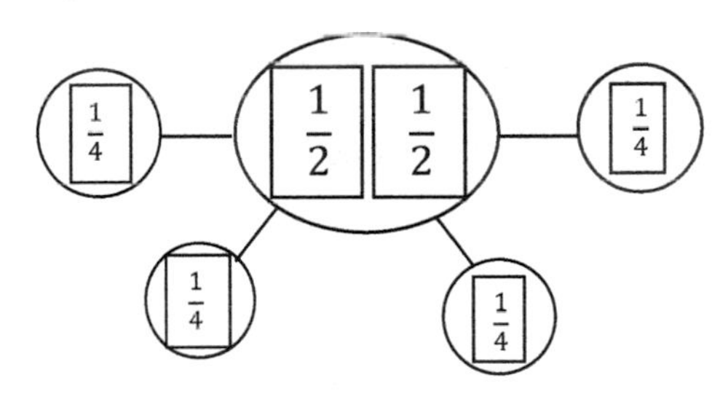

MP.7

T: Use your pieces to make this number bond on your board.

S: (Make the number bond.)

T: Discuss with your partner: Is this number bond true? Why or why not?

S: No, because the whole has only 2 pieces, but there are 4 parts! → But fourths are just halves cut in 2. So, they're the same pieces, but smaller now. → $\frac{2}{4}$ is equivalent to $\frac{1}{2}$. → So, $\frac{2}{2} = \frac{4}{4}$, just like what we wrote down on our *1 whole* rectangle.

T: I hear some of you saying that $\frac{2}{2}$ and $\frac{4}{4}$ both equal 1 whole. So, can we say that this is true? (Project or show Image 2, shown on the next page.)

S: No, because thirds aren't halves cut in 2. They look completely different. → But, when we put our thirds together and halves together, they make the same whole. → Before, we found with our pieces that 1 whole $= \frac{2}{2} = \frac{3}{3} = \frac{4}{4}$. → Then, it must be true!

NOTES ON MULTIPLE MEANS OF ENGAGEMENT:

Students working below grade level may appreciate tangibly proving that 2 halves is the same as 4 fourths. Encourage students to place the (paper) fourths on top of the halves to show equivalency.

Follow the same sequence with a variety of *wholes* and *parts* until students are comfortable with this representation of equivalence.

Image 2

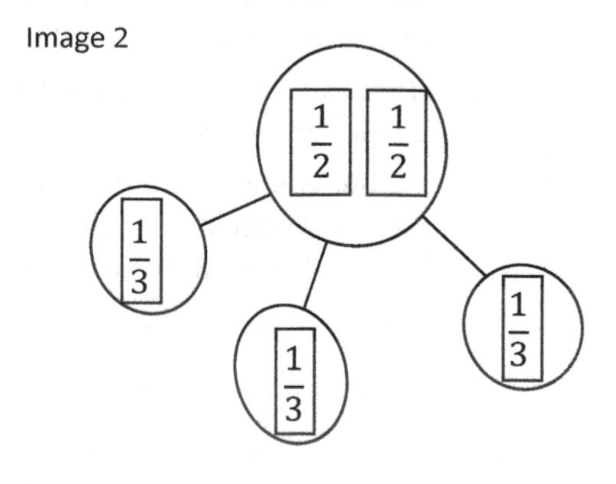

T: Now, let's place our different units on the same number line. Use your sentence strip to represent the interval from 0 to 1 on a number line. Mark the endpoints with your pencil now.

S: (Mark endpoints *0* and *1* below the number line.)

T: Go ahead and fold your sentence strip to partition one unit at a time into halves, fourths, thirds, and then sixths. Label each fraction above the number line. As you count, be sure to rename 0 and the whole. Use a different color crayon to mark and label the fraction for each unit.

S: (Fold the sentence strip and first label halves, then fourths, then thirds, and then sixths in different colors. Rename 0 and 1 in terms of each new unit.)

T: You should have a crowded number line! Compare it to your partner's.

S: (Compare.)

T: Before today, we've been noticing a lot of equivalent fractions between wholes on the number line. Today, notice the fractions you wrote at 0 and 1. Look first at the fractions for 0. What pattern do you notice?

S: They all have 0 copies of the unit! → The total number of equal parts changes. It shows you what unit you're going to count by. → Since our number line starts at 0, there is 0 of that unit in all of the fractions.

T: Even though the unit is different in each of our fractions at 0, are they equivalent? Think back to our work with shapes earlier.

S: We saw before that fractions with different units can still make the same whole. This time, the whole is just 0.

Follow the sequence to study the fractions written at 1. For both 0 and 1, students should see that every color they used is present.

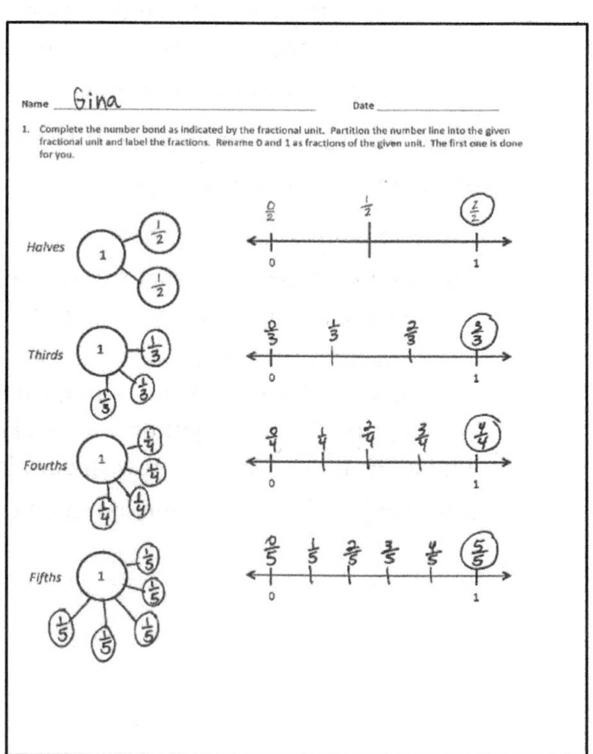

Problem Set (10 minutes)

Students should do their personal best to complete the Problem Set within the allotted 10 minutes. For some classes, it may be appropriate to modify the assignment by specifying which problems they work on first. Some problems do not specify a method for solving. Students should solve these problems using the RDW approach used for Application Problems.

Student Debrief (10 minutes)

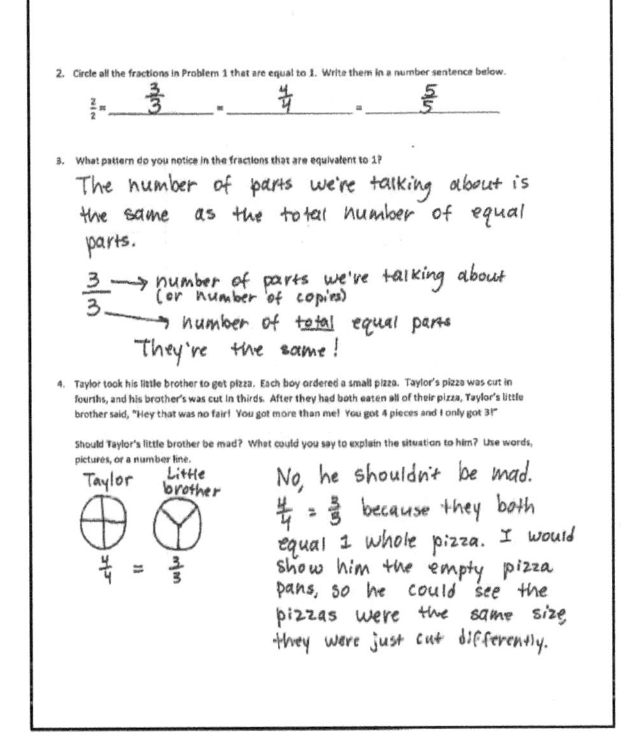

2. Circle all the fractions in Problem 1 that are equal to 1. Write them in a number sentence below.

$\frac{2}{2} = \frac{3}{3} = \frac{4}{4} = \frac{5}{5}$

3. What pattern do you notice in the fractions that are equivalent to 1?

The number of parts we're talking about is the same as the total number of equal parts.

3 → number of parts we're talking about (or number of copies)
3 → number of total equal parts
They're the same!

4. Taylor took his little brother to get pizza. Each boy ordered a small pizza. Taylor's pizza was cut in fourths, and his brother's was cut in thirds. After they had both eaten all of their pizza, Taylor's little brother said, "Hey that was no fair! You got more than me! You got 4 pieces and I only got 3!"

Should Taylor's little brother be mad? What could you say to explain the situation to him? Use words, pictures, or a number line.

Taylor / Little brother

$\frac{4}{4} = \frac{3}{3}$

No, he shouldn't be mad. $\frac{4}{4} = \frac{3}{3}$ because they both equal 1 whole pizza. I would show him the empty pizza pans, so he could see the pizzas were the same size they were just cut differently.

Lesson Objective: Express whole numbers as fractions and recognize equivalence with different units.

The Student Debrief is intended to invite reflection and active processing of the total lesson experience.

Invite students to review their solutions for the Problem Set. They should check work by comparing answers with a partner before going over answers as a class. Look for misconceptions or misunderstandings that can be addressed in the Student Debrief. Guide students in a conversation to debrief the Problem Set and process the lesson.

Any combination of the questions below may be used to lead the discussion.

- Invite students to share their thinking about Problem 3.
- Invite students to share their work on Problem 4.
- Have students use their fraction shapes from the lesson to model the number bonds in Problem 1.
- Ask students to generate other fractions equivalent to 1 whole. Provide the unit, and ask them to generate the fraction. The following is an example:

  T: The unit is millionths. What fraction is equivalent to 1 whole?

  S: Wow! 1,000,000 millionths!

Exit Ticket (3 minutes)

After the Student Debrief, instruct students to complete the Exit Ticket. A review of their work will help with assessing students' understanding of the concepts that were presented in today's lesson and planning more effectively for future lessons. The questions may be read aloud to the students.

A

Number Correct: _______

Add by Seven

1.	0 + 7 =	
2.	1 + 7 =	
3.	2 + 7 =	
4.	3 + 7 =	
5.	7 + 3 =	
6.	7 + 2 =	
7.	7 + 1 =	
8.	7 + 0 =	
9.	4 + 7 =	
10.	14 + 7 =	
11.	24 + 7 =	
12.	34 + 7 =	
13.	44 + 7 =	
14.	84 + 7 =	
15.	64 + 7 =	
16.	5 + 7 =	
17.	15 + 7 =	
18.	25 + 7 =	
19.	35 + 7 =	
20.	45 + 7 =	
21.	75 + 7 =	
22.	55 + 7 =	

23.	6 + 7 =	
24.	16 + 7 =	
25.	26 + 7 =	
26.	36 + 7 =	
27.	46 + 7 =	
28.	66 + 7 =	
29.	7 + 7 =	
30.	17 + 7 =	
31.	27 + 7 =	
32.	37 + 7 =	
33.	87 + 7 =	
34.	8 + 7 =	
35.	18 + 7 =	
36.	28 + 7 =	
37.	38 + 7 =	
38.	78 + 7 =	
39.	9 + 7 =	
40.	19 + 7 =	
41.	29 + 7 =	
42.	39 + 7 =	
43.	49 + 7 =	
44.	79 + 7 =	

B

Number Correct: _______

Improvement: _______

Add by Seven

1.	7 + 0 =	
2.	7 + 1 =	
3.	7 + 2 =	
4.	7 + 3 =	
5.	3 + 7 =	
6.	2 + 7 =	
7.	1 + 7 =	
8.	0 + 7 =	
9.	4 + 7 =	
10.	14 + 7 =	
11.	24 + 7 =	
12.	34 + 7 =	
13.	44 + 7 =	
14.	74 + 7 =	
15.	54 + 7 =	
16.	5 + 7 =	
17.	15 + 7 =	
18.	25 + 7 =	
19.	35 + 7 =	
20.	45 + 7 =	
21.	85 + 7 =	
22.	65 + 7 =	

23.	6 + 7 =	
24.	16 + 7 =	
25.	26 + 7 =	
26.	36 + 7 =	
27.	46 + 7 =	
28.	76 + 7 =	
29.	7 + 7 =	
30.	17 + 7 =	
31.	27 + 7 =	
32.	37 + 7 =	
33.	67 + 7 =	
34.	8 + 7 =	
35.	18 + 7 =	
36.	28 + 7 =	
37.	38 + 7 =	
38.	88 + 7 =	
39.	9 + 7 =	
40.	19 + 7 =	
41.	29 + 7 =	
42.	39 + 7 =	
43.	49 + 7 =	
44.	89 + 7 =	

Name ______________________________ Date ______________

1. Complete the number bond as indicated by the fractional unit. Partition the number line into the given fractional unit, and label the fractions. Rename 0 and 1 as fractions of the given unit. The first one is done for you.

Halves

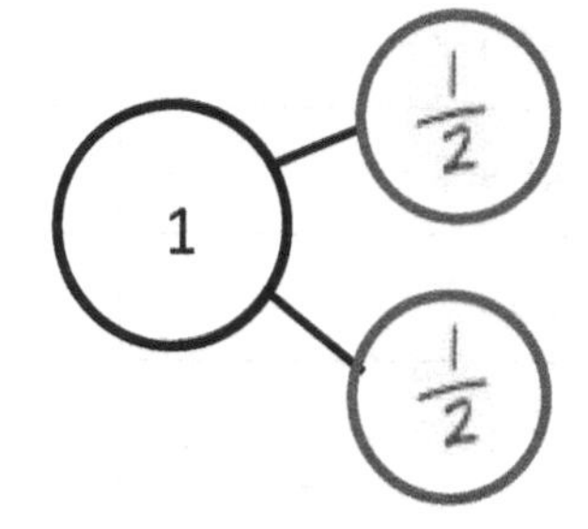

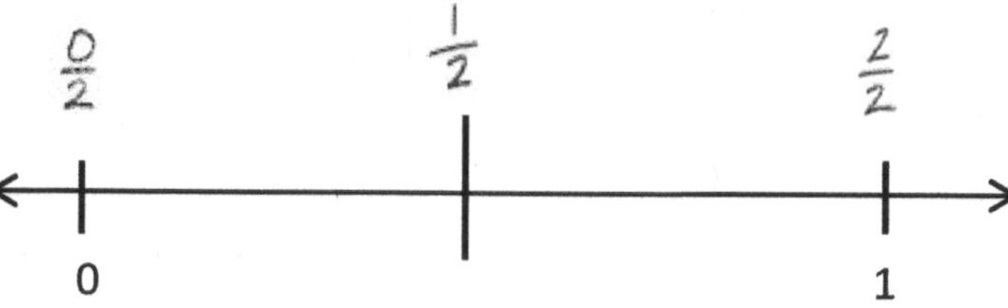

Thirds

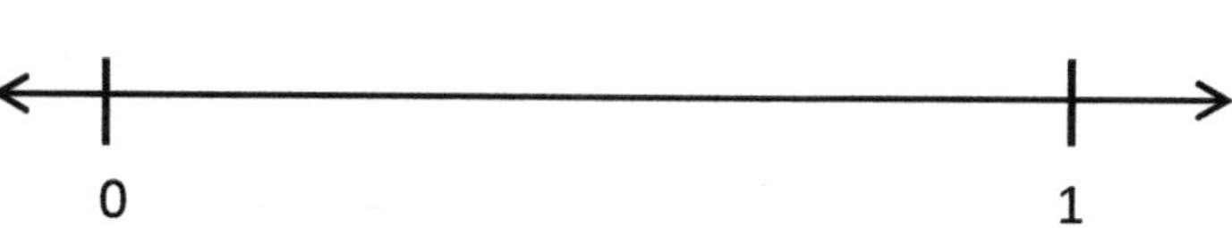

Fourths

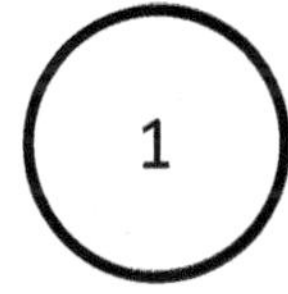

0

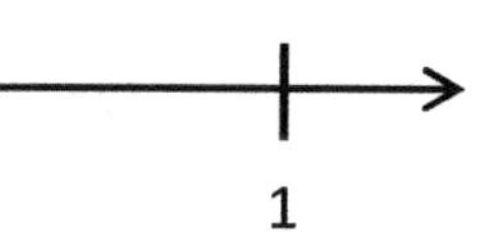

Fifths

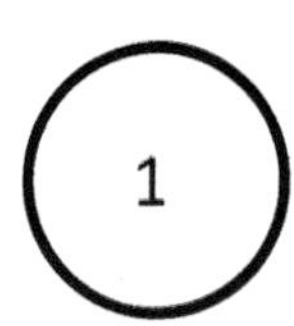

0 1

2. Circle all the fractions in Problem 1 that are equal to 1. Write them in a number sentence below.

$\frac{2}{2}$ = ____________________ = ________________________ = _________________________

3. What pattern do you notice in the fractions that are equivalent to 1?

4. Taylor took his little brother to get pizza. Each boy ordered a small pizza. Taylor's pizza was cut in fourths, and his brother's was cut in thirds. After they had both eaten all of their pizza, Taylor's little brother said, "Hey that was no fair! You got more than me! You got 4 pieces, and I only got 3."

   Should Taylor's little brother be mad? What could you say to explain the situation to him? Use words, pictures, or a number line.

Name ______________________________ Date ______________

1. Complete the number bond as indicated by the fractional unit. Partition the number line into the given fractional unit, and label the fractions. Rename 0 and 1 as fractions of the given unit.

Fourths (1)

0 1

2. How many copies of $\frac{1}{4}$ does it take to make 1 whole? What's the fraction for 1 whole in this case? Use the number line or the number bond in Problem 1 to help you explain.

Name ______________________________ Date ________________

1. Complete the number bond as indicated by the fractional unit. Partition the number line into the given fractional unit, and label the fractions. Rename 0 and 1 as fractions of the given unit.

Fifths

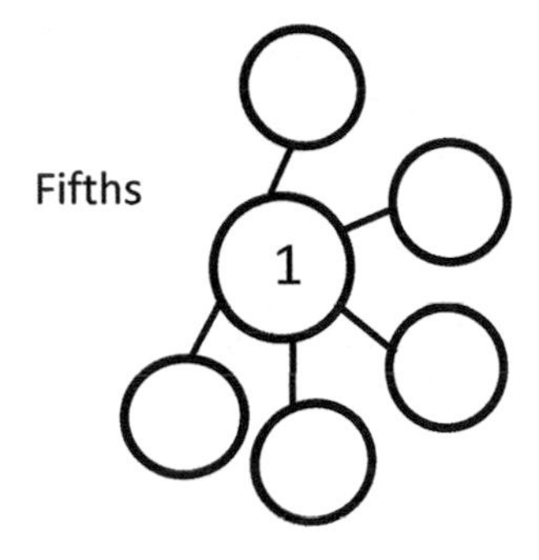

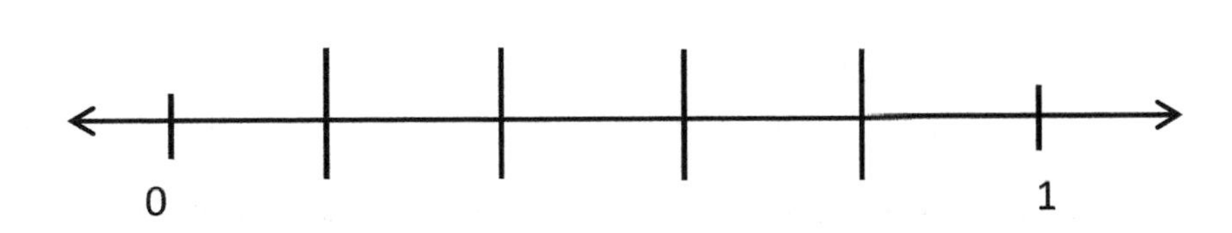

Sixths

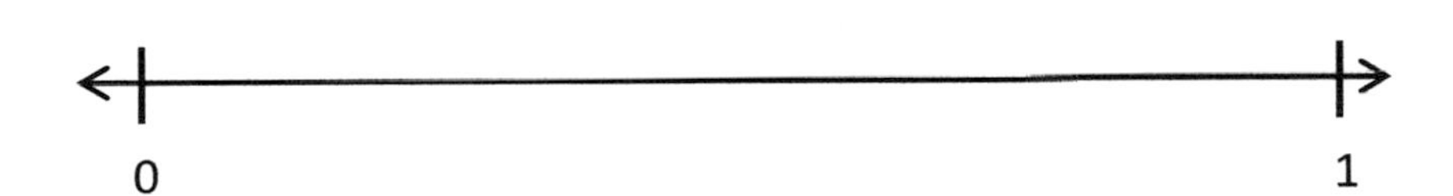

Sevenths

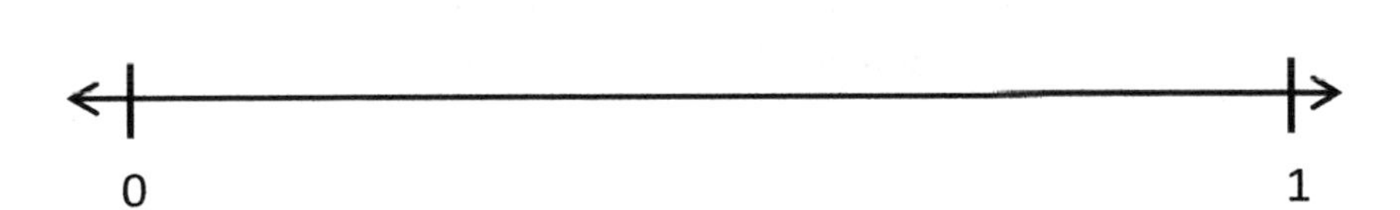

Eighths

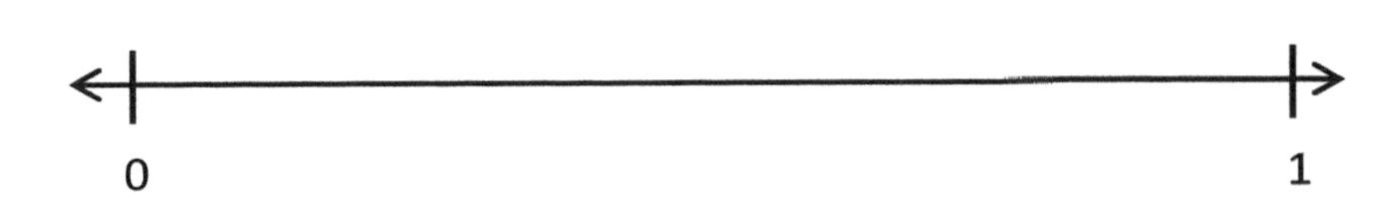

2. Circle all the fractions in Problem 1 that are equal to 1. Write them in a number sentence below.

$\frac{5}{5}$ = ____________________ = ________________________ = __________________________

3. What pattern do you notice in the fractions that are equivalent to 1? Following this pattern, how would you represent ninths as 1 whole?

4. In Art class, Mr. Joselyn gave everyone a 1-foot stick to measure and cut. Vivian measured and cut her stick into 5 equal pieces. Scott measured and cut his into 7 equal pieces. Scott said to Vivian, "The total length of my stick is longer than yours because I have 7 pieces, and you only have 5." Is Scott correct? Use words, pictures, or a number line to help you explain.

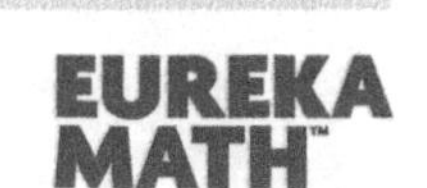

fourths

halves

sixths

1 whole

thirds

fraction pieces

Lesson 25

Objective: Express whole number fractions on the number line when the unit interval is 1.

Suggested Lesson Structure

■ Fluency Practice	(12 minutes)
■ Application Problem	(8 minutes)
■ Concept Development	(30 minutes)
■ Student Debrief	(10 minutes)
Total Time	**(60 minutes)**

Fluency Practice (12 minutes)

- Sprint: Subtract by Six **2.NBT.5** (8 minutes)
- Express Whole Numbers as Different Fractions **3.NF.3c** (4 minutes)

Sprint: Subtract by Six (8 minutes)

Materials: (S) Subtract by Six Sprint

Note: This Sprint supports fluency with subtraction by 6.

Express Whole Numbers as Different Fractions (4 minutes)

Materials: (S) Personal white board

Note: This activity reviews the concept of naming whole numbers as fractions from Lesson 24.

T: (Draw or project a number line from 0–4. Below the 0, write $0 = \frac{}{5}$.) 0 is how many fifths?

S: 0 fifths.

T: (Write $\frac{0}{5}$ below the 0 on the number line. Below the 1, write $1 = \frac{}{5}$.) 1 is how many fifths?

S: 5 fifths.

T: (Write $\frac{5}{5}$ below the 1 on the number line. Below the 2, write $2 = \frac{}{5}$.) On your personal white board, copy and fill in the number sentence.

S: (Write $2 = \frac{10}{5}$.)

T: (Write $\frac{10}{5}$ below the 2 on the number line. Write $3 = \frac{}{5}$.) On your board, copy and fill in the number sentence.

S: (Write $3 = \frac{15}{5}$.)

T: (Write $\frac{15}{5}$ below the 3 on the number line. Write $4 = \frac{}{5}$.) On your board, copy and fill in the number sentence.

S: (Write $4 = \frac{20}{5}$.)

T: (Write $\frac{20}{5}$ below the 4 on the number line.)

Continue the process for fourths.

Application Problem (8 minutes)

Lincoln drinks 1 eighth gallon of milk every morning.

a. How many days will it take Lincoln to drink 1 gallon of milk? Use a number line and words to explain your answer.

b. How many days will it take Lincoln to drink 2 gallons? Extend your number line to show 2 gallons, and use words to explain your answer.

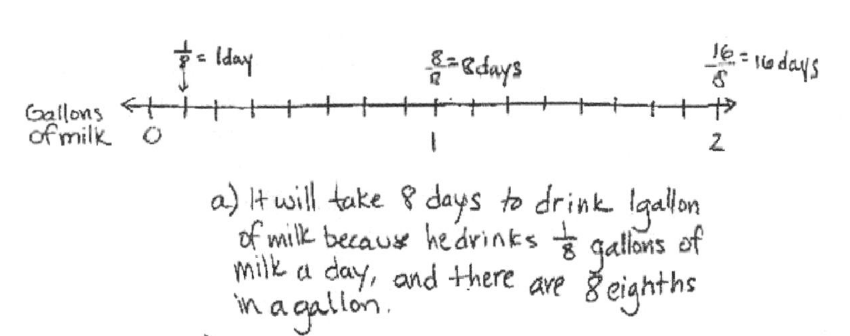

NOTES ON MULTIPLE MEANS OF ENGAGEMENT:

Scaffold the Application Problem for students working below grade level with step-by-step questioning. Ask (for example) the following:

- What is the unit fraction?
- Name the unit that we are partitioning.
- Count by eighths to reach 1 whole, labeling the number line as you count.
- How many eighths are in 1 gallon?
- How many days will it take to drink 1 gallon?
- Count by eighths to reach 2 wholes, labeling the number line as you count.
- How many eighths are in 2 gallons?
- How many days will it take to drink 2 gallons?

Note: This activity reviews the concept of naming whole numbers as fractions from Lesson 24. Invite students to discuss how their number line shows the gallons of milk and the number of days.

Concept Development (30 minutes)

Materials: (S) 3 wholes (Template 1), 6 wholes (Template 2), personal white board

Note on materials: Template 1 is used again in Lessons 27 and 29.

Begin with 3 wholes and 6 wholes in the personal white boards. 3 wholes should be faceup.

3 Wholes Template

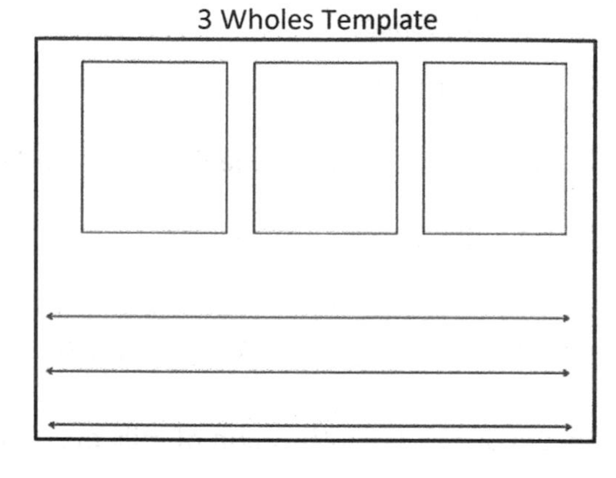

T: Each rectangle represents 1 whole. Partition the first rectangle into thirds. Write the whole as a fraction below it.

S: (Partition and label with $\frac{3}{3}$.)

T: $\frac{3}{3}$ is equivalent to how many wholes?

S: 1 whole!

T: Add that to your picture.

S: (Write $\frac{3}{3} = 1$.)

T: Now, partition the second rectangle into halves. Label the whole as a fraction below it.

S: (Partition and label with $\frac{2}{2}$.)

T: $\frac{2}{2}$ is equivalent to how many wholes?

S: 1 whole!

T: Add that to your picture.

S: (Write $\frac{2}{2} = 1$.)

T: Now, partition the third rectangle into wholes.

S: What do you mean? It is already a whole. → That means 0 partitions!

T: Talk with your partner about how we label this whole as a fraction.

S: 1. → That's not a fraction! It's $\frac{0}{1}$ because there are no parts. → No, it's $\frac{1}{0}$ because we didn't partition. → There's a pattern of the same number for the number of equal parts we're looking at and the total number of equal parts in whole number fractions. So, maybe this is $\frac{1}{1}$?

NOTES ON MULTIPLE MEANS OF REPRESENTATION:

For English language learners, increase wait time, speak clearly, use gestures, and make eye contact. When giving instructions for partitioning the 3 models, pause more frequently, and check for understanding.

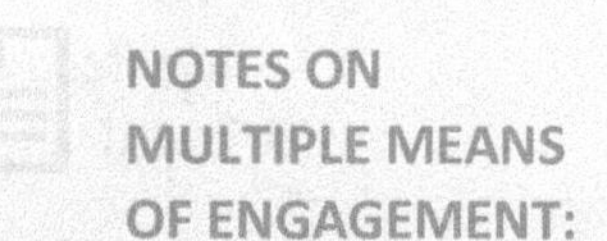

NOTES ON MULTIPLE MEANS OF ENGAGEMENT:

Students working above grade level might enjoy this discussion toward discovery. Guide students to respectfully respond to peers with sentence starters, such as, "That's an idea. I have a different way of seeing it. I think... because..." or "I see what you're saying, however..."

T: I hear some students noticing the pattern that whole number fractions have the same values for the number of equal parts we're counting and the total number of equal parts; therefore, an equivalent way of writing 1 whole as a fraction is to write it as $\frac{1}{1}$. We started with 1 whole. We didn't split it into more parts, so the whole is still in 1 piece, and we're counting that 1 piece. Let's look at the equivalent fractions we've written for 1 on the number line. At the bottom of 3 wholes, mark each of the 3 number lines with endpoints 0 and 1 above the line.

S: (Mark endpoints.)

T: Represent each rectangle on a different number line. Partition, label, and rename the wholes below the line.

S: (Partition and label number lines.)

T: What do you notice about the relationship between the partitioning on the rectangles and the number lines?

S: It's the same. → It goes 2 partitions, 1 partition, no partitions on the last number line!

T: Since we didn't partition because the unit is 1 whole, on the last number line, we renamed the whole...?

S: $\frac{1}{1}$.

6 Wholes Template

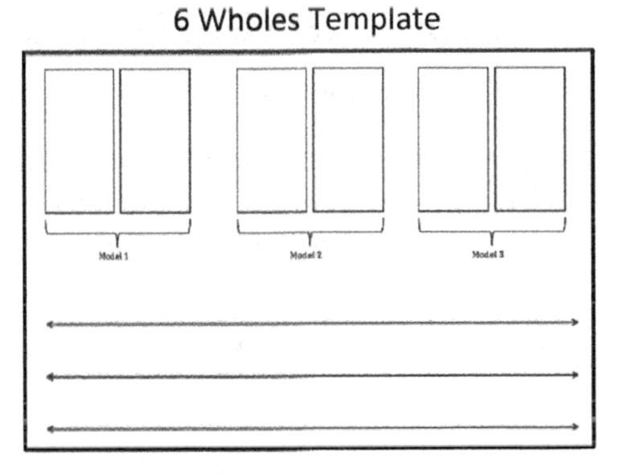

T: Flip your board over to 6 wholes. Each rectangle represents 1 whole. How many wholes are in each model?

S: 2 wholes.

T: Let's partition Model 1 into thirds, Model 2 into halves, and Model 3 into wholes. Use the completed 3 wholes to help if you need it.

S: (Partition.)

T: Now, work with your partner to label each model.

S: (Label Model 1: $\frac{6}{3}$ = 2. Label Model 2: $\frac{4}{2}$ = 2. Students may or may not label Model 3 correctly: $\frac{2}{1}$ = 2.)

T: Let's see how you labeled Model 3. How did you partition the model?

S: There are no partitions because they're both wholes.

T: How many copies of 1 whole does the model have?

S: 2 copies of 1 whole.

T: (Write $\frac{2}{1}$ on the board.) For Model 3, we write the fraction as $\frac{2}{1}$ because there are 2 copies (point to the 2 in the fraction) of the unit, 1 whole (point to the 1 in the fraction). Let's use our number lines again with these models. Label the endpoints on each number line 0 and 2.

Guide students through a similar sequence to the number line work they completed on 3 wholes.

T: I'd like you to circle $\frac{2}{2}$ on your second number line. Now, compare it to where you labeled $\frac{2}{1}$ on your third number line. Tell your partner the difference between $\frac{2}{1}$ and $\frac{2}{2}$.

S: $\frac{2}{2}$ means it's only 1 whole. There are 2 copies, and the unit is halves. → $\frac{2}{1}$ means there are 2 wholes, and the unit of each is 1 whole. → $\frac{2}{1}$ is much larger than $\frac{2}{2}$. It's another whole! You can see that right there on the number line.

If necessary, have students complete a similar sequence with fourths.

Problem Set (10 minutes)

Students should do their personal best to complete the Problem Set within the allotted 10 minutes. For some classes, it may be appropriate to modify the assignment by specifying which problems they work on first. Some problems do not specify a method for solving. Students should solve these problems using the RDW approach used for Application Problems.

Student Debrief (10 minutes)

Lesson Objective: Express whole number fractions on the number line when the unit interval is 1.

The Student Debrief is intended to invite reflection and active processing of the total lesson experience.

Invite students to review their solutions for the Problem Set. They should check work by comparing answers with a partner before going over answers as a class. Look for misconceptions or misunderstandings that can be addressed in the Student Debrief. Guide students in a conversation to debrief the Problem Set and process the lesson.

Any combination of the questions below may be used to lead the discussion.

- Problem 1 presents a slightly different sequence than the lesson. Invite students to share what they notice about the relationship between the models in Problem 1. Consider asking them to relate their work on that question to the guided practice in the lesson.
- Invite students to share their solutions to Problem 3. To solidify their understanding, ask them to apply their thinking to different fractions such as $\frac{3}{1}$ and $\frac{3}{3}$. Consider using a number line during this portion of the discussion to help students notice that the difference between these fractions is even greater and continues to grow as the numbers go higher.

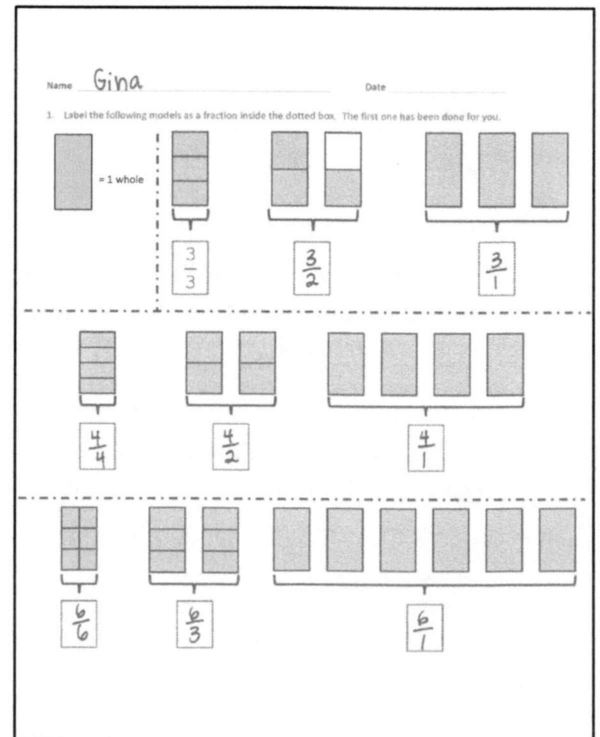

- Have students practice and articulate the lesson objective by closing with a series of pictures that can be quickly drawn on the board. For example, the teacher might make 10 circles and then say the following:

T: If each circle is 1 whole, how might you write the fraction for my total number of wholes?

S: $\frac{10}{1}$.

T: Explain to your partner how you know.

S: (Articulate understanding from the lesson.)

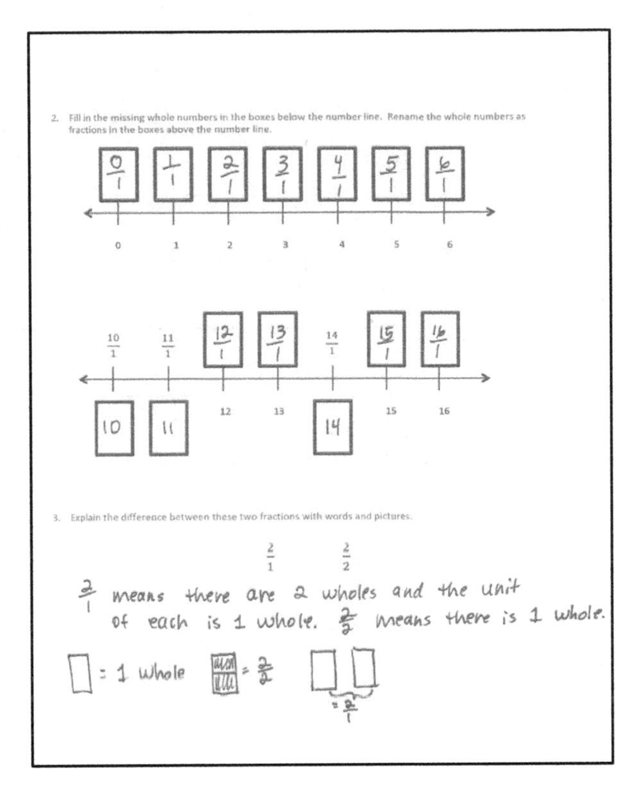

Exit Ticket (3 minutes)

After the Student Debrief, instruct students to complete the Exit Ticket. A review of their work will help with assessing students' understanding of the concepts that were presented in today's lesson and planning more effectively for future lessons. The questions may be read aloud to the students.

A

Number Correct: _______

Subtract by Six

1.	16 – 6 =	
2.	6 – 6 =	
3.	26 – 6 =	
4.	7 – 6 =	
5.	17 – 6 =	
6.	37 – 6 =	
7.	8 – 6 =	
8.	18 – 6 =	
9.	48 – 6 =	
10.	9 – 6 =	
11.	19 – 6 =	
12.	59 – 6 =	
13.	10 – 6 =	
14.	20 – 6 =	
15.	70 – 6 =	
16.	11 – 6 =	
17.	21 – 6 =	
18.	81 – 6 =	
19.	12 – 6 =	
20.	22 – 6 =	
21.	82 – 6 =	
22.	13 – 6 =	

23.	23 – 6 =	
24.	33 – 6 =	
25.	63 – 6 =	
26.	83 – 6 =	
27.	14 – 6 =	
28.	24 – 6 =	
29.	34 – 6 =	
30.	74 – 6 =	
31.	54 – 6 =	
32.	15 – 6 =	
33.	25 – 6 =	
34.	35 – 6 =	
35.	85 – 6 =	
36.	65 – 6 =	
37.	90 – 6 =	
38.	53 – 6 =	
39.	42 – 6 =	
40.	71 – 6 =	
41.	74 – 6 =	
42.	95 – 6 =	
43.	51 – 6 =	
44.	92 – 6 =	

B

Number Correct: _______

Improvement: _______

Subtract by Six

1.	6 – 6 =	
2.	16 – 6 =	
3.	26 – 6 =	
4.	7 – 6 =	
5.	17 – 6 =	
6.	67 – 6 =	
7.	8 – 6 =	
8.	18 – 6 =	
9.	78 – 6 =	
10.	9 – 6 =	
11.	19 – 6 =	
12.	89 – 6 =	
13.	10 – 6 =	
14.	20 – 6 =	
15.	90 – 6 =	
16.	11 – 6 =	
17.	21 – 6 =	
18.	41 – 6 =	
19.	12 – 6 =	
20.	22 – 6 =	
21.	42 – 6 =	
22.	13 – 6 =	

23.	23 – 6 =	
24.	33 – 6 =	
25.	53 – 6 =	
26.	73 – 6 =	
27.	14 – 6 =	
28.	24 – 6 =	
29.	34 – 6 =	
30.	64 – 6 =	
31.	44 – 6 =	
32.	15 – 6 =	
33.	25 – 6 =	
34.	35 – 6 =	
35.	75 – 6 =	
36.	55 – 6 =	
37.	70 – 6 =	
38.	63 – 6 =	
39.	52 – 6 =	
40.	81 – 6 =	
41.	64 – 6 =	
42.	85 – 6 =	
43.	91 – 6 =	
44.	52 – 6 =	

Name ______________________________ Date ______________

1. Label the following models as a fraction inside the dotted box. The first one has been done for you.

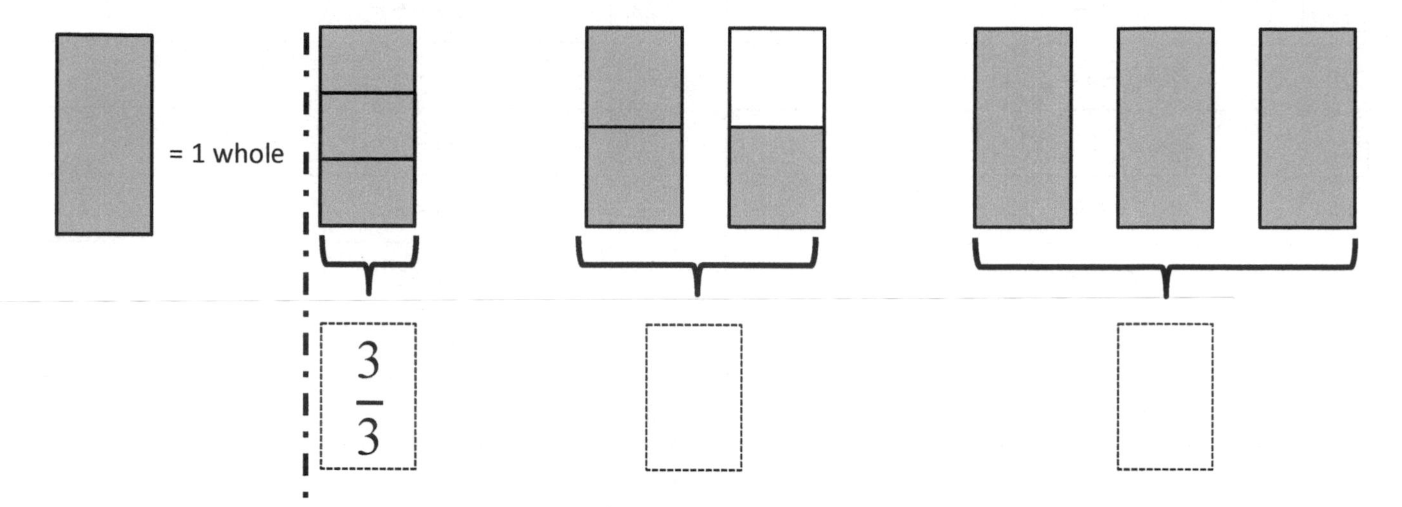

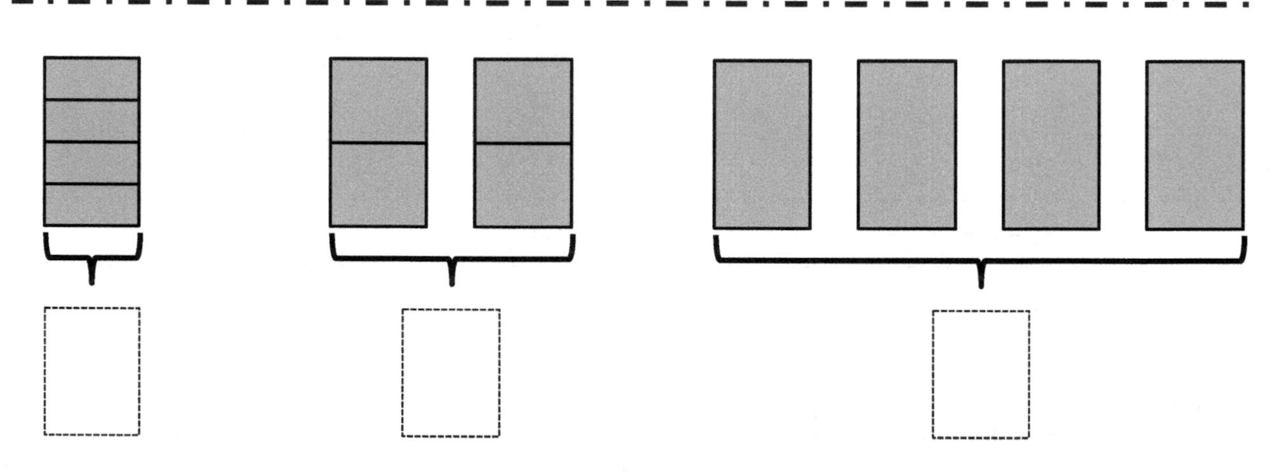

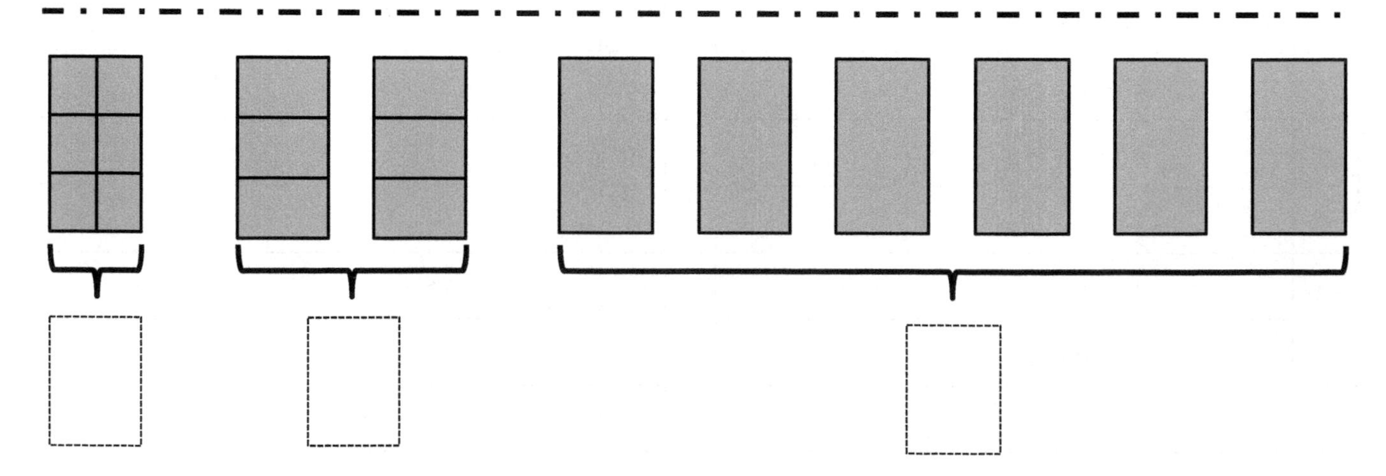

2. Fill in the missing whole numbers in the boxes below the number line. Rename the whole numbers as fractions in the boxes above the number line.

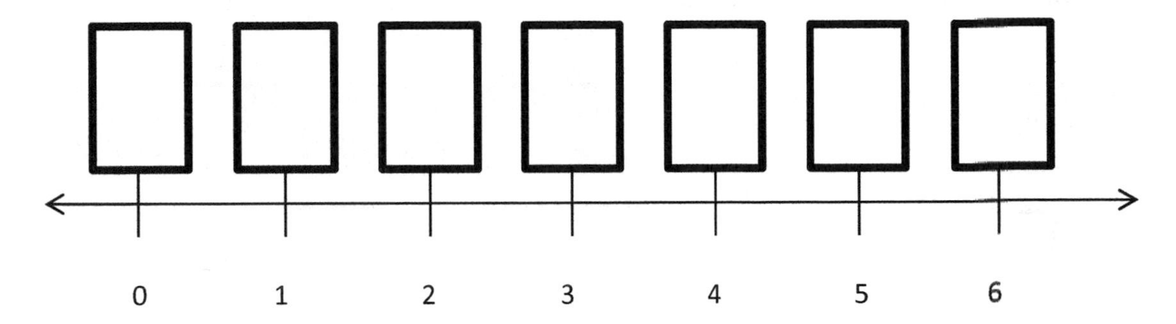

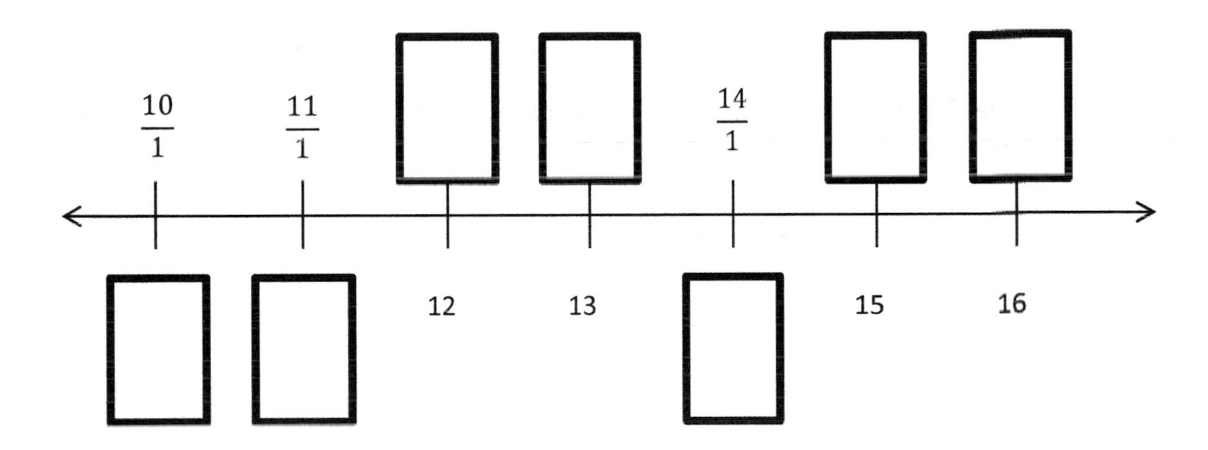

3. Explain the difference between these two fractions with words and pictures.

$$\frac{2}{1} \qquad \frac{2}{2}$$

Name ______________________________ Date ______________

1. Label the model as a fraction inside the box.

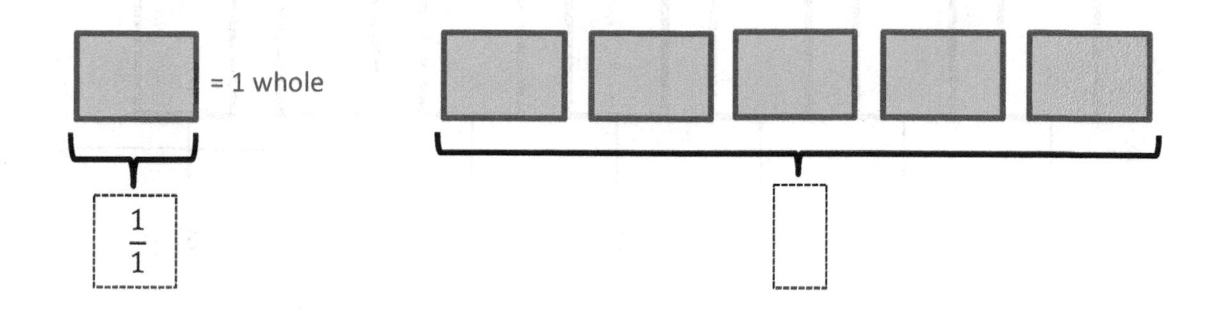

2. Partition the wholes into thirds. Rename the fraction for 3 wholes. Use the number line and words to explain your answer.

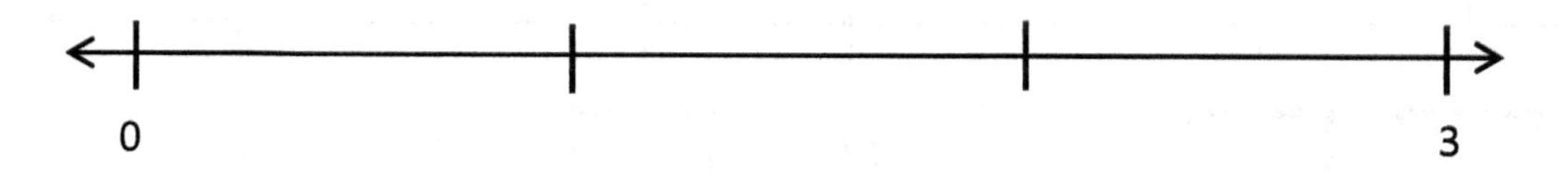

Name ______________________________ Date ______________

1. Label the following models as fractions inside the boxes.

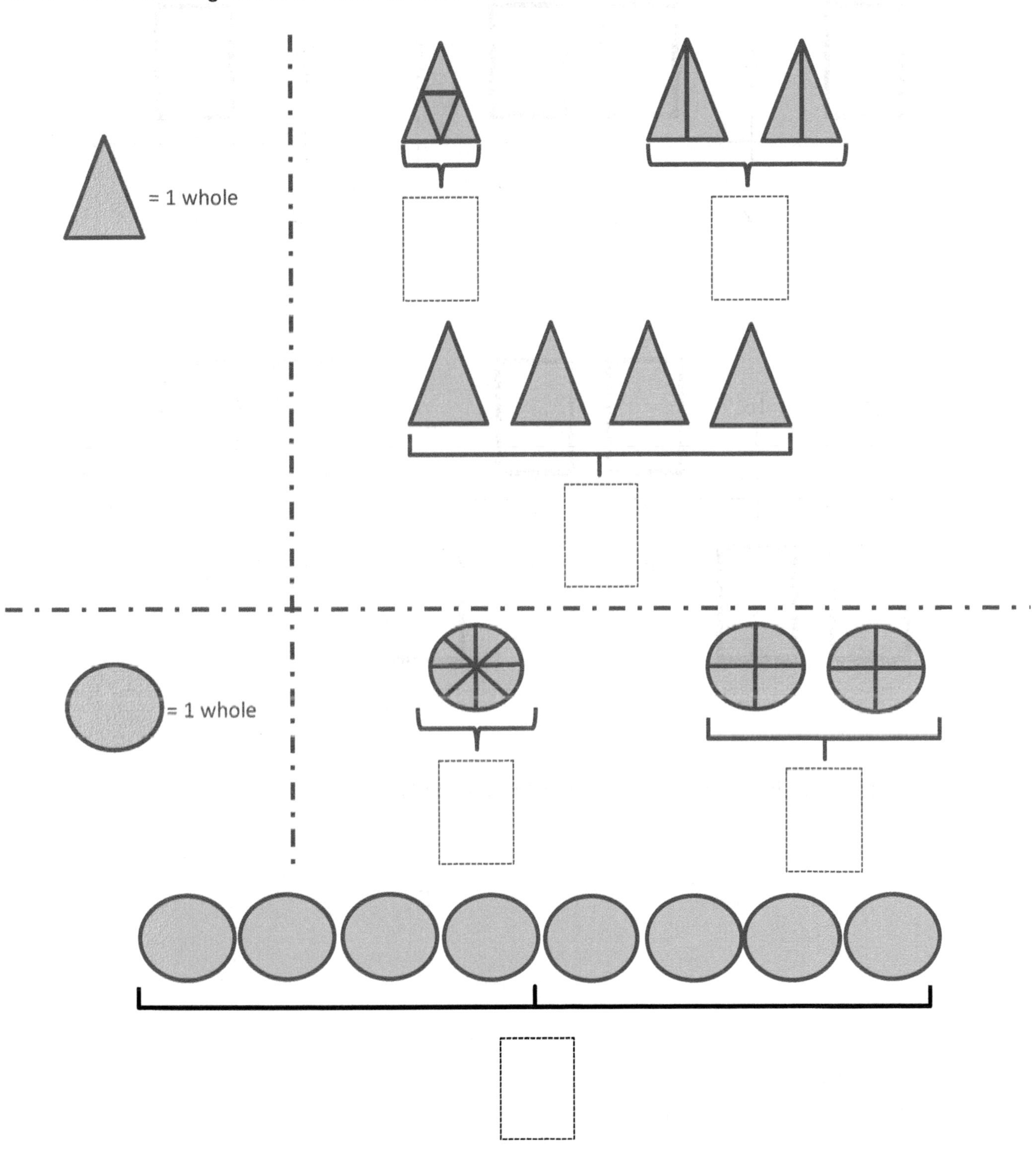

2. Fill in the missing whole numbers in the boxes below the number line. Rename the wholes as fractions in the boxes above the number line.

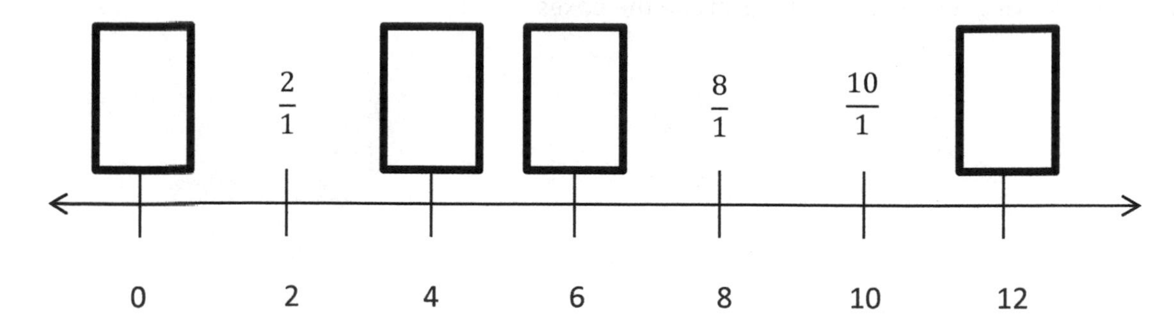

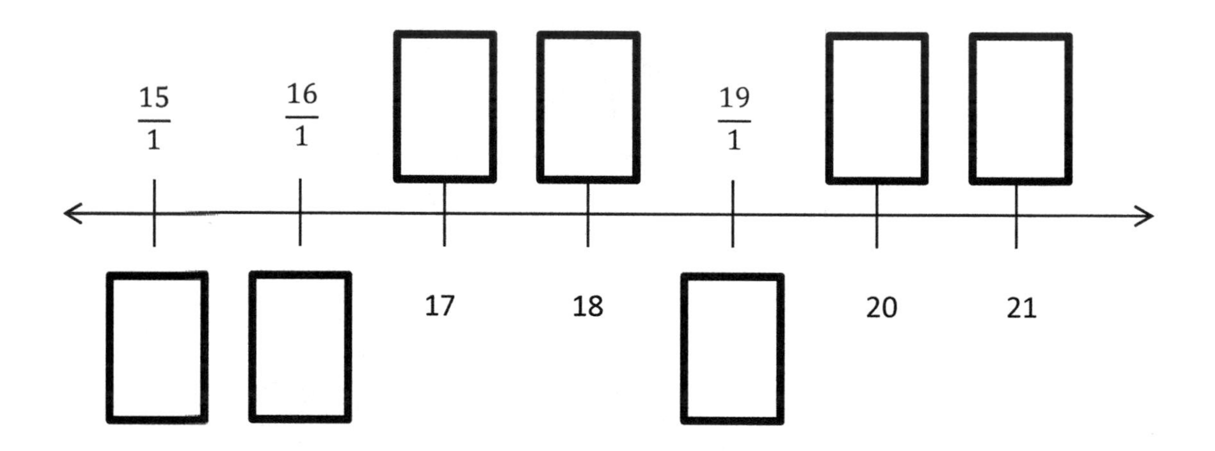

3. Explain the difference between these fractions with words and pictures.

$$\frac{5}{1} \qquad \frac{5}{5}$$

3 wholes

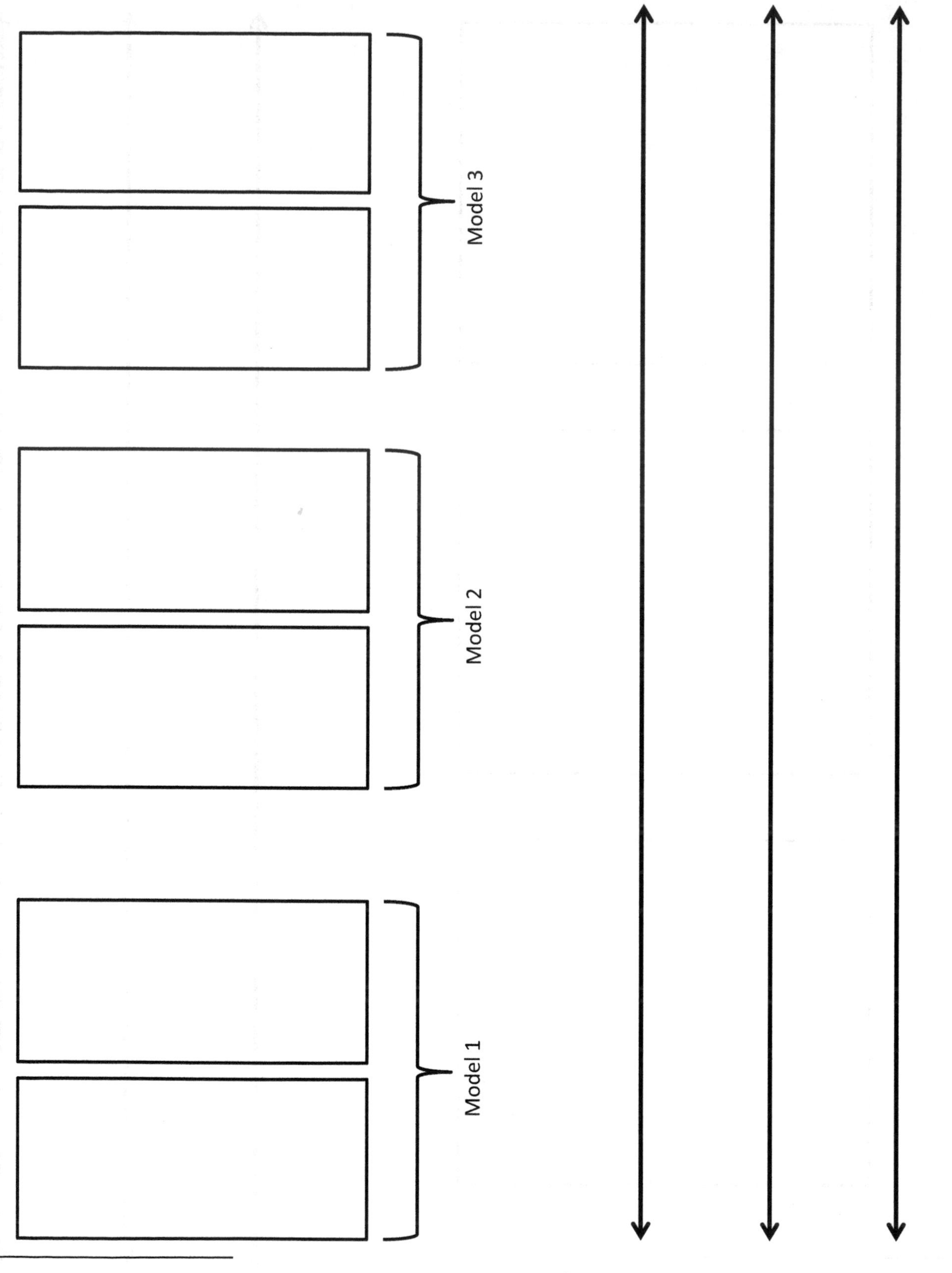

6 wholes

Lesson 26

Objective: Decompose whole number fractions greater than 1 using whole number equivalence with various models.

Suggested Lesson Structure

■ Fluency Practice	(14 minutes)
■ Application Problem	(6 minutes)
■ Concept Development	(30 minutes)
■ Student Debrief	(10 minutes)
Total Time	**(60 minutes)**

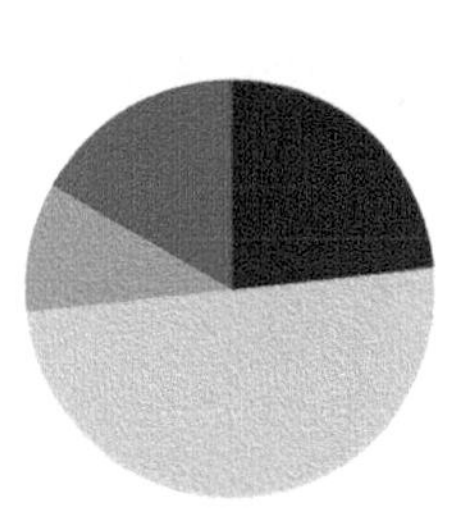

Fluency Practice (14 minutes)

- Sprint: Add by Eight **2.NBT.5** (8 minutes)
- Write Equal Fractions **3.NF.3d** (6 minutes)

Sprint: Add by Eight (8 minutes)

Materials: (S) Add by Eight Sprint

Note: This Sprint supports fluency with addition by 8.

Write Equal Fractions (6 minutes)

Materials: (S) Personal white board

Note: This activity reviews the skill of finding equivalent fractions with pictorial models from Lesson 20.

T: (Project $\frac{1}{2}$.) Say the fraction.

S: 1 half.

T: Draw a shape, shade 1 half, and write the fraction below it.

S: (Draw a shape partitioned into 2 equal parts with one part shaded. Write $\frac{1}{2}$ below the shape.)

T: (Write $\frac{1}{2} = \frac{}{4}$.) Draw the same shape, and partition it into fourths. Shade the fourths to show a fraction equivalent to $\frac{1}{2}$, and complete the number sentence.

S: (Draw the same shape partitioned into 4 equal parts with 2 parts shaded. Write $\frac{1}{2} = \frac{2}{4}$ below the shape.)

Repeat with the following possible sequence: $\frac{1}{3} = \frac{}{6}$, $\frac{1}{4} = \frac{}{8}$, and $\frac{1}{5} = \frac{2}{}$.

Application Problem (6 minutes)

Antonio works on his project for 4 thirds hours. His mom tells him that he must spend another 2 thirds of an hour on it. Draw a number bond and number line with copies of thirds to show how long Antonio needs to work altogether. Write the amount of time Antonio needs to work altogether as a whole number.

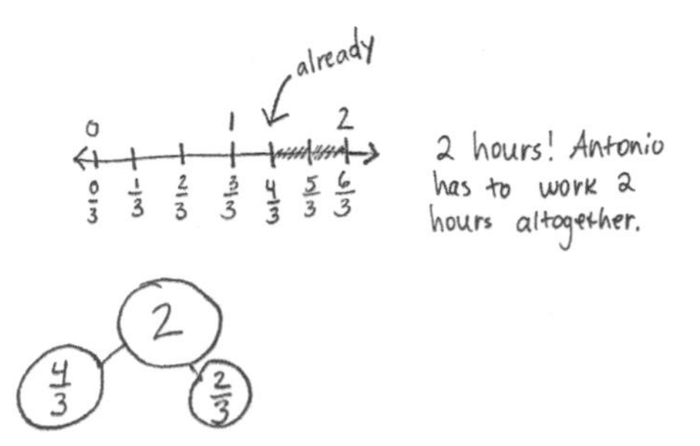

Note: This problem reviews placing fractions on a number line, using number bonds to compose fractions, and representing whole number fractions as whole numbers.

Concept Development (30 minutes)

Image 1

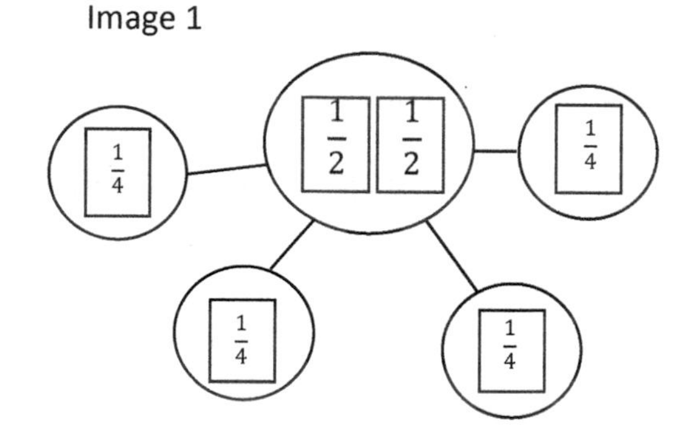

Materials: (S) Personal white board

Draw or project Image 1 on the right, which was also used in Lesson 24.

T: Turn and tell your partner why the number bond is true.

S: Because fourths come from cutting halves in 2 equal pieces. → Yeah, so $\frac{2}{2}$ and $\frac{4}{4}$ both equal 1 whole.

T: (Add 2 more halves to the whole, as shown in Image 2.) Talk to a partner: How do the parts change if we change the whole to look like this?

S: (Discuss.)

T: Work with a partner to draw the new model on your personal white board, and change the parts so that the number bond is true.

S: (Draw.)

T: (Draw or project Models 1 and 2, as shown on the next page.) As I look around the room, I see these two models. Discuss with your partner. Are they equivalent?

Image 2

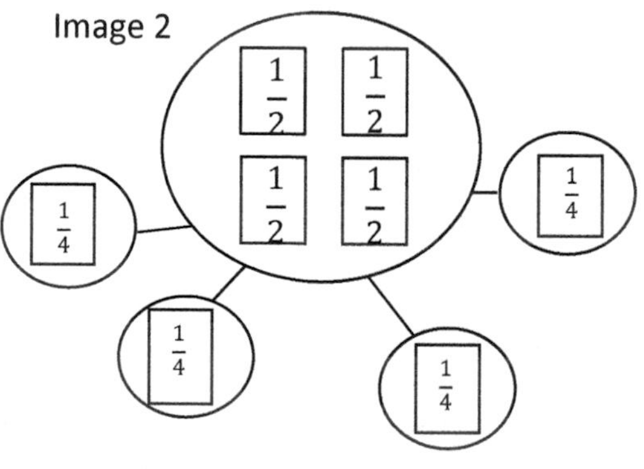

S: There are many more parts in the first model, so they aren't equal. → There are 8 total parts in both models. → 4 copies of $\frac{1}{4}$ makes $\frac{4}{4}$, and another 4 copies of $\frac{1}{4}$ makes another $\frac{4}{4}$. So, they are equivalent. → In the second model, they just made copies of 1 whole to show the total as 2 wholes.

Model 1

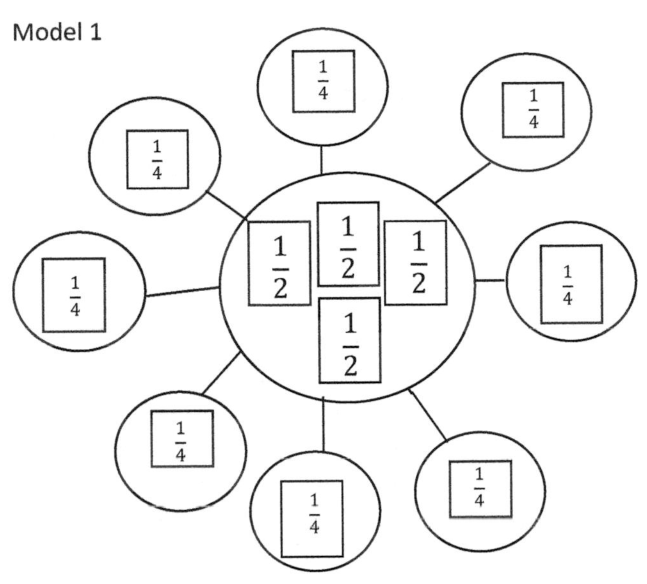

Model 2

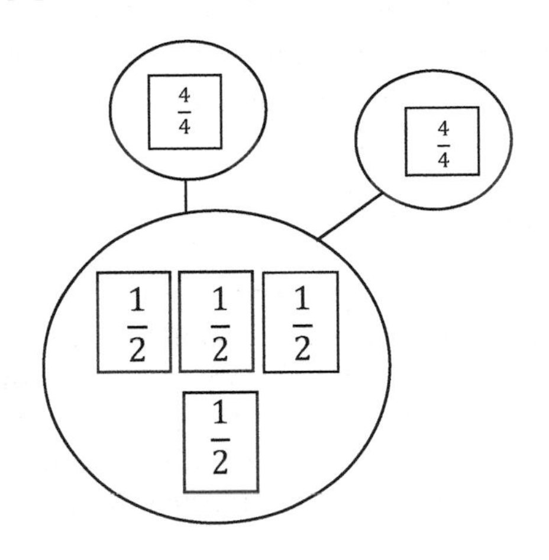

T: Model 2 does show a different way of writing the copies in Model 1. Instead of showing copies of unit fractions, the second model shows copies of 1 whole.

T: Let's see if we can show the equivalence of the number bonds on the number line. Draw a number line with endpoints 0 and 2. Label the wholes on top of the number line. Partition the number line into fourths, and label the fractions.

S: (Draw.)

T: How many fourths in 0?

S: 0 fourths!

T: How many fourths in 1?

S: 4 fourths!

T: How many fourths in 2?

S: 8 fourths!

T: Below each whole number on your number line, work with a partner to draw a number bond. As you draw number bonds, show copies of 1 whole instead of unit fractions if you can.

S: (Draw.)

T: What is the relationship between Models 1 and 2, as well as the number line and number bonds you just drew?

S: Our number bond for 2 on the number line looks just like Model 2. → But Model 2 has halves as the whole. → 4 halves make 2, so they're the same.

NOTES ON MULTIPLE MEANS OF ACTION AND EXPRESSION:

Partner talk is a valuable opportunity for English language learners to speak about their math ideas in English confidently and comfortably. Support limited English speakers with a sentence frame such as, "They are equivalent because Model 1 shows _____ fourths, and Model 2 shows _____ fourths."

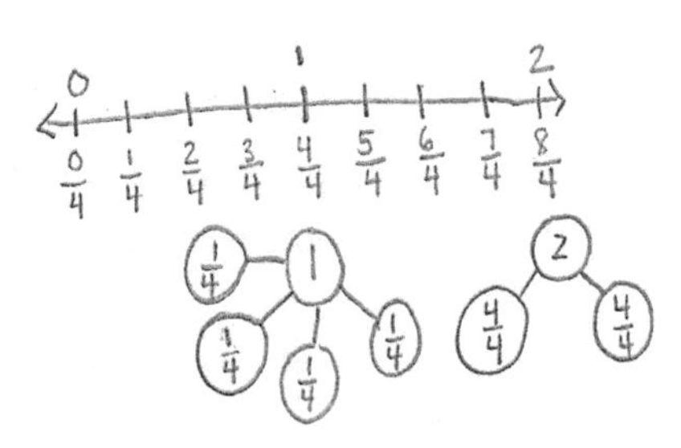

T: What about Model 1?

S: There are 8 fourths on the number line, just like Model 1 shows.

T: What is the difference between these 2 ways of showing the number bond?

S: One is way faster to write. → It's also easier to read because you can see the number of wholes inside of 2.

Problem Set (10 minutes)

Students should do their personal best to complete the Problem Set within the allotted 10 minutes. For some classes, it may be appropriate to modify the assignment by specifying which problems they work on first. Some problems do not specify a method for solving. Students should solve these problems using the RDW approach used for Application Problems.

NOTES ON MULTIPLE MEANS OF ENGAGEMENT:

As an alternative to the Problem Set, offer students working above grade level the option of drawing their own number lines with larger intervals (e.g., 6, 7, and 8) and their choice of fractional unit for partitioning (e.g., fifths).

Student Debrief (10 minutes)

Lesson Objective: Decompose whole number fractions greater than 1 using whole number equivalence with various models.

The Student Debrief is intended to invite reflection and active processing of the total lesson experience.

Invite students to review their solutions for the Problem Set. They should check work by comparing answers with a partner before going over answers as a class. Look for misconceptions or misunderstandings that can be addressed in the Student Debrief. Guide students in a conversation to debrief the Problem Set and process the lesson.

NOTES ON MULTIPLE MEANS OF ENGAGEMENT:

Use the chart on the Problem Set to help students working below grade level build understanding. After students have completed the halves and thirds, ask, "How is the whole related to the whole number fraction?" Discuss and verify predictions for sixths.

Any combination of the questions below may be used to lead the discussion.

- Compare the number lines and number bonds in Problem 1. What does each representation help you see?
- In Problem 2, what strategy did you use to find the whole number fractions without having to partition a number line again?
- Draw number bonds to demonstrate your answers in Problems 3 and 4 using copies of wholes.
- How is the way that we expressed whole number fractions today different from the way we've been doing it?
- Why is it helpful to know how to rename wholes to make number bonds with larger whole numbers?

Exit Ticket (3 minutes)

After the Student Debrief, instruct students to complete the Exit Ticket. A review of their work will help with assessing students' understanding of the concepts that were presented in today's lesson and planning more effectively for future lessons. The questions may be read aloud to the students.

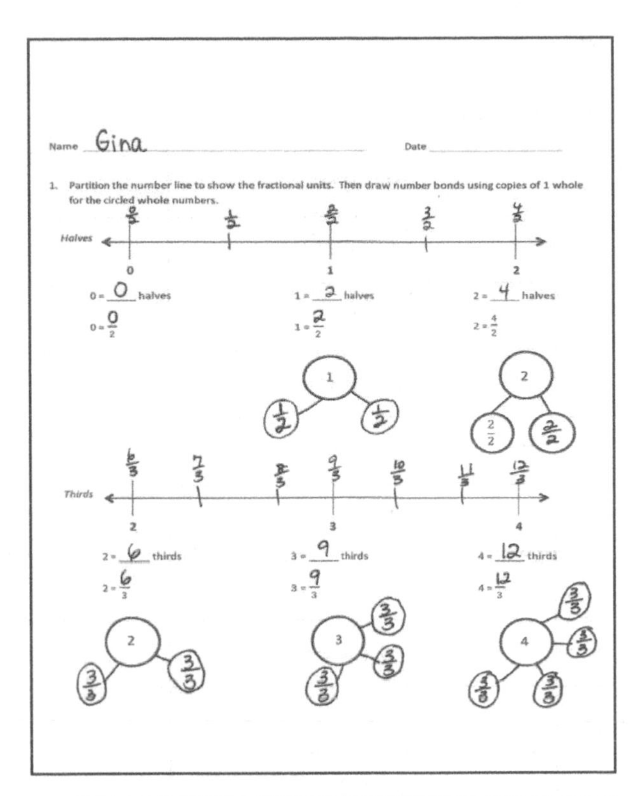

Name Gina Date

1. Partition the number line to show the fractional units. Then draw number bonds using copies of 1 whole for the circled whole numbers.

Halves: $\frac{0}{2}$, $\frac{1}{2}$, $\frac{2}{2}$, $\frac{3}{2}$, $\frac{4}{2}$ (0, 1, 2)

0 = 0 halves 1 = 2 halves 2 = 4 halves

$0 = \frac{0}{2}$ $1 = \frac{2}{2}$ $2 = \frac{4}{2}$

Thirds: $\frac{6}{3}$, $\frac{7}{3}$, $\frac{8}{3}$, $\frac{9}{3}$, $\frac{10}{3}$, $\frac{11}{3}$, $\frac{12}{3}$ (2, 3, 4)

2 = 6 thirds 3 = 9 thirds 4 = 12 thirds

$2 = \frac{6}{3}$ $3 = \frac{9}{3}$ $4 = \frac{12}{3}$

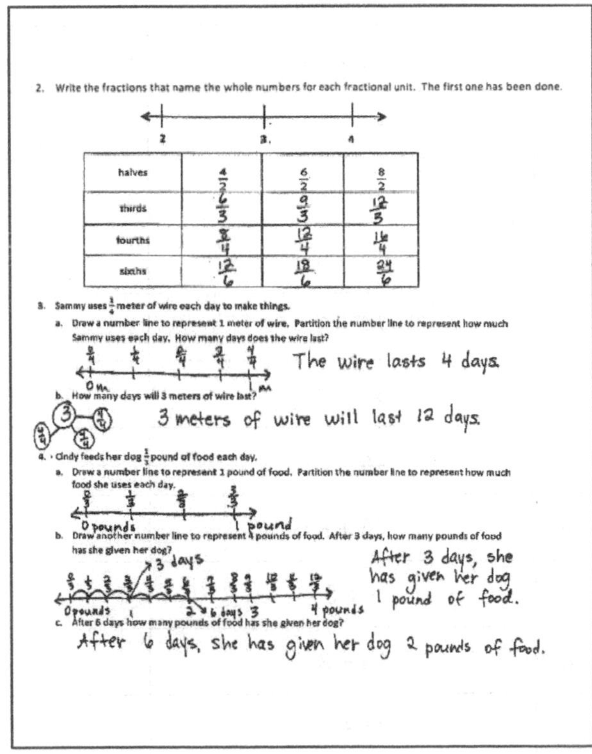

2. Write the fractions that name the whole numbers for each fractional unit. The first one has been done.

	2	3	4
halves	$\frac{4}{2}$	$\frac{6}{2}$	$\frac{8}{2}$
thirds	$\frac{6}{3}$	$\frac{9}{3}$	$\frac{12}{3}$
fourths	$\frac{8}{4}$	$\frac{12}{4}$	$\frac{16}{4}$
sixths	$\frac{12}{6}$	$\frac{18}{6}$	$\frac{24}{6}$

3. Sammy uses $\frac{1}{4}$ meter of wire each day to make things.
   a. Draw a number line to represent 1 meter of wire. Partition the number line to represent how much Sammy uses each day. How many days does the wire last?

   The wire lasts 4 days.

   b. How many days will 3 meters of wire last?

   3 meters of wire will last 12 days.

4. Cindy feeds her dog $\frac{1}{3}$ pound of food each day.
   a. Draw a number line to represent 1 pound of food. Partition the number line to represent how much food she uses each day.
   b. Draw another number line to represent 4 pounds of food. After 3 days, how many pounds of food has she given her dog?

   After 3 days, she has given her dog 1 pound of food.

   c. After 6 days how many pounds of food has she given her dog?

   After 6 days, she has given her dog 2 pounds of food.

A

Number Correct: _______

Add by Eight

1.	0 + 8 =	
2.	1 + 8 =	
3.	2 + 8 =	
4.	8 + 2 =	
5.	1 + 8 =	
6.	0 + 8 =	
7.	3 + 8 =	
8.	13 + 8 =	
9.	23 + 8 =	
10.	33 + 8 =	
11.	43 + 8 =	
12.	83 + 8 =	
13.	4 + 8 =	
14.	14 + 8 =	
15.	24 + 8 =	
16.	34 + 8 =	
17.	44 + 8 =	
18.	74 + 8 =	
19.	5 + 8 =	
20.	15 + 8 =	
21.	25 + 8 =	
22.	35 + 8 =	

23.	65 + 8 =	
24.	6 + 8 =	
25.	16 + 8 =	
26.	26 + 8 =	
27.	36 + 8 =	
28.	86 + 8 =	
29.	46 + 8 =	
30.	7 + 8 =	
31.	17 + 8 =	
32.	27 + 8 =	
33.	37 + 8 =	
34.	77 + 8 =	
35.	8 + 8 =	
36.	18 + 8 =	
37.	28 + 8 =	
38.	38 + 8 =	
39.	68 + 8 =	
40.	9 + 8 =	
41.	19 + 8 =	
42.	29 + 8 =	
43.	39 + 8 =	
44.	89 + 8 =	

B

Number Correct: _______

Improvement: _______

Add by Eight

1.	8 + 0 =	
2.	8 + 1 =	
3.	8 + 2 =	
4.	2 + 8 =	
5.	1 + 8 =	
6.	0 + 8 =	
7.	3 + 8 =	
8.	13 + 8 =	
9.	23 + 8 =	
10.	33 + 8 =	
11.	43 + 8 =	
12.	73 + 8 =	
13.	4 + 8 =	
14.	14 + 8 =	
15.	24 + 8 =	
16.	34 + 8 =	
17.	44 + 8 =	
18.	84 + 8 =	
19.	5 + 8 =	
20.	15 + 8 =	
21.	25 + 8 =	
22.	35 + 8 =	
23.	55 + 8 =	
24.	6 + 8 =	
25.	16 + 8 =	
26.	26 + 8 =	
27.	36 + 8 =	
28.	66 + 8 =	
29.	56 + 8 =	
30.	7 + 8 =	
31.	17 + 8 =	
32.	27 + 8 =	
33.	37 + 8 =	
34.	67 + 8 =	
35.	8 + 8 =	
36.	18 + 8 =	
37.	28 + 8 =	
38.	38 + 8 =	
39.	78 + 8 =	
40.	9 + 8 =	
41.	19 + 8 =	
42.	29 + 8 =	
43.	39 + 8 =	
44.	89 + 8 =	

Name ______________________________ Date ______________

1. Partition the number line to show the fractional units. Then, draw number bonds using copies of 1 whole for the circled whole numbers.

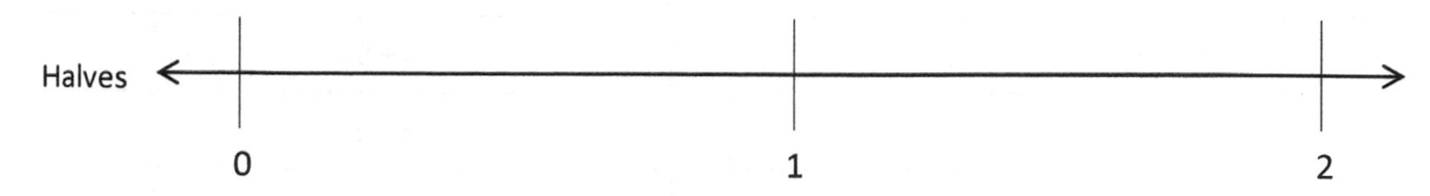

0 = _____ halves 1 = _____ halves 2 = _____ halves

$0 = \frac{\square}{2}$ $1 = \frac{\square}{2}$ $2 = \frac{4}{2}$

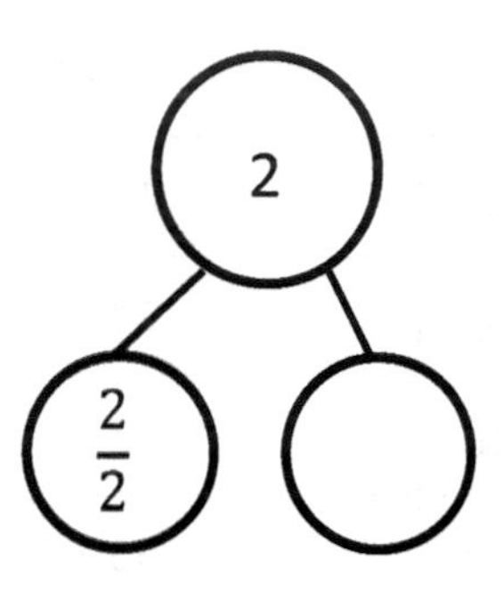

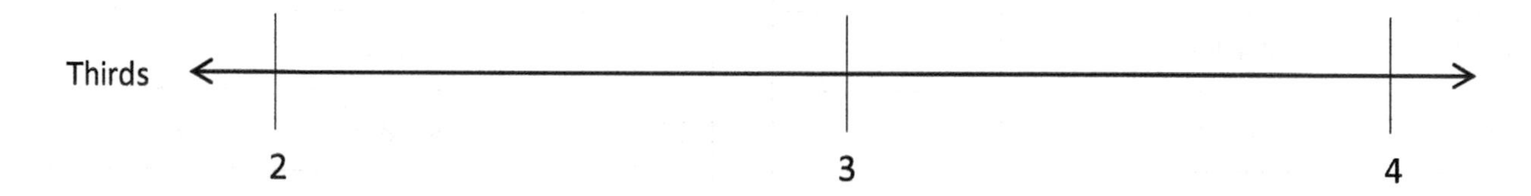

2 = _____ thirds 3 = _____ thirds 4 = _____ thirds

$2 = \frac{\square}{3}$ $3 = \frac{\square}{3}$ $4 = \frac{\square}{3}$

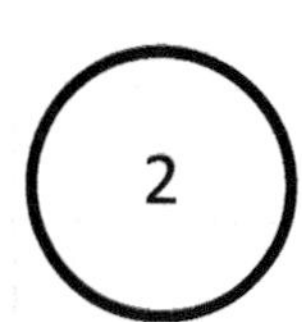

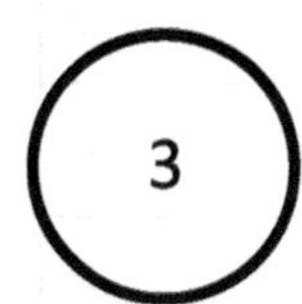

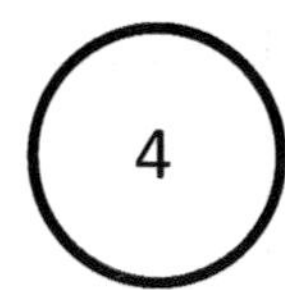

2. Write the fractions that name the whole numbers for each fractional unit. The first one has been done.

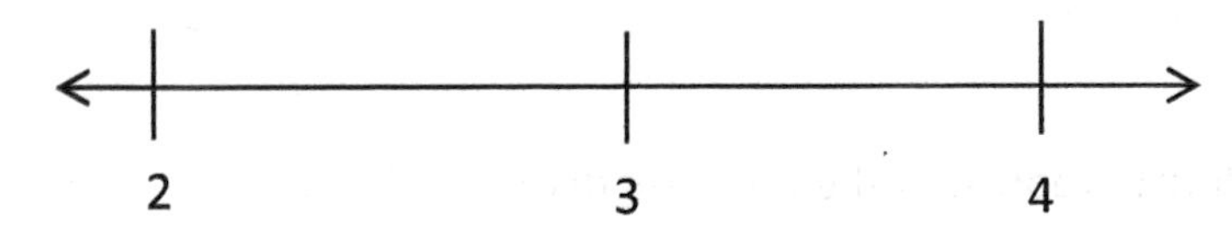

Halves	$\frac{4}{2}$	$\frac{6}{2}$	$\frac{8}{2}$
Thirds			
Fourths			
Sixths			

3. Sammy uses $\frac{1}{4}$ meter of wire each day to make things.

   a. Draw a number line to represent 1 meter of wire. Partition the number line to represent how much Sammy uses each day. How many days does the wire last?

   b. How many days will 3 meters of wire last?

4. Cindy feeds her dog $\frac{1}{3}$ pound of food each day.

   a. Draw a number line to represent 1 pound of food. Partition the number line to represent how much food she uses each day.

   b. Draw another number line to represent 4 pounds of food. After 3 days, how many pounds of food has she given her dog?

   c. After 6 days, how many pounds of food has she given her dog?

Name ______________________________ Date ________________

Irene has 2 yards of fabric.

a. Draw a number line to represent the total length of Irene's fabric.

b. Irene cuts her fabric into pieces of $\frac{1}{5}$ yard in length. Partition the number line to show her cuts.

c. How many $\frac{1}{5}$-yard pieces does she cut altogether? Use number bonds with copies of wholes to help you explain.

Name ______________________________ Date ____________________

1. Partition the number line to show the fractional units. Then, draw number bonds with copies of 1 whole for the circled whole numbers.

Sixths

0 1 2

$0 =$ ______ sixths $1 =$ ______ sixths $2 =$ ______ sixths

$0 = \frac{\square}{6}$ $1 = \frac{\square}{6}$ $2 = \frac{12}{6}$

1

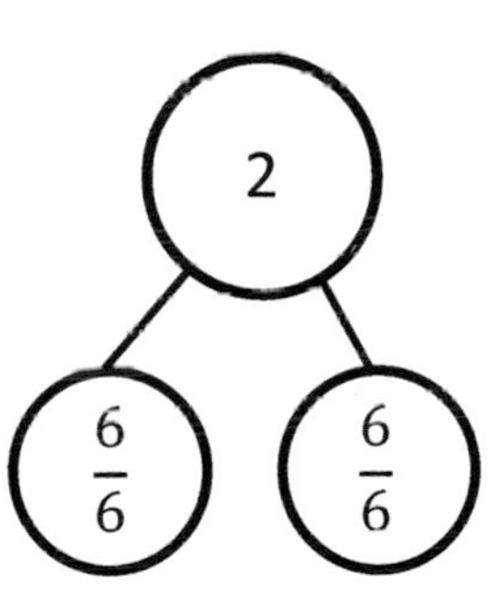

Fifths

2 3 4

$2 =$ ______ fifths $3 =$ ______ fifths $4 =$ ______ fifths

$2 = \frac{\square}{5}$ $3 = \frac{\square}{5}$ $4 = \frac{\square}{5}$

2

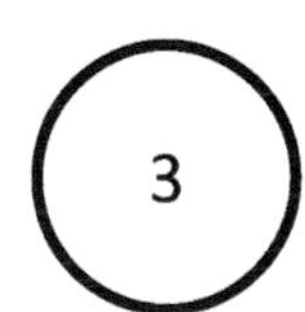

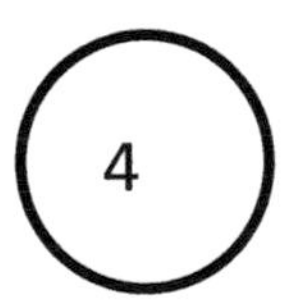

2. Write the fractions that name the whole numbers for each fractional unit. The first one has been done for you.

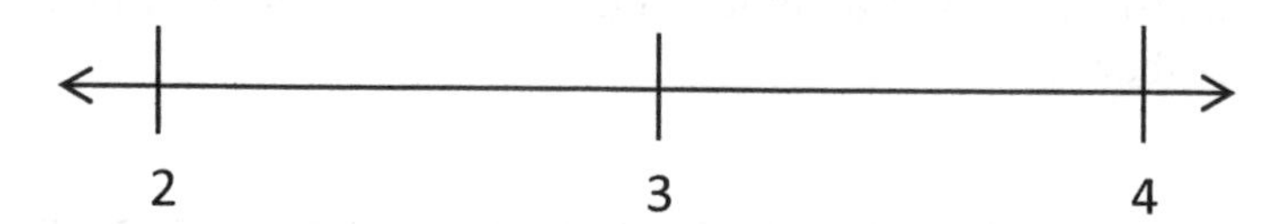

Thirds	$\frac{6}{3}$	$\frac{9}{3}$	$\frac{12}{3}$
Sevenths			
Eighths			
Tenths			

3. Rider dribbles the ball down $\frac{1}{3}$ of the basketball court on the first day of practice. Each day after that, he dribbles $\frac{1}{3}$ of the way more than he did the day before. Draw a number line to represent the court. Partition the number line to represent how far Rider dribbles on Day 1, Day 2, and Day 3 of practice. What fraction of the way does he dribble on Day 3?

Lesson 27

Objective: Explain equivalence by manipulating units and reasoning about their size.

Suggested Lesson Structure

■ Fluency Practice	(12 minutes)
■ Application Problem	(8 minutes)
■ Concept Development	(30 minutes)
■ Student Debrief	(10 minutes)
Total Time	**(60 minutes)**

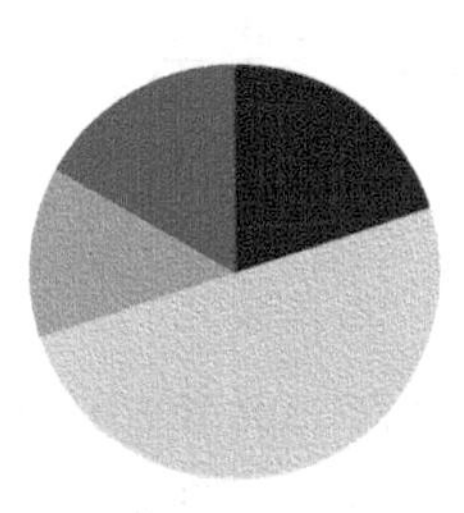

Fluency Practice (12 minutes)

- Sprint: Subtract by Seven **2.NBT.5** (8 minutes)
- Recognize the Fraction **3.G.2** (4 minutes)

Sprint: Subtract by Seven (8 minutes)

Materials: (S) Subtract by Seven Sprint

Note: This Sprint supports fluency with subtraction by 7.

Recognize the Fraction (4 minutes)

Materials: (S) Personal white board

Note: This activity reviews the concept of naming various fractions, depending on the designation of the whole.

T: (Project or draw a shaded rectangular model.) This equals 1 whole. (Project or draw 1 whole partitioned into 3 equal shaded units.) On your personal white board, write the fraction.

S: (Write $\frac{3}{3}$.)

T: (Project or draw 2 wholes, each partitioned into 3 equal shaded units.) On your board, write the fraction.

S: (Write $\frac{6}{3}$.)

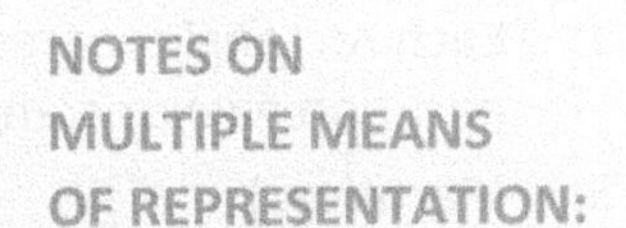

NOTES ON MULTIPLE MEANS OF REPRESENTATION:

Have English language learners practice *Recognize the Fraction* orally to practice speaking math language in English with the support of a model.

T: (Project or draw 3 wholes, each partitioned into 3 shaded parts.) On your board, write the fraction.

S: (Write $\frac{9}{3}$.)

T: (Project or draw 3 wholes, each partitioned into 3 parts. 3 parts in the first 2 wholes are shaded. 1 part of the third whole is shaded.) On your board, write the fraction.

S: (Write $\frac{7}{3}$.)

Continue with the following possible sequence: $\frac{4}{4}, \frac{8}{4}, \frac{12}{4}, \frac{9}{4}, \frac{6}{5}$, and $\frac{9}{8}$.

Application Problem (8 minutes)

The branch of a tree is 2 meters long. Monica chops the branch for firewood. She cuts pieces that are $\frac{1}{6}$ meter long. Draw a number line to show the total length of the branch. Partition and label each of Monica's cuts.

a. How many pieces does Monica have altogether?

b. Write 2 equivalent fractions to describe the total length of Monica's branch.

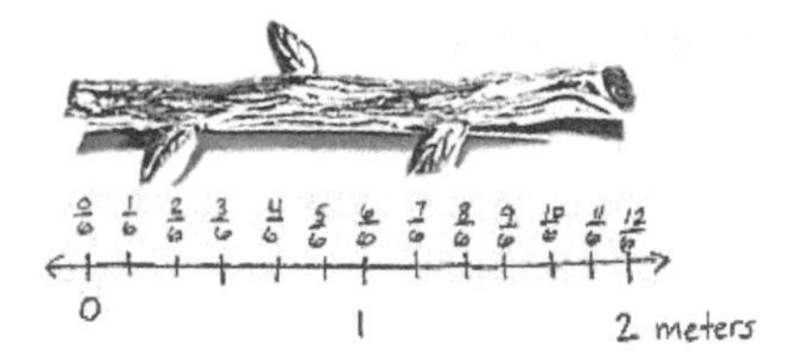

a) Monica has 12 pieces.

b) $\frac{12}{6} = \frac{2}{1}$

Note: This problem reviews partitioning wholes on the number line, labeling fractions on a number line, and naming equivalent fractions.

NOTES ON MULTIPLE MEANS OF ENGAGEMENT:

For English language learners, demonstrate that words can have multiple meanings. Here, *cut* means to draw a line (or lines) that divides the unit into smaller equal parts.

Students working below grade level may benefit from revisiting the discussion of *doubling, tripling, halving*, and cutting unit fractions as presented in Lesson 22.

Concept Development (30 minutes)

Materials: (S) 3 wholes (Lesson 25 Template 1), personal white board, fraction strips (3 per student), math journal

Pass out 3 wholes, and have students slip it into their personal white boards.

T: Each rectangle represents 1 whole. Estimate to partition each rectangle into thirds.

S: (Partition.)

T: How can we double the number of units in the second rectangle?

S: We cut each third in 2.

T: Go ahead and partition.

S: (Partition.)

T: What's our new unit?

S: Sixths!

3 wholes (Lesson 25 Template 1)

Repeat this process for the third rectangle. Instead of having students double, have them triple the original thirds.

MP.3

T: Label the fractions in each model.

S: (Label.)

T: What is different about these models?

S: They all started as thirds, but then we cut them into different parts. → The parts are different sizes. → Yes, they're different units.

T: What is the same about these models?

S: The whole.

T: Talk to your partner about the relationship between the number of parts and the size of parts in each model.

S: 3 is the smallest number, but thirds have the biggest size. → As I drew more lines to partition, the size of the parts got smaller. → That's because the whole is cut into more pieces when there are ninths than when there are thirds.

T: (Give each student 3 fraction strips.) Fold all 3 fraction strips into halves.

S: (Fold.)

T: Fold your second and third fraction strips to double the number of units.

S: (Fold.)

T: What's the new unit on these fraction strips?

S: Fourths!

T: Fold your third fraction strip to double the number of units again.

S: (Fold.)

T: What's the new unit on your third fraction strip?

S: Eighths!

T: Compare the number of parts and the size of the parts with the number of times you folded the strip. What happens to the size of the parts when you fold the strip more times?

S: The more I folded, the smaller the parts got. → Yeah, that's because you folded the whole to make more units.

T: Open your math journal to a new page, and glue your strips in a column, making sure the ends line up. Glue them from the largest unit to the smallest.

S: (Glue.)

T: Use your fraction strips to find the fractions equivalent to $\frac{4}{8}$. Shade them.

S: (Shade $\frac{4}{8}$, $\frac{2}{4}$, and $\frac{1}{2}$.)

T: Talk with your partner: What do you notice about the size of parts and number of parts in equivalent fractions?

S: You can see that there are more eighths than halves or fourths shaded to cover the same amount of the strip. → It's the same as before then. As the number of parts gets larger, the size of them gets smaller. → That's because the shaded area in equivalent fractions doesn't change, even though the number of parts gets larger.

If necessary, reinforce the concept with other examples using these fraction strips.

Image 1

T: (Show Image 1.) Let's practice this idea a bit more on our personal white boards. Draw my shape on your board. The entire figure represents 1 whole.

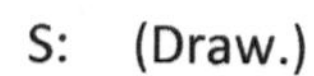

S: (Draw.)

T: Write the shaded fraction.

S: (Write $\frac{1}{4}$.)

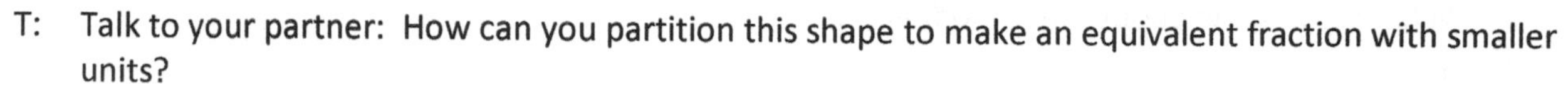

T: Talk to your partner: How can you partition this shape to make an equivalent fraction with smaller units?

S: We can cut each small rectangle in 2 pieces from top to bottom to make eighths. → Or we can make 2 horizontal cuts to make twelfths.

T: Use one of these strategies now. (Circulate as students work to select a few different examples to share with the class.)

S: (Partition.)

T: Let's look at our classmates' work. (Show examples of $\frac{2}{8}$, $\frac{3}{12}$, $\frac{4}{16}$, etc.) As we partitioned with more parts, what happens to the shaded area and number of parts needed to make them equivalent?

S: The size of the parts gets smaller, but the number of them gets larger.

T: Even though the parts changed, did the area covered by the shaded region change?

S: No.

Consider having students practice independently. The shape to the right is more challenging because triangles are more difficult to make into equal parts.

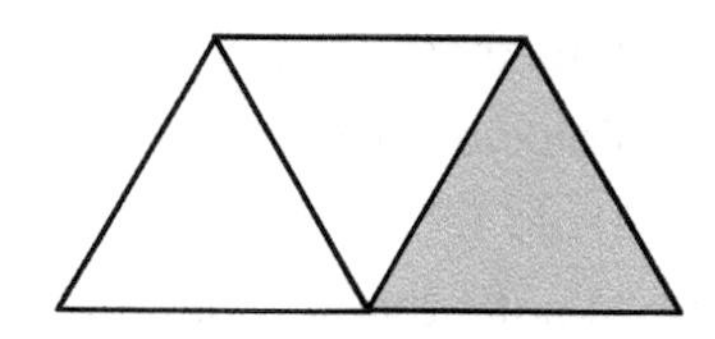

Problem Set (10 minutes)

Students should do their personal best to complete the Problem Set within the allotted 10 minutes. For some classes, it may be appropriate to modify the assignment by specifying which problems they work on first. Some problems do not specify a method for solving. Students should solve these problems using the RDW approach used for Application Problems.

NOTES ON MULTIPLE MEANS OF ENGAGEMENT:

Extend Problem 5 on the Problem Set for students working above grade level. Instead of *doubling,* have students *triple* or *quadruple.* Let students choose the fractional unit into which the rectangle is partitioned.

Student Debrief (10 minutes)

Lesson Objective: Explain equivalence by manipulating units and reasoning about their size.

The Student Debrief is intended to invite reflection and active processing of the total lesson experience.

Invite students to review their solutions for the Problem Set. They should check work by comparing answers with a partner before going over answers as a class. Look for misconceptions or misunderstandings that can be addressed in the Student Debrief. Guide students in a conversation to debrief the Problem Set and process the lesson.

Any combination of the questions below may be used to lead the discussion.

- How did using the fraction strips help you with Problem 2? Talk about the relationship between them.
- What was your strategy for Problems 3 and 4? How did it change or stay the same?
- Why is it important that the magic wand in Problem 5 keeps the whole the same?
- How does the magic wand in Problem 5 make it easy to create equivalent fractions?

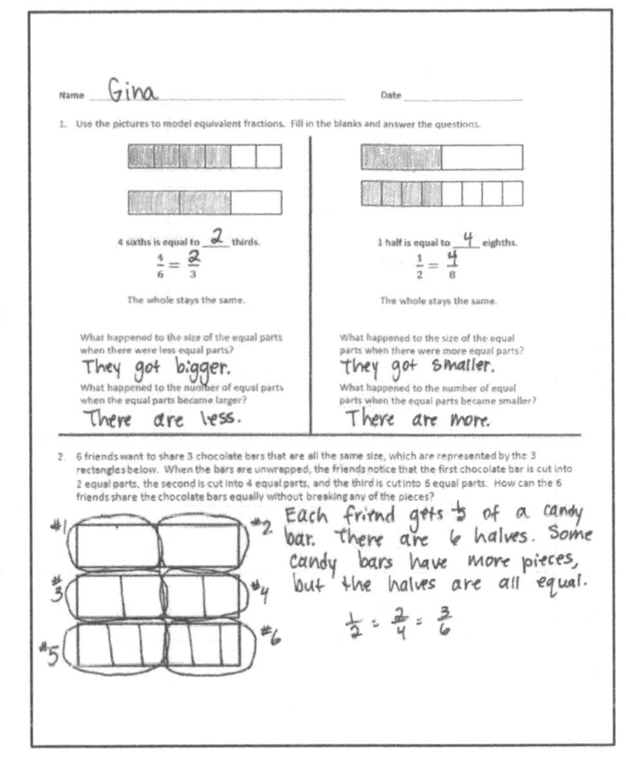

Name Gina Date

1. Use the pictures to model equivalent fractions. Fill in the blanks and answer the questions.

4 sixths is equal to 2 thirds.

$\frac{4}{6} = \frac{2}{3}$

The whole stays the same.

What happened to the size of the equal parts when there were less equal parts?

They got bigger.

What happened to the number of equal parts when the equal parts became larger?

There are less.

1 half is equal to 4 eighths.

$\frac{1}{2} = \frac{4}{8}$

The whole stays the same.

What happened to the size of the equal parts when there were more equal parts?

They got smaller.

What happened to the number of equal parts when the equal parts became smaller?

There are more.

2. 6 friends want to share 3 chocolate bars that are all the same size, which are represented by the 3 rectangles below. When the bars are unwrapped, the friends notice that the first chocolate bar is cut into 2 equal parts, the second is cut into 4 equal parts, and the third is cut into 6 equal parts. How can the 6 friends share the chocolate bars equally without breaking any of the pieces?

Each friend gets $\frac{1}{2}$ of a candy bar. There are 6 halves. Some candy bars have more pieces, but the halves are all equal.

$\frac{1}{2} = \frac{2}{4} = \frac{3}{6}$

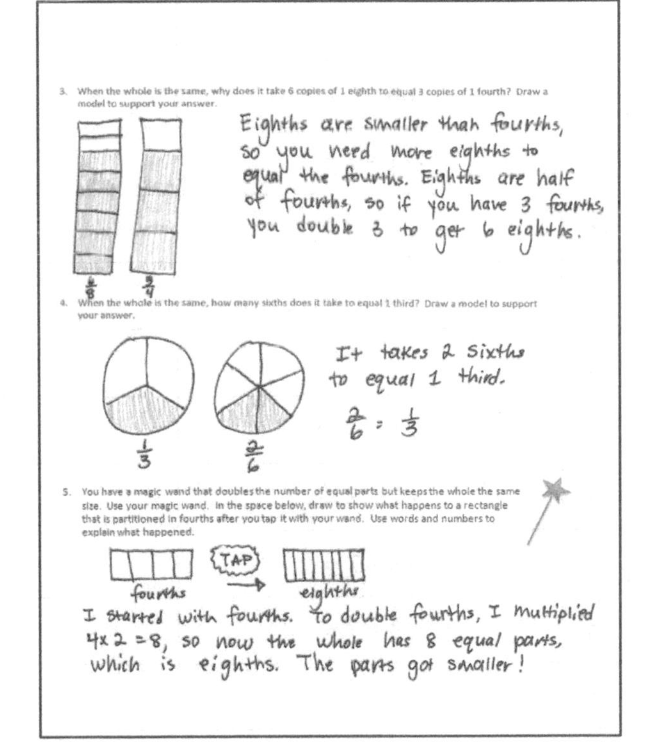

3. When the whole is the same, why does it take 6 copies of 1 eighth to equal 3 copies of 1 fourth? Draw a model to support your answer.

Eighths are smaller than fourths, so you need more eighths to equal the fourths. Eighths are half of fourths, so if you have 3 fourths, you double 3 to get 6 eighths.

4. When the whole is the same, how many sixths does it take to equal 1 third? Draw a model to support your answer.

It takes 2 sixths to equal 1 third.

$\frac{2}{6} = \frac{1}{3}$

5. You have a magic wand that doubles the number of equal parts but keeps the whole the same size. Use your magic wand. In the space below, draw to show what happens to a rectangle that is partitioned in fourths after you tap it with your wand. Use words and numbers to explain what happened.

I started with fourths. To double fourths, I multiplied $4 \times 2 = 8$, so now the whole has 8 equal parts, which is eighths. The parts got smaller!

Exit Ticket (3 minutes)

After the Student Debrief, instruct students to complete the Exit Ticket. A review of their work will help with assessing students' understanding of the concepts that were presented in today's lesson and planning more effectively for future lessons. The questions may be read aloud to the students.

A

Number Correct: ______

Subtract by Seven

1.	17 – 7 =	
2.	7 – 7 =	
3.	27 – 7 =	
4.	8 – 7 =	
5.	18 – 7 =	
6.	38 – 7 =	
7.	9 – 7 =	
8.	19 – 7 =	
9.	49 – 7 =	
10.	10 – 7 =	
11.	20 – 7 =	
12.	60 – 7 =	
13.	11 – 7 =	
14.	21 – 7 =	
15.	71 – 7 =	
16.	12 – 7 =	
17.	22 – 7 =	
18.	82 – 7 =	
19.	13 – 7 =	
20.	23 – 7 =	
21.	83 – 7 =	
22.	14 – 7 =	

23.	24 – 7 =	
24.	34 – 7 =	
25.	64 – 7 =	
26.	84 – 7 =	
27.	15 – 7 =	
28.	25 – 7 =	
29.	35 – 7 =	
30.	75 – 7 =	
31.	55 – 7 =	
32.	16 – 7 =	
33.	26 – 7 =	
34.	36 – 7 =	
35.	86 – 7 =	
36.	66 – 7 =	
37.	90 – 7 =	
38.	53 – 7 =	
39.	42 – 7 =	
40.	71 – 7 =	
41.	74 – 7 =	
42.	56 – 7 =	
43.	95 – 7 =	
44.	92 – 7 =	

B

Number Correct: _______

Improvement: _______

Subtract by Seven

1.	7 – 7 =	
2.	17 – 7 =	
3.	27 – 7 =	
4.	8 – 7 =	
5.	18 – 7 =	
6.	68 – 7 =	
7.	9 – 7 =	
8.	19 – 7 =	
9.	79 – 7 =	
10.	10 – 7 =	
11.	20 – 7 =	
12.	90 – 7 =	
13.	11 – 7 =	
14.	21 – 7 =	
15.	91 – 7 =	
16.	12 – 7 =	
17.	22 – 7 =	
18.	42 – 7 =	
19.	13 – 7 =	
20.	23 – 7 =	
21.	43 – 7 =	
22.	14 – 7 =	

23.	24 – 7 =	
24.	34 – 7 =	
25.	54 – 7 =	
26.	74 – 7 =	
27.	15 – 7 =	
28.	25 – 7 =	
29.	35 – 7 =	
30.	65 – 7 =	
31.	45 – 7 =	
32.	16 – 7 =	
33.	26 – 7 =	
34.	36 – 7 =	
35.	76 – 7 =	
36.	56 – 7 =	
37.	70 – 7 =	
38.	63 – 7 =	
39.	52 – 7 =	
40.	81 – 7 =	
41.	74 – 7 =	
42.	66 – 7 =	
43.	85 – 7 =	
44.	52 – 7 =	

Name ______________________________ Date ____________

1. Use the pictures to model equivalent fractions. Fill in the blanks, and answer the questions.

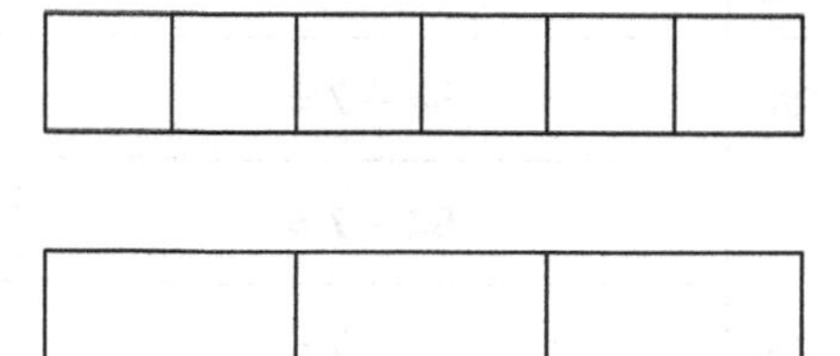

4 sixths is equal to _____ thirds.

$\frac{4}{6} = \frac{\square}{3}$

The whole stays the same.

What happened to the size of the equal parts when there were fewer equal parts?

What happened to the number of equal parts when the equal parts became larger?

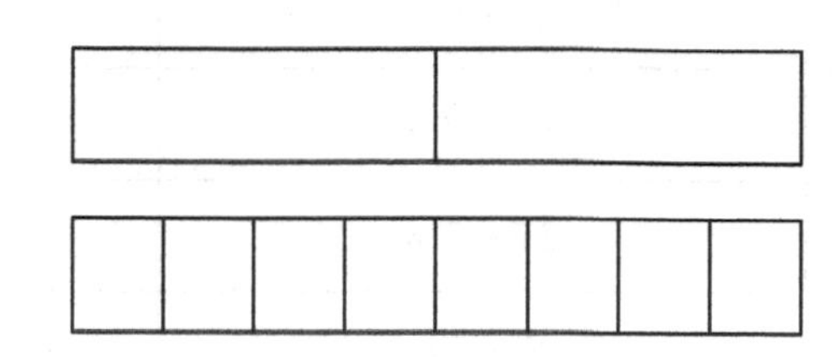

1 half is equal to _____ eighths.

$\frac{1}{2} = \frac{\square}{8}$

The whole stays the same.

What happened to the size of the equal parts when there were more equal parts?

What happened to the number of equal parts when the equal parts became smaller?

2. 6 friends want to share 3 chocolate bars that are all the same size, which are represented by the 3 rectangles below. When the bars are unwrapped, the friends notice that the first chocolate bar is cut into 2 equal parts, the second is cut into 4 equal parts, and the third is cut into 6 equal parts. How can the 6 friends share the chocolate bars equally without breaking any of the pieces?

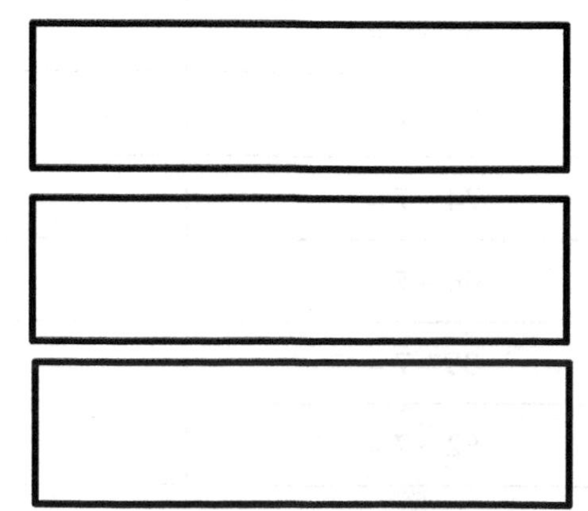

3. When the whole is the same, why does it take 6 copies of 1 eighth to equal 3 copies of 1 fourth? Draw a model to support your answer.

4. When the whole is the same, how many sixths does it take to equal 1 third? Draw a model to support your answer.

5. You have a magic wand that doubles the number of equal parts but keeps the whole the same size. Use your magic wand. In the space below, draw to show what happens to a rectangle that is partitioned in fourths after you tap it with your wand. Use words and numbers to explain what happened.

Name ______________________________ Date ______________

1. Solve.

2 thirds is equal to _____ twelfths.

$$\frac{2}{3} = \frac{}{12}$$

2. Draw and label two models that show fractions equivalent to those in Problem 1.

3. Use words to explain why the two fractions in Problem 1 are equal.

Name ______________________________ Date ____________

1. Use the pictures to model equivalent fractions. Fill in the blanks, and answer the questions.

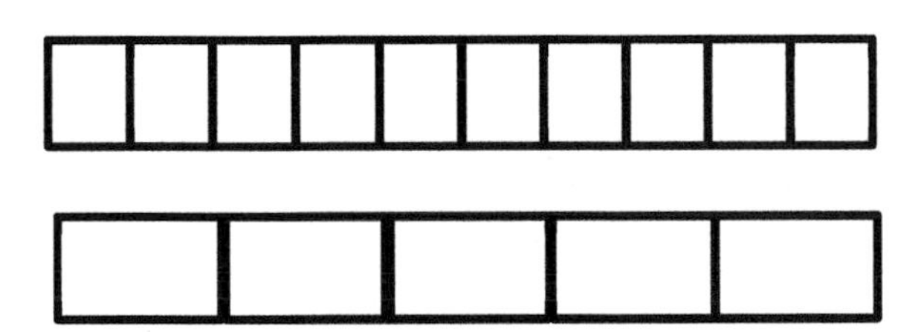

2 tenths is equal to _____ fifths.

$$\frac{2}{10} = \frac{\ }{5}$$

The whole stays the same.

What happened to the size of the equal parts when there were fewer equal parts?

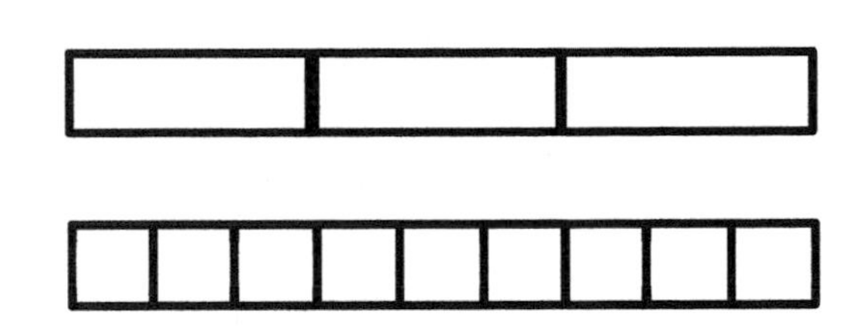

1 third is equal to _____ ninths.

$$\frac{1}{3} = \frac{\ }{9}$$

The whole stays the same.

What happened to the size of the equal parts when there were more equal parts?

2. 8 students share 2 pizzas that are the same size, which are represented by the 2 circles below. They notice that the first pizza is cut into 4 equal slices, and the second is cut into 8 equal slices. How can the 8 students share the pizzas equally without cutting any of the pieces?

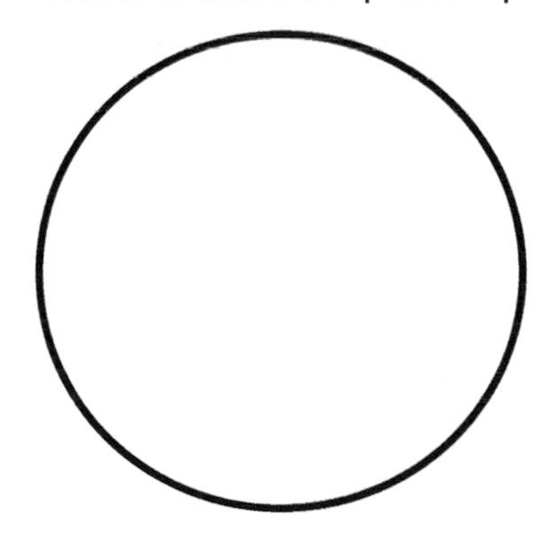

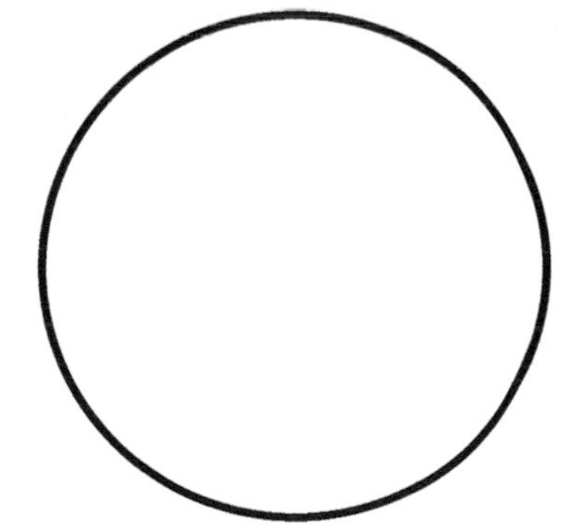

3. When the whole is the same, why does it take 4 copies of 1 tenth to equal 2 copies of 1 fifth? Draw a model to support your answer.

4. When the whole is the same, how many eighths does it take to equal 1 fourth? Draw a model to support your answer.

5. Mr. Pham cuts a cake into 8 equal slices. Then, he cuts every slice in half. How many of the smaller slices does he have? Use words and numbers to explain your answer.

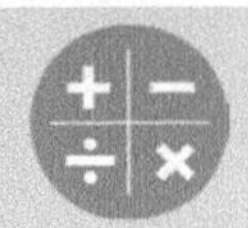

Topic F

Comparison, Order, and Size of Fractions

3.NF.3d

Focus Standard:		3.NF.3	Explain equivalence of fractions in special cases, and compare fractions by reasoning about their size. d. Compare two fractions with the same numerator or the same denominator by reasoning about their size. Recognize that comparisons are valid only when the two fractions refer to the same whole. Record the results of comparisons with the symbols >, =, or <, and justify the conclusions, e.g., by using a visual fraction model.
Instructional Days:		3	
Coherence	**-Links from:**	G2–M8	Time, Shapes, and Fractions as Equal Parts of Shapes
	-Links to:	G4–M5	Fraction Equivalence, Ordering, and Operations

Fraction strips and the number line carry into Topic F as students compare fractions with the same numerator. As they study and compare different fractions, students continue to reason about their size. They develop the understanding that the numerator or number of copies of the fractional unit (shaded parts) does not necessarily determine the size of the fraction. The module closes with an exploration in which students are guided to develop a method for precisely partitioning various wholes into any fractional unit, using the number line as a measurement tool.

A Teaching Sequence Toward Mastery of Comparison, Order, and Size of Fractions

Objective 1: **Compare fractions with the same numerator pictorially.**
(Lesson 28)

Objective 2: **Compare fractions with the same numerator using <, >, or =, and use a model to reason about their size.**
(Lesson 29)

Objective 3: **Partition various wholes precisely into equal parts using a number line method.**
(Lesson 30)

Lesson 28

Objective: Compare fractions with the same numerator pictorially.

Suggested Lesson Structure

Fluency Practice	(12 minutes)
Application Problem	(8 minutes)
Concept Development	(30 minutes)
Student Debrief	(10 minutes)
Total Time	**(60 minutes)**

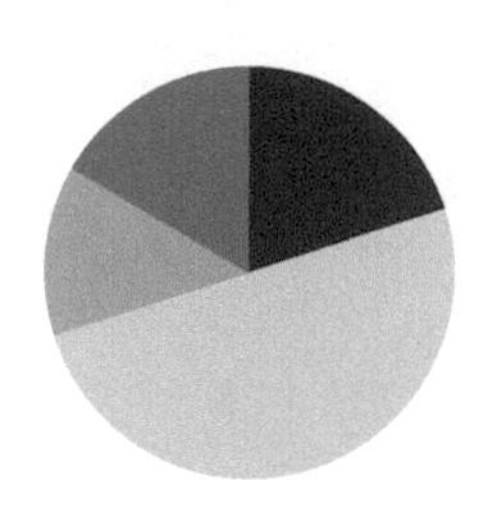

Fluency Practice (12 minutes)

- Sprint: Subtract by Eight **2.NBT.5** (8 minutes)
- Recognize Equal Fractions **3.NF.3b** (4 minutes)

Sprint: Subtract by Eight (8 minutes)

Materials: (S) Subtract by Eight Sprint

Note: This Sprint supports fluency with subtraction by 8.

Recognize Equal Fractions (4 minutes)

Materials: (S) Personal white board

Note: This activity reviews the concepts of representing and naming equivalent fractions.

T: (Project or draw a rectangle partitioned into 2 equal units with the first unit shaded.) Say the fraction that's shaded.

S: 1 half.

T: (Write $\frac{1}{2}$ to the side of the rectangle. Project or draw a rectangle partitioned into 4 equal, unshaded units directly below the first rectangle.) Say the fractional unit of this shape.

S: Fourths.

T: I'm going to start shading in fourths. Tell me to stop when I've shaded enough fourths to equal 1 half. (Shade 2 fourths.)

S: Stop!

T: (Write $\frac{1}{2} = \frac{}{4}$ to the side of the rectangle.) 1 half is the same as how many fourths?

S: 2 fourths.

T: (Write $\frac{1}{2} = \frac{2}{4}$.)

Continue with the following possible sequence: $\frac{1}{3} = \frac{\ }{9}$ and $\frac{6}{8} = \frac{\ }{4}$.

Application Problem (8 minutes)

LaTonya has 2 equal-sized hotdogs. She cut the first one into thirds at lunch. Later, she cut the second hotdog to make double the number of pieces. Draw a model of LaTonya's hotdogs.

a. How many pieces is the second hotdog cut into?

b. If she wants to eat $\frac{2}{3}$ of the second hotdog, how many pieces should she eat?

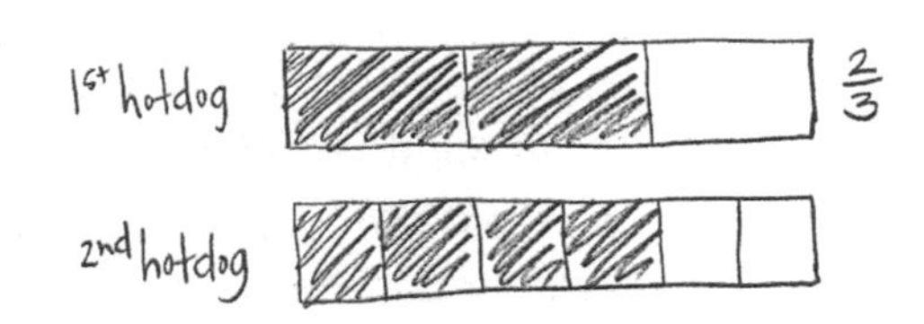

a) The 2nd hot dog is cut into 6 pieces.

b) She should eat $\frac{4}{6}$ of it. That means 4 pieces.

Note: This problem reviews the concept of equivalent fractions from Topic E. Encourage students to find other equivalent fractions based on their models. This problem is used in the Concept Development to provide a context in which students can compare fractions with the same numerators.

Concept Development (30 minutes)

Materials: (S) Work from Application Problem, personal white board

T: Look again at your models of LaTonya's hotdogs. Let's change the problem slightly. What if LaTonya eats 2 pieces of each hotdog? Figure out what fraction of each hotdog she eats.

S: (Work.) She eats $\frac{2}{3}$ of the first one and $\frac{2}{6}$ of the second one.

T: Did LaTonya eat the same amount of the first hotdog and second hotdog?

S: (Use models for help.) No.

T: But she ate 2 pieces of each hotdog. Why is the amount she ate different?

S: The number of pieces she ate is the same, but the size of each piece is different. → Just like we saw yesterday, the more you cut up a whole, the smaller the pieces get. → So, eating 2 pieces of thirds is more hotdog than 2 pieces of sixths.

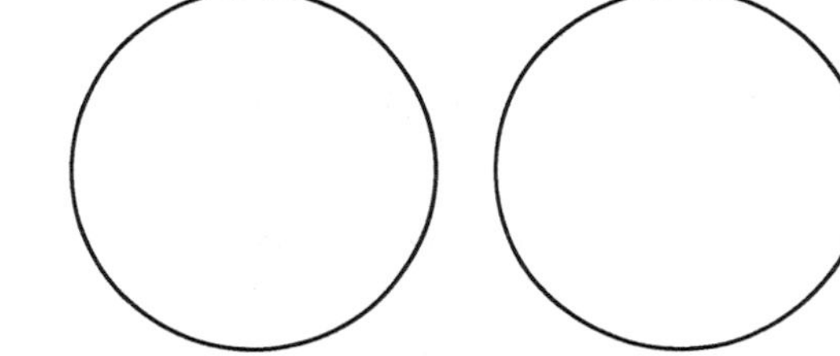

MP.2

T: (Project or draw the circles on the right.) Draw my pizzas on your personal white board.

S: (Draw shapes.)

T: Estimate to partition both pizzas into fourths.

S: (Partition.)

T: Partition the second pizza to double the number of units.

S: (Partition.)

T: What units do we have?

S: Fourths and eighths.

T: Shade in 3 fourths and 3 eighths.

S: (Shade.)

MP.2

T: Which shaded portion would you rather eat? The fourths or eighths? Why?

S: I'd rather eat the fourths because it's way more pizza. → I'd rather eat the eighths because I'm not that hungry, and it's less.

T: But both choices are 3 pieces. Aren't they equivalent?

S: No. You can see fourths are larger. → We know because the more times you cut the whole, the smaller the pieces get. → So, eighths are tiny compared to fourths! → The number of pieces we shaded is the same, but the sizes of the pieces are different, so the shaded amounts are not equivalent.

If necessary, continue with other examples varying the pictorial models.

T: Let's work in pairs to play a comparison game. Partner A, draw a whole and shade a fraction of the whole. Label the shaded part.

S: (Partner A draws and labels.)

T: Partner B, draw a fraction that is less than Partner A's fraction. Use the same whole and same number of shaded parts, but choose a different fractional unit. Label the shaded parts.

S: (Partner B draws and labels.)

T: Partner A, check your friend's work to be sure the fraction is less than yours.

S: (Partner A checks and helps make any corrections necessary.)

T: Partner B, draw a whole, and shade a fraction. I will say *less than* or *greater than* for Partner A to draw another fraction.

Play several rounds.

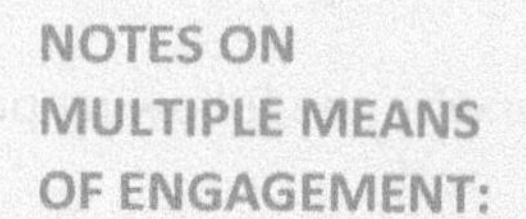

NOTES ON MULTIPLE MEANS OF ENGAGEMENT:

Give students working below grade level the option of rectangular pizzas (rather than circles) to ease the task of partitioning.

NOTES ON MULTIPLE MEANS OF ACTION AND EXPRESSION:

As students play a comparison game, facilitate peer-to-peer talk for English language learners with sentence frames, such as the following:

- "I partitioned into ____ (fractional unit). I shaded ___ (number of) ____ (fractional unit)."
- "I drew ___ (fractional unit), too. I shaded ___ (number of) ____ (fractional unit). ____ is less than ____."

NOTES ON MULTIPLE MEANS OF ENGAGEMENT:

Extend Page 1 of the Problem Set for students working above grade level so they can use their knowledge of equivalencies. Say, "If 2 thirds is greater than 2 fifths, use equivalent fractions to name the same comparison. For example, 4 sixths is greater than 2 fifths."

Problem Set (10 minutes)

Students should do their personal best to complete the Problem Set within the allotted 10 minutes. For some classes, it may be appropriate to modify the assignment by specifying which problems they work on first. Some problems do not specify a method for solving. Students should solve these problems using the RDW approach used for Application Problems.

Name Gina Date

Shade the models to compare the fractions. Circle the larger fraction for each problem.

1. 2 fifths
   2 thirds

2. 2 tenths
   2 eighths

3. 3 fourths
   3 eighths

4. 4 eighths
   4 sixths

5. 3 thirds
   3 sixths

Student Debrief (10 minutes)

Lesson Objective: Compare fractions with the same numerator pictorially.

The Student Debrief is intended to invite reflection and active processing of the total lesson experience.

Invite students to review their solutions for the Problem Set. They should check work by comparing answers with a partner before going over answers as a class. Look for misconceptions or misunderstandings that can be addressed in the Student Debrief. Guide students in a conversation to debrief the Problem Set and process the lesson.

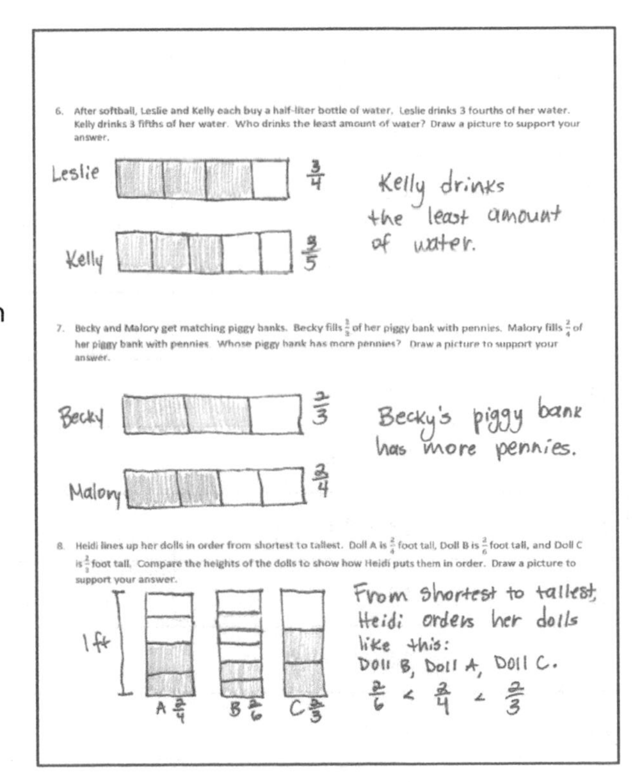

6. After softball, Leslie and Kelly each buy a half-liter bottle of water. Leslie drinks 3 fourths of her water. Kelly drinks 3 fifths of her water. Who drinks the least amount of water? Draw a picture to support your answer.

   Leslie $\frac{3}{4}$
   Kelly $\frac{3}{5}$
   Kelly drinks the least amount of water.

7. Becky and Malory get matching piggy banks. Becky fills $\frac{2}{3}$ of her piggy bank with pennies. Malory fills $\frac{2}{4}$ of her piggy bank with pennies. Whose piggy bank has more pennies? Draw a picture to support your answer.

   Becky $\frac{2}{3}$
   Malory $\frac{2}{4}$
   Becky's piggy bank has more pennies.

8. Heidi lines up her dolls in order from shortest to tallest. Doll A is $\frac{2}{4}$ foot tall, Doll B is $\frac{2}{6}$ foot tall, and Doll C is $\frac{2}{3}$ foot tall. Compare the heights of the dolls to show how Heidi puts them in order. Draw a picture to support your answer.

   1 ft
   A $\frac{2}{4}$ B $\frac{2}{6}$ C $\frac{2}{3}$
   From shortest to tallest, Heidi orders her dolls like this:
   Doll B, Doll A, Doll C.
   $\frac{2}{6} < \frac{2}{4} < \frac{2}{3}$

Any combination of the questions below may be used to lead the discussion.

- Look at your answers for Problems 7 and 8. Is 2 parts always equal to 2 parts? Why or why not?
- If you only know the number of shaded parts, can you tell if fractions are equivalent? Why or why not?

Exit Ticket (3 minutes)

After the Student Debrief, instruct students to complete the Exit Ticket. A review of their work will help with assessing students' understanding of the concepts that were presented in today's lesson and planning more effectively for future lessons. The questions may be read aloud to the students.

A

Number Correct: ______

Subtract by Eight

1.	18 – 8 =	
2.	8 – 8 =	
3.	28 – 8 =	
4.	9 – 8 =	
5.	19 – 8 =	
6.	39 – 8 =	
7.	10 – 8 =	
8.	20 – 8 =	
9.	50 – 8 =	
10.	11 – 8 =	
11.	21 – 8 =	
12.	71 – 8 =	
13.	12 – 8 =	
14.	22 – 8 =	
15.	82 – 8 =	
16.	13 – 8 =	
17.	23 – 8 =	
18.	83 – 8 =	
19.	14 – 8 =	
20.	24 – 8 =	
21.	34 – 8 =	
22.	54 – 8 =	

23.	74 – 8 =	
24.	15 – 8 =	
25.	25 – 8 =	
26.	35 – 8 =	
27.	85 – 8 =	
28.	65 – 8 =	
29.	16 – 8 =	
30.	26 – 8 =	
31.	36 – 8 =	
32.	96 – 8 =	
33.	76 – 8 =	
34.	17 – 8 =	
35.	27 – 8 =	
36.	37 – 8 =	
37.	87 – 8 =	
38.	67 – 8 =	
39.	70 – 8 =	
40.	62 – 8 =	
41.	84 – 8 =	
42.	66 – 8 =	
43.	91 – 8 =	
44.	75 – 8 =	

B

Number Correct: _______

Improvement: _______

Subtract by Eight

1.	8 – 8 =	
2.	18 – 8 =	
3.	28 – 8 =	
4.	9 – 8 =	
5.	19 – 8 =	
6.	69 – 8 =	
7.	10 – 8 =	
8.	20 – 8 =	
9.	60 – 8 =	
10.	11 – 8 =	
11.	21 – 8 =	
12.	81 – 8 =	
13.	12 – 8 =	
14.	22 – 8 =	
15.	52 – 8 =	
16.	13 – 8 =	
17.	23 – 8 =	
18.	93 – 8 =	
19.	14 – 8 =	
20.	24 – 8 =	
21.	34 – 8 =	
22.	74 – 8 =	

23.	94 – 8 =	
24.	15 – 8 =	
25.	25 – 8 =	
26.	35 – 8 =	
27.	95 – 8 =	
28.	75 – 8 =	
29.	16 – 8 =	
30.	26 – 8 =	
31.	36 – 8 =	
32.	66 – 8 =	
33.	46 – 8 =	
34.	17 – 8 =	
35.	27 – 8 =	
36.	37 – 8 =	
37.	97 – 8 =	
38.	77 – 8 =	
39.	80 – 8 =	
40.	71 – 8 =	
41.	53 – 8 =	
42.	45 – 8 =	
43.	87 – 8 =	
44.	54 – 8 =	

Name ______________________________ Date ______________

Shade the models to compare the fractions. Circle the larger fraction for each problem.

1. 2 fifths

   2 thirds

2. 2 tenths

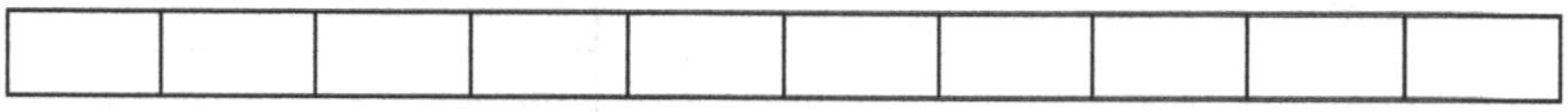

   2 eighths

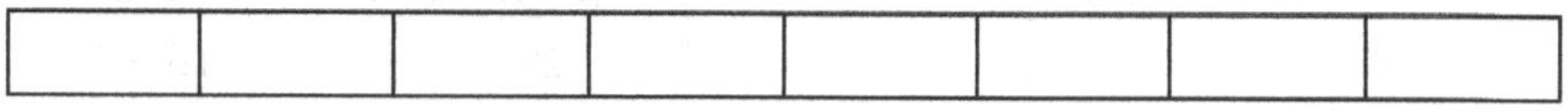

3. 3 fourths

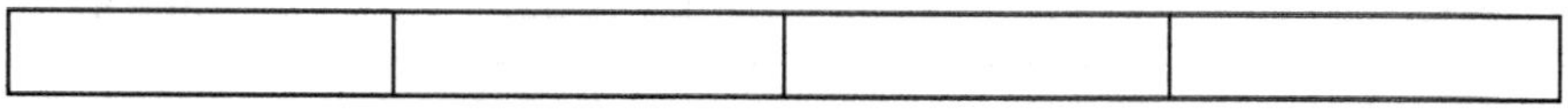

   3 eighths

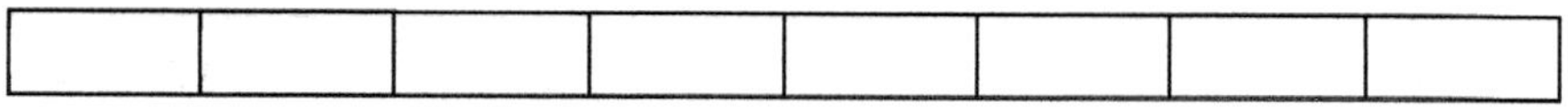

4. 4 eighths

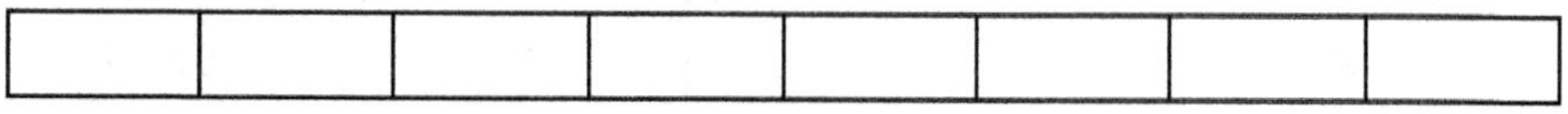

   4 sixths

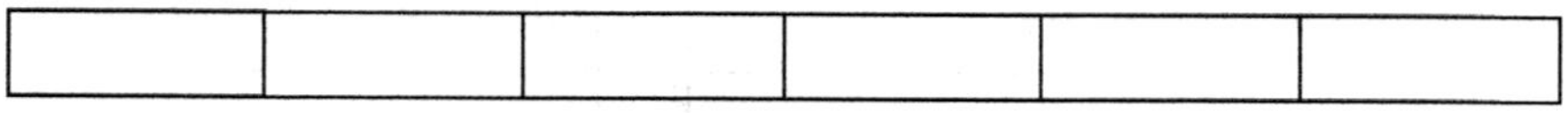

5. 3 thirds

   3 sixths

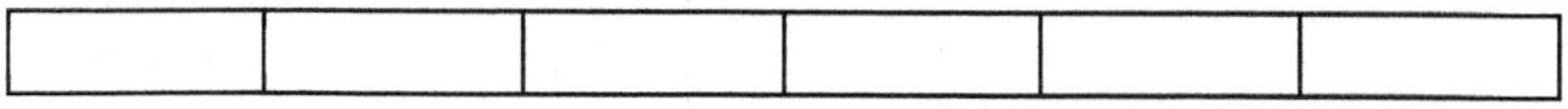

6. After softball, Leslie and Kelly each buy a half-liter bottle of water. Leslie drinks 3 fourths of her water. Kelly drinks 3 fifths of her water. Who drinks the least amount of water? Draw a picture to support your answer.

7. Becky and Malory get matching piggy banks. Becky fills $\frac{2}{3}$ of her piggy bank with pennies. Malory fills $\frac{2}{4}$ of her piggy bank with pennies. Whose piggy bank has more pennies? Draw a picture to support your answer.

8. Heidi lines up her dolls in order from shortest to tallest. Doll A is $\frac{2}{4}$ foot tall, Doll B is $\frac{2}{6}$ foot tall, and Doll C is $\frac{2}{3}$ foot tall. Compare the heights of the dolls to show how Heidi puts them in order. Draw a picture to support your answer.

Name ______________________________ Date ______________

1. Shade the models to compare the fractions.

2 thirds

2 eighths

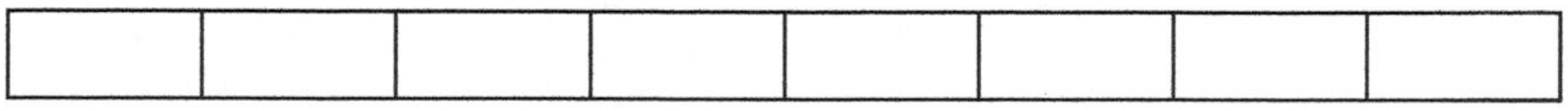

Which is larger, 2 thirds or 2 eighths? Why? Use words to explain.

2. Draw a model for each fraction. Circle the smaller fraction.

3 sevenths

3 fourths

Name ______________________________ Date ______________

Shade the models to compare the fractions. Circle the larger fraction for each problem.

1. 1 half

   1 fifth

2. 2 sevenths

   2 fourths

3. 4 fifths

   4 ninths

4. 5 sevenths

   5 tenths

5. 4 sixths

   4 fourths

6. Saleem and Edwin use inch rulers to measure the lengths of their caterpillars. Saleem's caterpillar measures 3 fourths of an inch. Edwin's caterpillar measures 3 eighths of an inch. Whose caterpillar is longer? Draw a picture to support your answer.

7. Lily and Jasmine each bake the same-sized chocolate cake. Lily puts $\frac{5}{10}$ of a cup of sugar into her cake. Jasmine puts $\frac{5}{6}$ of a cup of sugar into her cake. Who uses less sugar? Draw a picture to support your answer.

Lesson 29

Objective: Compare fractions with the same numerator using <, >, or =, and use a model to reason about their size.

Suggested Lesson Structure

■ Fluency Practice	(12 minutes)
■ Application Problem	(8 minutes)
■ Concept Development	(30 minutes)
■ Student Debrief	(10 minutes)
Total Time	**(60 minutes)**

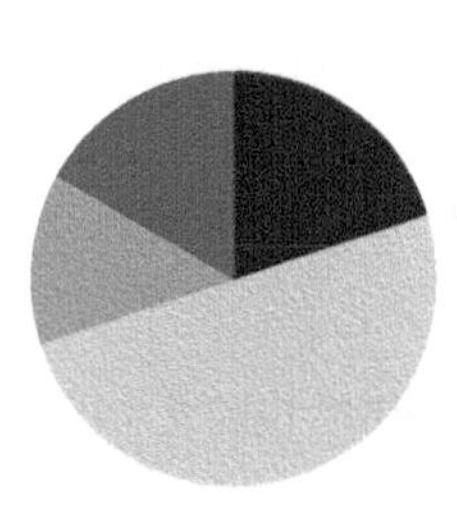

Fluency Practice (12 minutes)

- Multiply by 8 **3.OA.4** (8 minutes)
- Compare Fractions with the Same Numerator **3.NF.3d** (4 minutes)

Multiply by 8 (8 minutes)

Materials: (S) Multiply by 8 (5–9) Pattern Sheet

Note: This Pattern Sheet supports fluency with multiplication using units of 8.

T: Skip-count by eights. (Write multiples horizontally as students count.)
S: 8, 16, 24, 32, 40, 48, 56, 64, 72, 80.
T: (Write 5 × 8 = ____.) Let's skip-count by eights to find the answer. (Count with fingers to 5 as students count.)
S: 8, 16, 24, 32, 40.
T: (Circle 40, and write 5 × 8 = 40 above it. Write 3 × 8 = ____.) Let's skip-count up by eights again. (Count with fingers to 3 as students count.)
S: 8, 16, 24.
T: Let's see how we can skip-count down to find the answer, too. Start at 40. (Count down with your fingers as students say numbers.)
S: 40, 32, 24.
T: (Write 7 × 8 = ____.) Let's skip-count up by eights. (Count with fingers to 7 as students count.)
S: 8, 16, 24, 32, 40, 48, 56.
T: (Write 9 × 8 = ____.) Let's skip-count up by eights. (Count with fingers to 9 as students count.)
S: 8, 16, 24, 32, 40, 48, 56, 64, 72.

T: Let's see how we can skip-count down to find the answer, too. Start at 80. (Count down with your fingers as students say the numbers.)

S: 80, 72.

T: Let's practice multiplying by 8. Be sure to work left to right across the page. (Distribute Multiply by 8 Pattern Sheet.)

Compare Fractions with the Same Numerator (4 minutes)

Materials: (S) Personal white board

Note: This fluency activity reviews the concept of pictorially comparing fractions with the same numerators from Lesson 28.

T: (Project or draw a rectangle partitioned into 3 equal units with the first 2 units shaded.) Say the fraction that is shaded.

S: 2 thirds.

T: (Write $\frac{2}{3}$ to the left of the rectangle. Project or draw a rectangle of 6 equal, unshaded units directly below the first rectangle. Next to the second rectangle, write $\frac{2}{6}$.) How many units should I shade to show 2 sixths?

S: 2.

T: (Shade the first 2 units in the second rectangle.) On your personal white board, write the larger fraction.

S: (Write $\frac{2}{3}$.)

Continue with the following possible sequence: 3 tenths and 3 fourths, 5 sixths and 5 eighths, and 7 eighths and 7 tenths.

Application Problem (8 minutes)

Catherine and Diana buy matching scrapbooks. Catherine decorates $\frac{5}{9}$ of the pages in her book. Diana decorates $\frac{5}{6}$ of the pages in her book. Who has decorated more pages of her scrapbook? Draw a picture to support your answer.

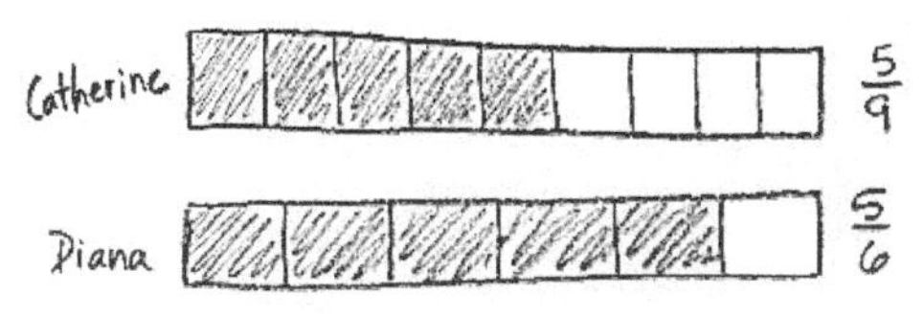

Diana has decorated more of her scrapbook than Catherine.

> **NOTES ON MULTIPLE MEANS OF ENGAGEMENT:**
>
> Challenge students working above grade level to model the comparison on a number line (or two). Have students evaluate and compare the models. Ask (for example), "How might you decide when to use a rectangular model rather than a number line to solve?"

Note: This problem reviews the concept of pictorially comparing fractions with the same numerators from Lesson 28.

Concept Development (30 minutes)

Materials: (S) Personal white board, 3 wholes (Lesson 25 Template 1)

Seat students in pairs facing each other in a large circle around the room. 3 wholes should be in their personal white boards.

3 wholes (Lesson 25 Template 1)

T: Today, we'll only use the first rectangle. At my signal, draw and shade a fraction less than $\frac{1}{2}$, and label it below the rectangle. (Signal.)

S: (Draw and label.)

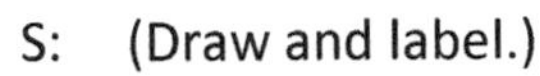

T: Check your partner's work to make sure it's less than $\frac{1}{2}$.

S: (Check.)

T: This is how we're going to play a game today. For the next round, we'll see which partner is quicker but still accurate. As soon as you finish drawing, raise your personal white board. If you are quicker, then you are the winner of the round. If you are the winner of the round, you will stand up, and your partner will stay seated. If you are standing, you will then move to partner with the person on your right, who is still seated. Ready? Erase your boards. At my signal, draw and label a fraction that is greater than $\frac{1}{2}$. (Signal.)

S: (Draw and label.)

The student who goes around the entire circle and arrives back at his original place faster than the other students wins the game. The winner can also be the student who has moved the furthest if it takes too long to play all the way around. Move the game at a brisk pace. Use a variety of fractions, and mix it up between greater than and less than so that students constantly need to update their drawings and feel challenged. If preferred, mix it up by calling out *equal to*.

T: (Draw or show the images on the right.) Draw my shapes on your board. Make sure they match in size like mine.

S: (Draw.)

T: Partition both shapes into sixths.

S: (Partition.)

T: Partition the second shape to show double the number of units in the same whole.

S: (Partition.)

T: What fractional units do we have?

S: Sixths and twelfths.

T: Shade in 4 units of each shape, and label the shaded fraction below each shape.

S: (Shade and label.)

T: Whispering to your partner, say a sentence comparing the fractions using the words *greater than, less than*, or *equal to*.

S: $\frac{4}{6}$ is greater than $\frac{4}{12}$.

T: Now, write the comparison as a number sentence with the correct symbol between the fractions.

S: (Write $\frac{4}{6} > \frac{4}{12}$.)

T: (Draw or show the images on the right.) Draw my rectangles on your board. Make sure they match in size like mine.

S: (Draw.)

T: Partition the first rectangle into sevenths and the second one into fifths.

S: (Partition.)

T: Shade in 3 units of each rectangle, and label the shaded fraction below each rectangle.

S: (Shade and label.)

T: Whispering to your partner, say a sentence comparing the fractions using the words *greater than*, *less than*, or *equal to*.

S: $\frac{3}{7}$ is less than $\frac{3}{5}$.

T: Now, write the comparison as a number sentence with the correct symbol between the fractions.

S: (Write $\frac{3}{7} < \frac{3}{5}$.)

Do other examples, if necessary, using a variety of shapes and units.

T: Draw 2 number lines on your board, and label the endpoints 0 and 1.

S: (Draw and label.)

T: Partition the first number line into eighths and the second one into tenths.

S: (Partition.)

T: On the first number line, label $\frac{8}{8}$.

S: (Label.)

T: On the second number line, label 2 copies of $\frac{5}{10}$.

S: (Label.)

T: Whispering to your partner, say a sentence comparing the fractions using the words *greater than*, *less than*, or *equal to*.

S: Wait, they're the same! $\frac{8}{8}$ is equal to $\frac{10}{10}$.

T: How do you know?

S: Because they have the same point on the number line. That means they're equivalent.

T: Now, write the comparison as a number sentence with the correct symbol between the fractions.

S: (Write $\frac{8}{8} = \frac{10}{10}$.)

Do other examples with the number line. In subsequent examples that use smaller units or units that are farther apart, move to using a single number line.

Problem Set (10 minutes)

Students should do their personal best to complete the Problem Set within the allotted 10 minutes. For some classes, it may be appropriate to modify the assignment by specifying which problems they work on first. Some problems do not specify a method for solving. Students should solve these problems using the RDW approach used for Application Problems.

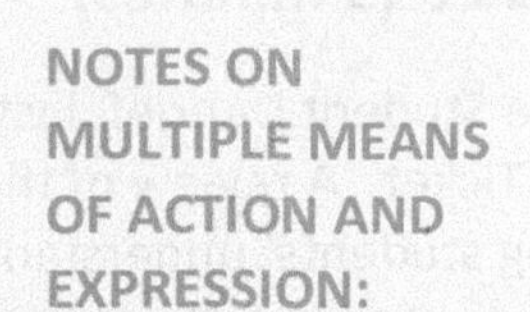

NOTES ON MULTIPLE MEANS OF ACTION AND EXPRESSION:

English language learners and students working below grade level may benefit from math (and English) fluency practice using the Problem Set. For Problems 1 through 4, encourage learners to whisper the unit fraction, whisper count the shaded units (e.g., 1 sixth, 2 sixths), and whisper the shaded fraction as they write.

Student Debrief (10 minutes)

Lesson Objective: Compare fractions with the same numerator using <, >, or =, and use a model to reason about their size.

The Student Debrief is intended to invite reflection and active processing of the total lesson experience.

Invite students to review their solutions for the Problem Set. They should check work by comparing answers with a partner before going over answers as a class. Look for misconceptions or misunderstandings that can be addressed in the Student Debrief. Guide students in a conversation to debrief the Problem Set and process the lesson.

Any combination of the questions below may be used to lead the discussion.

- Look at the models in Problems 1–4. When comparing fractions, why is it so important that the wholes are the same size?
- Tell a partner how you used the models in Problems 1–4 to determine *greater than, less than,* or *equal to.*
- What if you didn't have the models for these problems? How could you compare the fractions? (Write pairs of fractions with the same numerators on the board, and have students compare them without using a model.)
- To extend the lesson, draw fraction models greater than 1, and guide students to compare. For example, use $\frac{12}{9}$ and $\frac{12}{7}$.

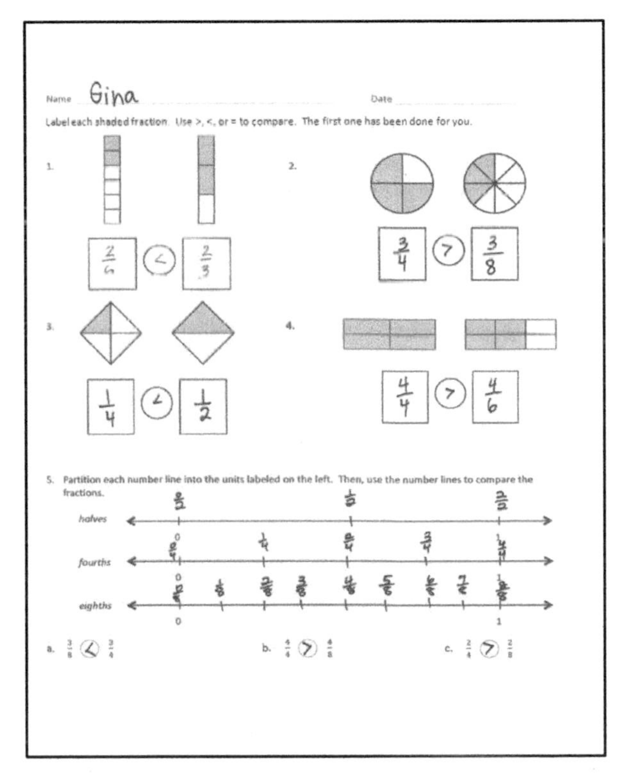

Exit Ticket (3 minutes)

After the Student Debrief, instruct students to complete the Exit Ticket. A review of their work will help with assessing students' understanding of the concepts that were presented in today's lesson and planning more effectively for future lessons. The questions may be read aloud to the students.

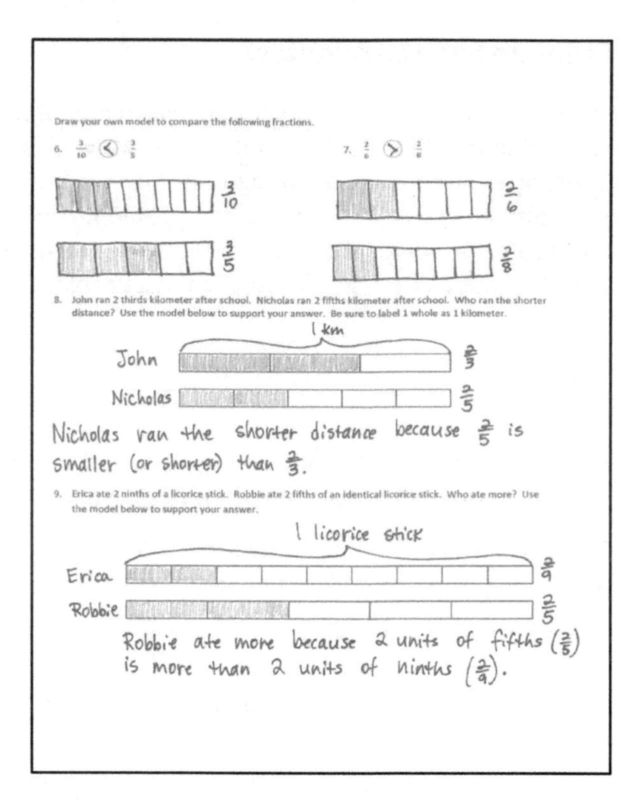

Draw your own model to compare the following fractions.

6. $\frac{3}{10}$ (<) $\frac{3}{5}$

7. $\frac{2}{6}$ (>) $\frac{2}{8}$

8. John ran 2 thirds kilometer after school. Nicholas ran 2 fifths kilometer after school. Who ran the shorter distance? Use the model below to support your answer. Be sure to label 1 whole as 1 kilometer.

9. Erica ate 2 ninths of a licorice stick. Robbie ate 2 fifths of an identical licorice stick. Who ate more? Use the model below to support your answer.

Multiply.

8 x 1 = ______	8 x 2 = ______	8 x 3 = ______	8 x 4 = ______
8 x 5 = ______	8 x 6 = ______	8 x 7 = ______	8 x 8 = ______
8 x 9 = ______	8 x 10 = ______	8 x 5 = ______	8 x 6 = ______
8 x 5 = ______	8 x 7 = ______	8 x 5 = ______	8 x 8 = ______
8 x 5 = ______	8 x 9 = ______	8 x 5 = ______	8 x 10 = ______
8 x 6 = ______	8 x 5 = ______	8 x 6 = ______	8 x 7 = ______
8 x 6 = ______	8 x 8 = ______	8 x 6 = ______	8 x 9 = ______
8 x 6 = ______	8 x 7 = ______	8 x 6 = ______	8 x 7 = ______
8 x 8 = ______	8 x 7 = ______	8 x 9 = ______	8 x 7 = ______
8 x 8 = ______	8 x 6 = ______	8 x 8 = ______	8 x 7 = ______
8 x 8 = ______	8 x 9 = ______	8 x 9 = ______	8 x 6 = ______
8 x 9 = ______	8 x 7 = ______	8 x 9 = ______	8 x 8 = ______
8 x 9 = ______	8 x 8 = ______	8 x 6 = ______	8 x 9 = ______
8 x 7 = ______	8 x 9 = ______	8 x 6 = ______	8 x 8 = ______
8 x 9 = ______	8 x 7 = ______	8 x 6 = ______	8 x 8 = ______

multiply by 8 (5–9)

Name ______________________________ Date ________________

Label each shaded fraction. Use >, <, or = to compare. The first one has been done for you.

1.

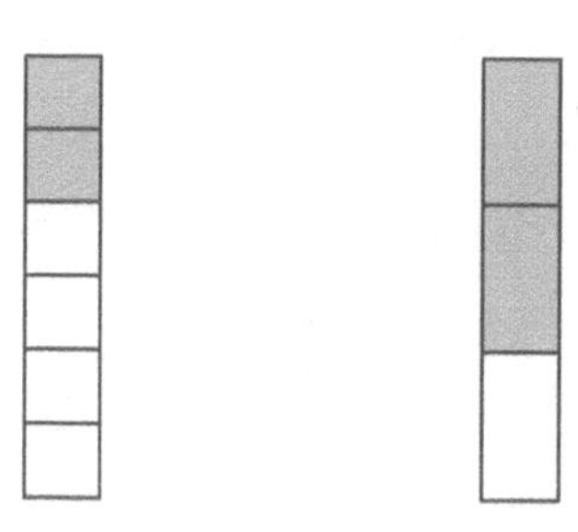

$\frac{2}{6}$ < $\frac{2}{3}$

2.

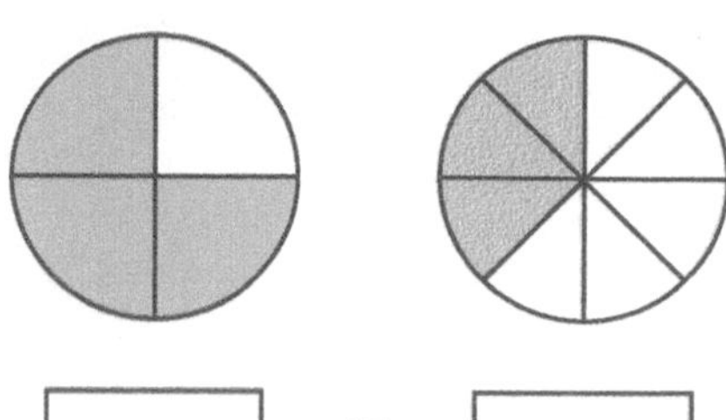

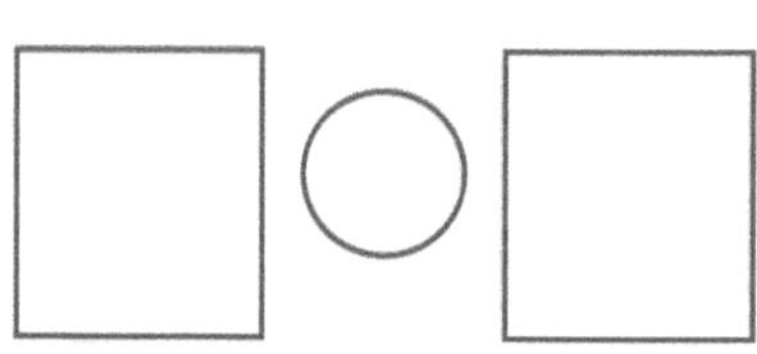

3.

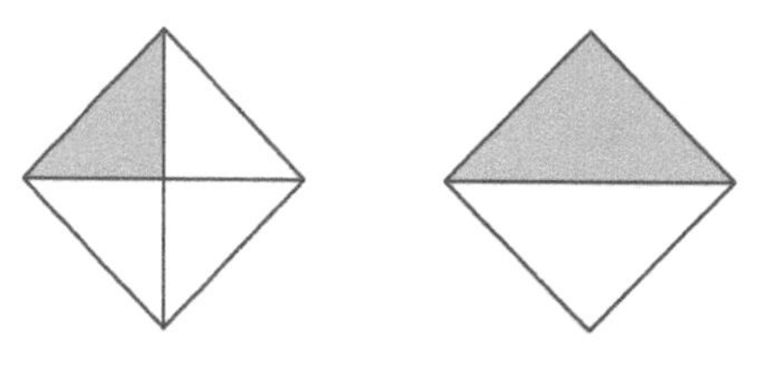

4.

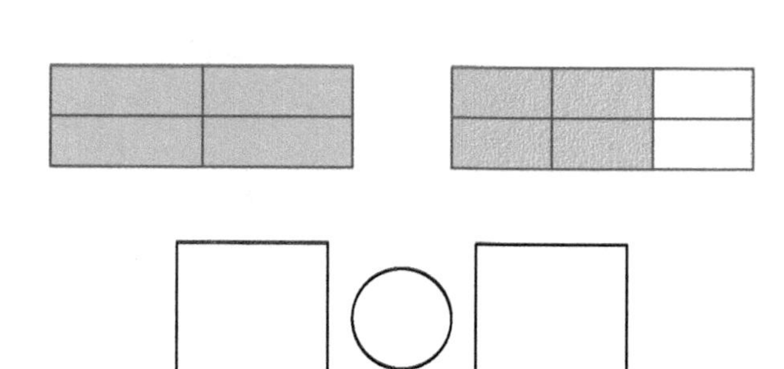

5. Partition each number line into the units labeled on the left. Then, use the number lines to compare the fractions.

halves 0 1

fourths 0 1

eighths 0 1

a. $\frac{3}{8}$ ◯ $\frac{3}{4}$

b. $\frac{4}{4}$ ◯ $\frac{4}{8}$

c. $\frac{2}{4}$ ◯ $\frac{2}{8}$

Draw your own model to compare the following fractions.

6. $\frac{3}{10}$ ◯ $\frac{3}{5}$

7. $\frac{2}{6}$ ◯ $\frac{2}{8}$

8. John ran 2 thirds of a kilometer after school. Nicholas ran 2 fifths of a kilometer after school. Who ran the shorter distance? Use the model below to support your answer. Be sure to label 1 whole as 1 kilometer.

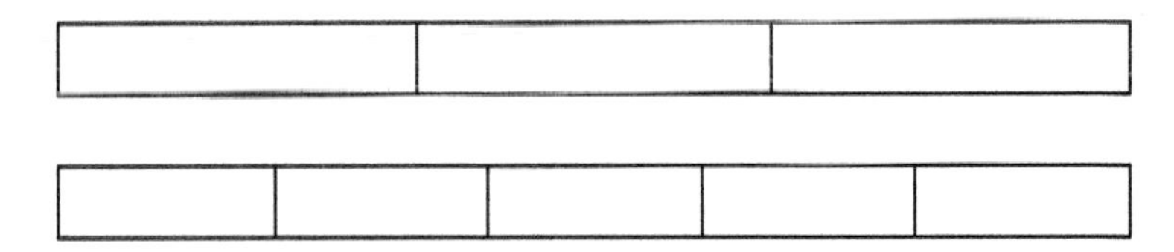

9. Erica ate 2 ninths of a licorice stick. Robbie ate 2 fifths of an identical licorice stick. Who ate more? Use the model below to support your answer.

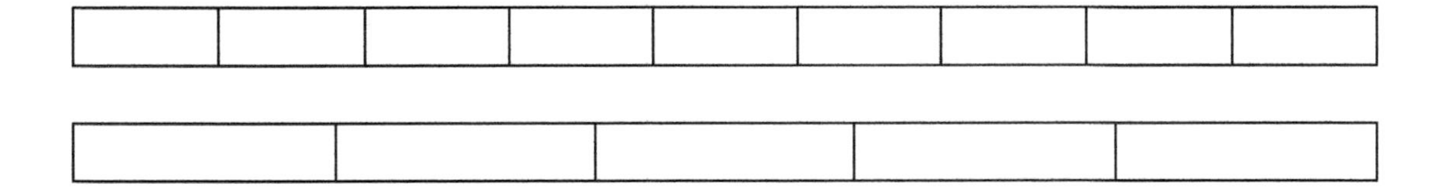

Name ______________________________ Date ______________

1. Complete the number sentence by writing >, <, or =.

   $\frac{3}{5}$ ________ $\frac{3}{9}$

2. Draw 2 number lines with endpoints 0 and 1 to show each fraction in Problem 1. Use the number lines to explain how you know your comparison in Problem 1 is correct.

Name ______________________________ Date ______________

Label each shaded fraction. Use >, <, or = to compare.

1.

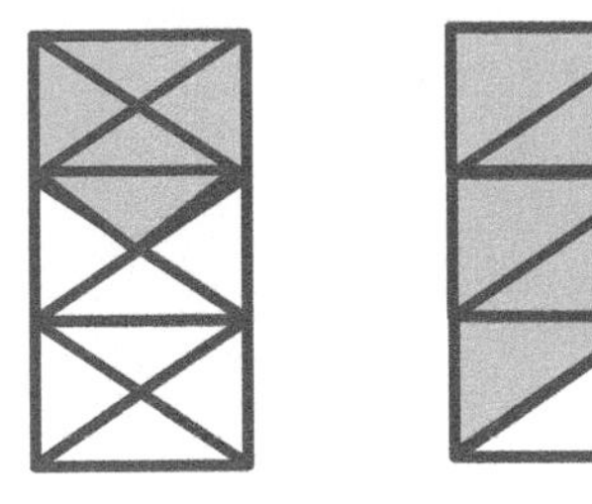

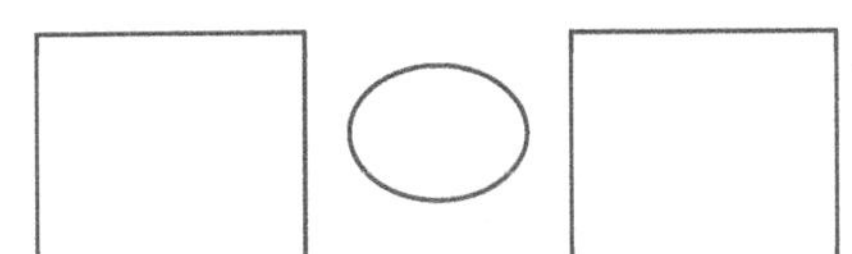

2.

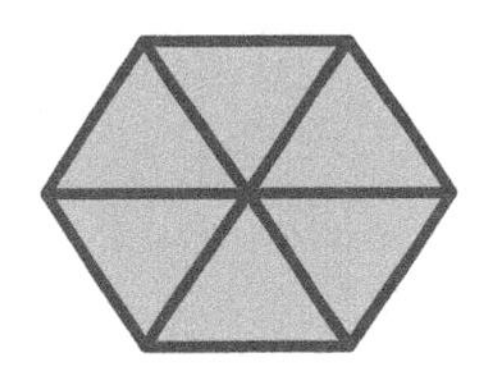

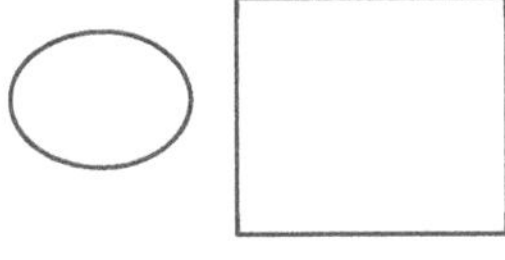

3.

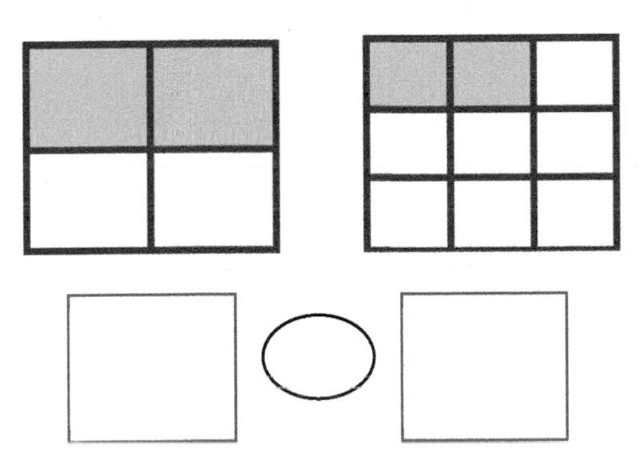

4.

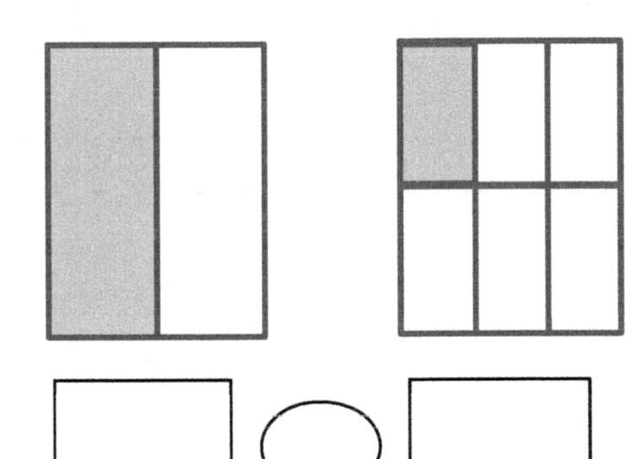

5. Partition each number line into the units labeled on the left. Then, use the number lines to compare the fractions.

thirds
0 1

sixths
0 1

ninths
0 1

a. $\frac{2}{6}$ ◯ $\frac{2}{3}$

b. $\frac{5}{9}$ ◯ $\frac{5}{6}$

c. $\frac{3}{3}$ ◯ $\frac{3}{9}$

Draw your own models to compare the following fractions.

6. $\frac{7}{10}$ ◯ $\frac{7}{8}$

7. $\frac{4}{6}$ ◯ $\frac{4}{9}$

8. For an art project, Michello used $\frac{3}{4}$ of a glue stick. Yamin used $\frac{3}{6}$ of an identical glue stick. Who used more of the glue stick? Use the model below to support your answer. Be sure to label 1 whole as 1 glue stick.

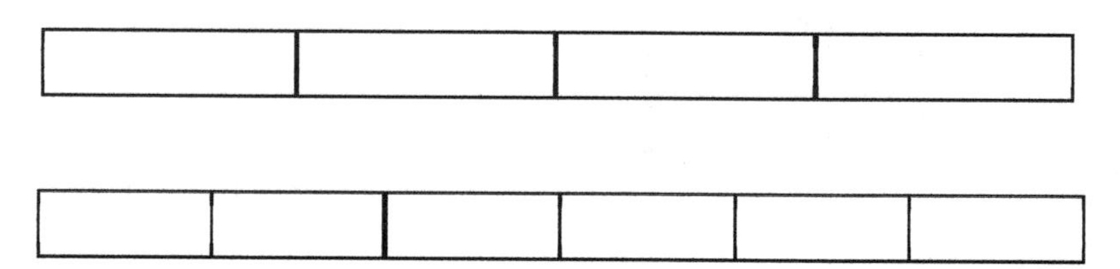

9. After gym class, Jahsir drank 2 eighths of a bottle of water. Jade drank 2 fifths of an identical bottle of water. Who drank less water? Use the model below to support your answer.

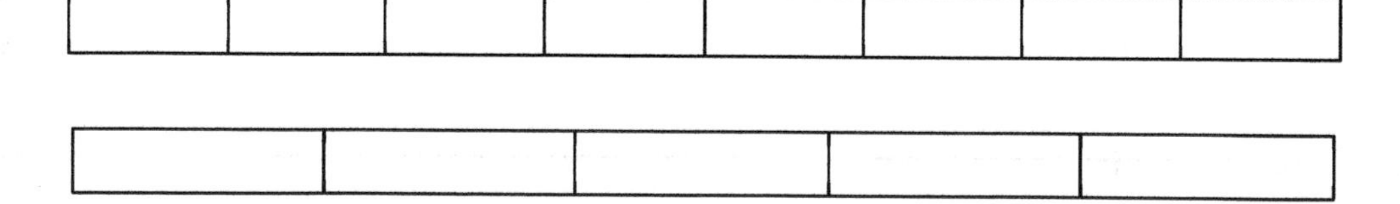

Lesson 30

Objective: Partition various wholes precisely into equal parts using a number line method.

Suggested Lesson Structure

■	Fluency Practice	(12 minutes)
■	Concept Development	(40 minutes)
■	Student Debrief	(8 minutes)
	Total Time	**(60 minutes)**

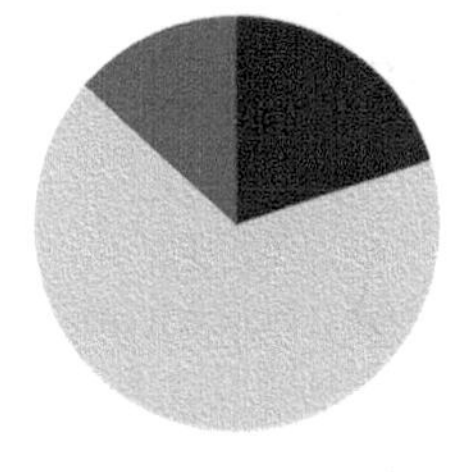

Fluency Practice (12 minutes)

- Multiply by 9 **3.OA.4** (8 minutes)
- Compare Fractions with the Same Numerator **3.NF.3d** (4 minutes)

Multiply by 9 (8 minutes)

Materials: (S) Multiply by 9 (1–5) Pattern Sheet

Note: This Pattern Sheet supports fluency with multiplication using units of 9.

T: Skip-count by nines. (Write multiples horizontally as students count.)

S: 9, 18, 27, 36, 45, 54, 63, 72, 81, 90.

T: (Write 5 × 9 = _____.) Let's skip-count by nines to find the answer. (Count with fingers to 5 as students count.)

S: 9, 18, 27, 36, 45.

T: (Circle 45, and write 5 × 9 = 45 above it. Write 4 × 9 = _____.) Skip-count by nines. (Count with fingers to 4 as students count.)

S: 9, 18, 27, 36.

T: Let's arrive at the answer by skip-counting down, starting at 45. (Hold up 5 fingers as students say 45, and take away 1 finger as students count.)

S: 45, 36.

T: (Write 7 × 9 = _____.) Skip-count by nines. (Count with fingers to 7 as students count.)

S: 9, 18, 27, 36, 45, 54, 63.

T: Let's skip-count, starting at 45. (Hold up 5 fingers as students say 45, and count up with fingers as students count.)

S: 45, 54, 63.

T: (Write 9 × 9 = _____.) Skip-count by nines. (Count with fingers to 9 as students count.)
S: 9, 18, 27, 36, 45, 54, 63, 72, 81.
T: Let's skip-count down starting at 90. (Hold up 10 fingers as students say 90 and remove 1 finger as students count.)
S: 90, 81.
T: Let's practice multiplying by 9. Be sure to work left to right across the page. (Distribute Multiply by 9 Pattern Sheet.)

Compare Fractions with the Same Numerator (4 minutes)

Materials: (S) Personal white board

Note: This fluency activity reviews the concept of pictorially comparing fractions with the same numerators from Lessons 28 and 29.

T: (Project a figure showing 3 fourths shaded.) Say the fraction of the figure that is shaded.
S: 3 fourths.
T: (Write $\frac{3}{4}$ directly below the figure. To the right of the first figure, project one that is the same size and shape that is 3 eighths shaded.) Say the fraction of the figure that is shaded.
S: 3 eighths.
T: (Write $\frac{3}{8}$ directly below the second figure.) On your personal white board, write each fraction. Between the fractions, use the greater than or less than symbol (write > and <) to show which fraction is larger.
S: (Write $\frac{3}{4} > \frac{3}{8}$.)

Continue with the following possible sequence: $\frac{5}{10}$ and $\frac{5}{8}$, $\frac{2}{5}$ and $\frac{2}{3}$, and $\frac{4}{5}$ and $\frac{4}{6}$.

NOTES ON MULTIPLE MEANS OF ACTION AND EXPRESSION:

Students working below grade level may benefit from naming the fractional unit (e.g., eighths) before naming the shaded fraction. Solidify understanding of greater than and less than symbols by soliciting a simultaneous oral response (e.g., "3 fourths is greater than 3 eighths")

Provide sentence frames for English language learners, such as "___ is greater than ___."

Concept Development (40 minutes)

Materials: (S) 9-inch × 1-inch strips of red construction paper (at least 5 per student), lined paper (Template) or wide-ruled notebook paper (several pieces per student), 12-inch ruler

Note: Please read the directions for the Exit Ticket before beginning.

NOTES ON MATERIALS:

It is highly recommended to try the activity with the prepared materials before presenting it to students. Even small variations in the width of spaces on wide-ruled notebook paper or in the 9-inch × 1-inch paper strips may result in adjusting the directions slightly to obtain the desired result.

T: Think back on our lessons. Talk to your partner about how to partition a number line into thirds.

S: Draw the line, and then estimate 3 equal parts. → Use your folded fraction strip to measure. → Measure a 3-inch line with a ruler, and then mark off each inch. → Or on a 6-inch line, 1 mark would be at each 2 inches. → Don't forget to mark 0. → Yes, you always have to start measuring from 0.

MP.6

T: Let's explore a method to mark off any fractional unit precisely without the use of a ruler, just with lined paper.

Step 1: Draw a number line and mark the 0 endpoint.

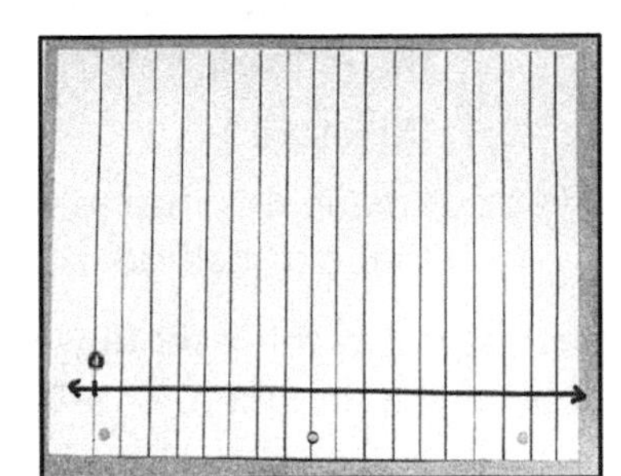

T: (Give students the lined paper or notebook paper.) Turn your paper so the margin is horizontal. Draw a number line on top of the margin.

T: Mark 0 on the point where I did. (Demonstrate.) Talk to your partner: How can we equally and precisely partition this number line into thirds?

S: We can use the vertical lines. → Each line can be an equal part. → We can count 2 lines for each third. → Or 3 spaces or 4 to make an equal part, just so long as each part has the same number. → Oh, I see; this is the answer. → But the teacher said any piece of paper. If we make thirds on this paper, it won't help us make thirds on every paper.

Step 2: Measure equal units using the paper's lines.

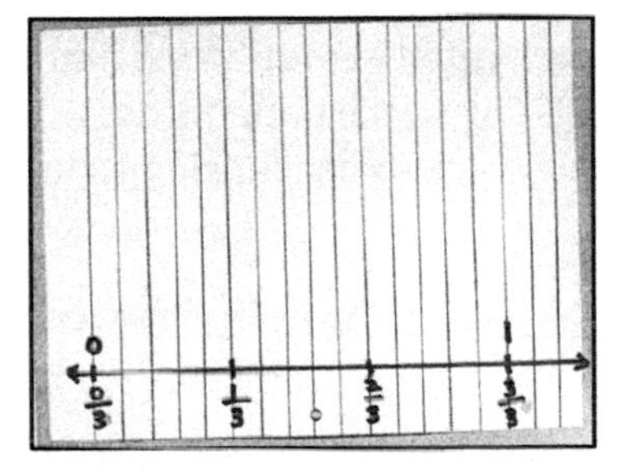

T: Use the paper's vertical lines to measure. Let's make each part 5 spaces long. Label the number line from 0 to 1 using 5 spaces for each third. Discuss in pairs how you know these are precise thirds.

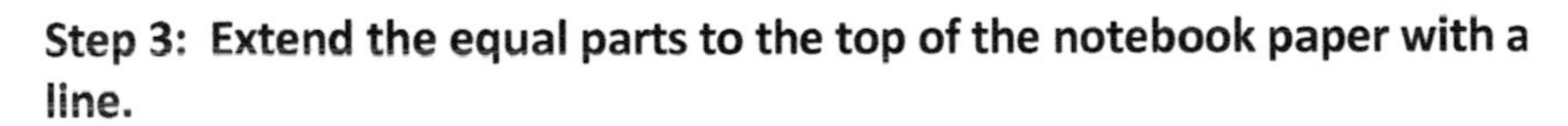

Step 3: Extend the equal parts to the top of the notebook paper with a line.

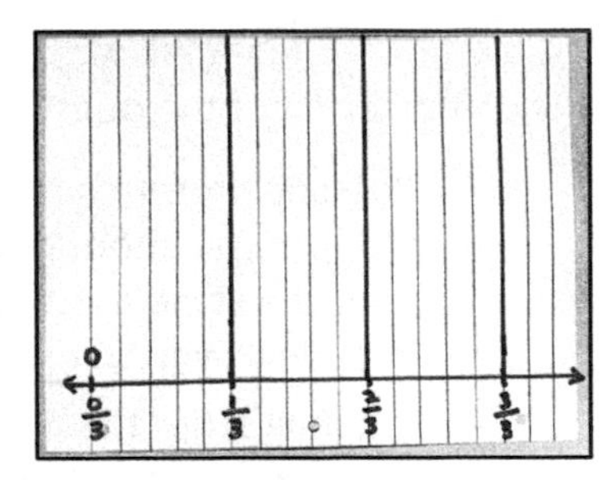

T: Draw vertical lines up from your number line to the top of the paper at each third. (Hold up 1 red strip of paper.) Talk to your partner about how we might use these lines to partition this red strip into thirds.

S: (Discuss.)

T: (Pass out 1 red strip to each student.) The challenge is to partition the red strip precisely into thirds. Let the left end of the strip be 0. The right end of the strip is 1.

S: The strip is too long. → We can't cut it? → No. The teacher said no. How can we do this? (Circulate and listen, but don't give an answer.)

Step 4: Angle the red strip so that the left end touches the 0 endpoint on the original number line. The right end touches the line at 1.

Step 5: Mark off equal units, which are indicated by the vertical extensions of the points on the original number line.

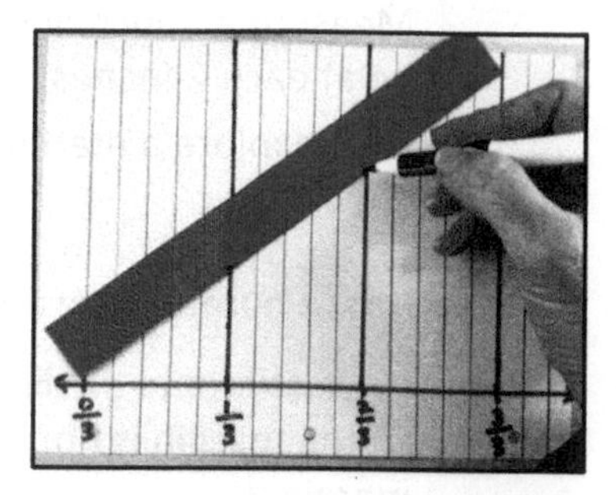

MP.6

T: Do your units look equal?

S: I'm not sure. → They look equal. → I think they're equal because we used the spaces on the paper to make equal units of thirds.

T: Verify that they are equal with your ruler. Measure the full length of the red strip in inches. Measure the equal parts.

S: (Measure.)

T: I made this strip 9 inches long just so you could verify that our method partitions precisely.

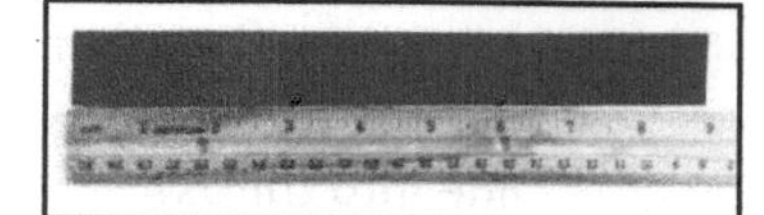

Have students think about why this method works. Have them review the process step by step.

Problem Set (10 minutes)

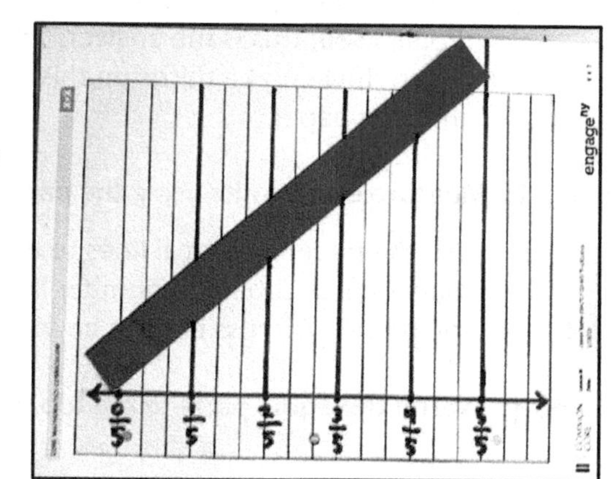

There is no Problem Set sheet for this lesson. In cooperative groups, challenge students to use the same process to precisely mark off other red strips into halves, fourths, etc. It is particularly exciting to partition fifths, sevenths, ninths, and tenths since those are so challenging to fold.

Student Debrief (8 minutes)

Lesson Objective: Partition various wholes precisely into equal parts using a number line method.

The Student Debrief is intended to invite reflection and active processing of the total lesson experience. Look for misconceptions or misunderstandings that can be addressed in the Student Debrief.

Any combination of the questions below may be used to lead the discussion.

- (Possibly present a meter strip.) Could we use this method to partition strips of any length? Talk to your partner about how we could partition this longer strip. Model partitioning the meter strip by using the same method. Simply tape additional lined papers above the lined paper with the thirds. This allows you to make a sharper angle with the meter strip.
- This long strip (the meter strip), shorter strip (the red strip), and number line (the one at the base of the paper) were all partitioned during our work. What is the same and different about them?
- Why do you think this method works? Why are the fractional units still equal when we angle the paper? Do you need to measure to check that they are?
- How might having this skill be helpful in your lives or math class?

- Explain to students that this lesson will be very important in their high school mathematics. Also explain that a mathematician invented it to prepare them for success later in their math journey.

Exit Ticket

There is no Exit Ticket sheet for this lesson. Instead, assess students by circulating and taking notes. Consider the following:

- Is the student able to generalize the method to partition into other fractional units?
- The quality of the new efforts and what mistakes a student made either conceptually (not understanding the angling of the strip) or at a skill level (such as not using the paper's lines properly to partition equal units).
- The role students take within cooperative groups for the Problem Set. Which students articulate directions? Explanations? Which students execute well but silently?

Multiply.

9 x 1 = ______	9 x 2 = ______	9 x 3 = ______	9 x 4 = ______
9 x 5 = ______	9 x 1 = ______	9 x 2 = ______	9 x 1 = ______
9 x 3 = ______	9 x 1 = ______	9 x 4 = ______	9 x 1 = ______
9 x 5 = ______	9 x 1 = ______	9 x 2 = ______	9 x 3 = ______
9 x 2 = ______	9 x 4 = ______	9 x 2 = ______	9 x 5 = ______
9 x 2 = ______	9 x 1 = ______	9 x 2 = ______	9 x 3 = ______
9 x 1 = ______	9 x 3 = ______	9 x 2 = ______	9 x 3 = ______
9 x 4 = ______	9 x 3 = ______	9 x 5 = ______	9 x 3 = ______
9 x 4 = ______	9 x 1 = ______	9 x 4 = ______	9 x 2 = ______
9 x 4 = ______	9 x 3 = ______	9 x 4 = ______	9 x 5 = ______
9 x 4 = ______	9 x 5 = ______	9 x 1 = ______	9 x 5 = ______
9 x 2 = ______	9 x 5 = ______	9 x 3 = ______	9 x 5 = ______
9 x 4 = ______	9 x 2 = ______	9 x 4 = ______	9 x 3 = ______
9 x 5 = ______	9 x 3 = ______	9 x 2 = ______	9 x 4 = ______
9 x 3 = ______	9 x 5 = ______	9 x 2 = ______	9 x 4 = ______

---

multiply by 9 (1–5)

Name ____________________________________ Date ________________

Describe step by step the experience you had of partitioning a length into equal units by simply using a piece of notebook paper and a straight edge. Illustrate the process.

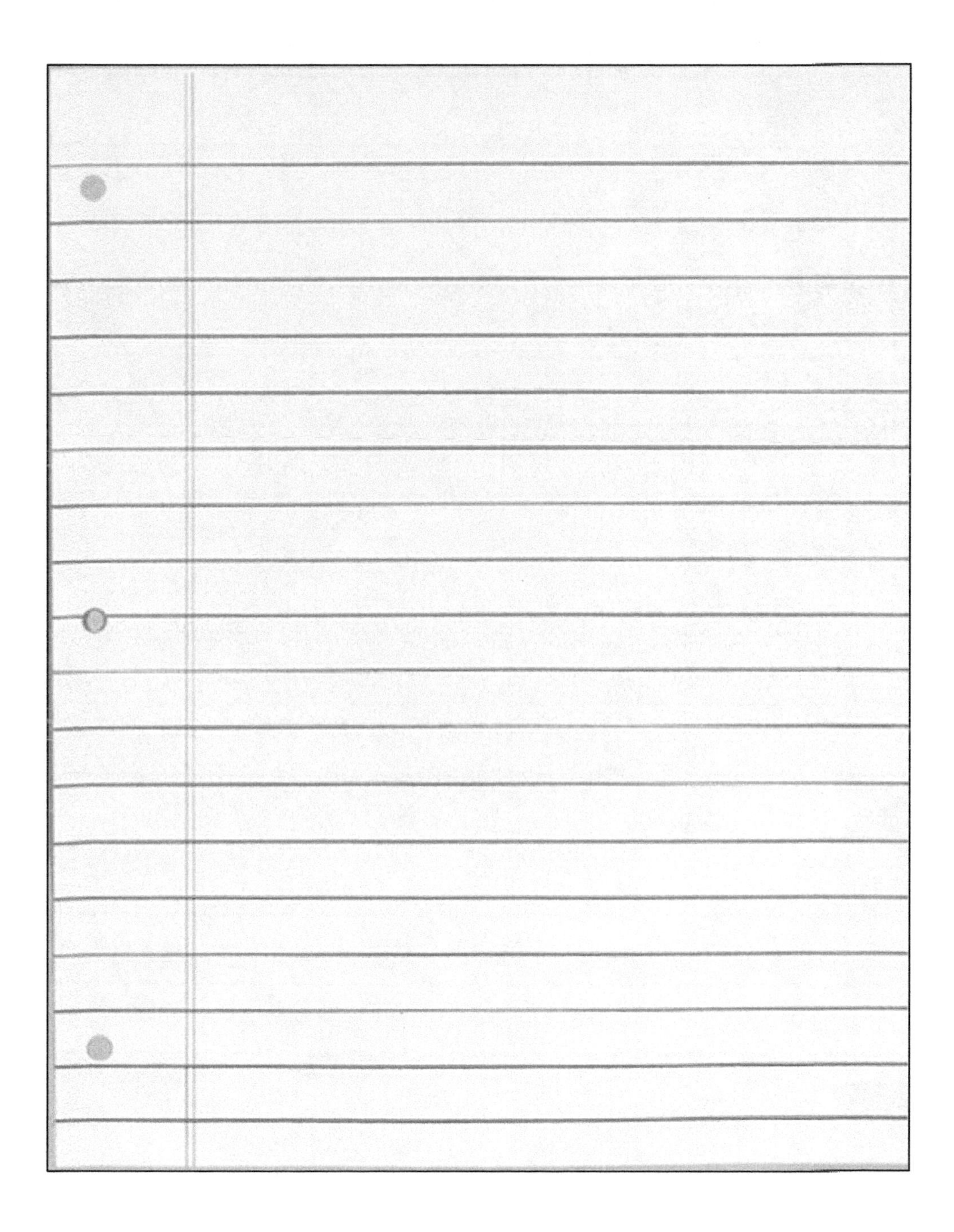

lined paper

Name ______________________________ Date ________________

1. Jerry put 7 equally spaced hooks on a straight wire so students could hang up their coats. The whole length is from the first hook to the last hook.

   a. On the picture below, label the fraction of the wire's length where each hook is located.

   b. At what fraction is Betsy's coat if she hangs it at the halfway point?

   c. Write a fraction that is equivalent to your answer for Part (b).

2. Jerry used the picture below to show his son how to find a fraction equal to $\frac{2}{3}$. Explain what Jerry might have said and done using words, pictures, and numbers.

3. Jerry and his son have the exact same granola bars. Jerry has eaten $\frac{3}{6}$ of his granola bar. His son has eaten $\frac{3}{8}$ of his own granola bar. Who has eaten more? Explain your answer using words, pictures, and numbers.

4. Jerry has a fruit roll that is 4 feet long.

   a. Label the number line to show how Jerry might cut his fruit roll into pieces $\frac{1}{3}$ of a foot long. Label every fraction on the number line, including renaming the wholes.

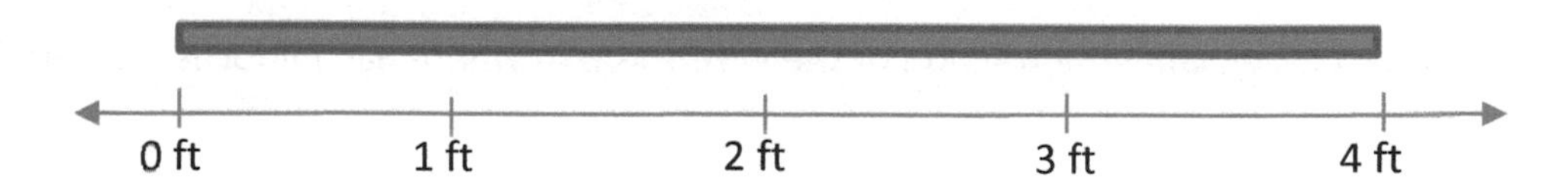

   b. Jerry cut his fruit roll into pieces that are $\frac{1}{3}$ of a foot long. Jerry and his 2 sons each eat one piece. What fraction of the whole fruit roll is eaten? Explain your answer using words, pictures, and numbers.

   c. Jerry's son says that 1 third is the same as 2 sixths. Do you agree? Why or why not? Use words, pictures, and numbers to explain your answer.

End-of-Module Assessment Task Standards Addressed	Topics A–F

Develop understanding of fractions as numbers.

3.NF.1 Understand a fraction $1/b$ as the quantity formed by 1 part when a whole is partitioned into b equal parts; understand a fraction a/b as the quantity formed by a parts of size $1/b$.

3.NF.2 Understand a fraction as a number on the number line; represent fractions on a number line diagram.

a. Represent a fraction $1/b$ on a number line diagram by defining the interval from 0 to 1 as the whole and partitioning it into b equal parts. Recognize that each part has size $1/b$ and that the endpoint of the part based at 0 locates the number $1/b$ on the number line.

b. Represent a fraction a/b on a number line diagram by marking off a lengths $1/b$ from 0. Recognize that the resulting interval has size a/b and that its endpoint locates the number a/b on the number line.

3.NF.3 Explain equivalence of fractions in special cases, and compare fractions by reasoning about their size.

a. Understand two fractions as equivalent (equal) if they are the same size, or the same point on a number line.

b. Recognize and generate simple equivalent fractions, e.g., $1/2 = 2/4$, $4/6 = 2/3$. Explain why the fractions are equivalent, e.g., by using a visual fraction model.

c. Express whole numbers as fractions, and recognize fractions that are equivalent to whole numbers. *Examples: Express 3 in the form $3 = 3/1$; recognize that $6/1 = 6$; locate 4/4 and 1 at the same point of a number line.*

d. Compare two fractions with the same numerator or the same denominator by reasoning about their size. Recognize that comparisons are valid only when the two fractions refer to the same whole. Record the results of comparisons with the symbols >, =, or <, and justify the conclusions, e.g., by using a visual fraction model.

Reason with shapes and their attributes.

3.G.2 Partition shapes into parts with equal areas. Express the area of each part as a unit fraction of the whole. *For example, partition a shape into 4 parts with equal area, and describe the area of each part as 1/4 of the area of the shape.*

Evaluating Student Learning Outcomes

A Progression Toward Mastery is provided to describe steps that illuminate the gradually increasing understandings that students develop *on their way to proficiency.* In this chart, this progress is presented from left (Step 1) to right (Step 4). The learning goal for students is to achieve Step 4 mastery. These steps are meant to help teachers and students identify and celebrate what the students CAN do now and what they need to work on next.

A Progression Toward Mastery				
Assessment Task Item and Standards Assessed	**STEP 1 Little evidence of reasoning without a correct answer.** **(1 Point)**	**STEP 2 Evidence of some reasoning without a correct answer.** **(2 Points)**	**STEP 3 Evidence of some reasoning with a correct answer or evidence of solid reasoning with an incorrect answer.** **(3 Points)**	**STEP 4 Evidence of solid reasoning with a correct answer.** **(4 Points)**
1 **3.NF.2a** **3.NF.3a**	The student is unable to label the number line.	The student labels the number line but thinks $\frac{2}{6}$ is $\frac{1}{2}$ because of the 2 in the numerator. Clear flaws in understanding are visible.	The student shows good reasoning and makes one small mistake, such as failing to correctly label $\frac{0}{6}$ or failing to identify the fraction equal to $\frac{1}{2}$.	The student correctly: ▪ Labels the number line with sixths. ▪ Identifies $\frac{3}{6}$ as the halfway point for Betsy's coat. ▪ Writes any fraction equivalent to $\frac{3}{6}$, such as $\frac{1}{2}$.
2 **3.NF.3b** **3.G.2** **3.NF.1**	The student does not demonstrate understanding.	The student may partition the strip correctly but does not give a clear explanation.	The student's explanation lacks clarity, but the drawing shows understanding. The strip is labeled.	The student uses words, pictures, and numbers to: ▪ Explain how Jerry would make smaller equal parts. ▪ Name a fraction equal to $\frac{2}{3}$, such as $\frac{4}{6}$, $\frac{6}{9}$, or $\frac{8}{12}$.
3 **3.NF.3d** **3.NF.1**	The student does not demonstrate understanding of the meaning of the question and does not produce meaningful work.	The student may say that the son has eaten more but does show some understanding. This is possibly evidenced by two fraction strips correctly partitioned but perhaps not the same size.	The student shows that Jerry has eaten more and correctly compares $\frac{3}{6}$ to $\frac{3}{8}$; the explanation includes some reasoning.	The student clearly explains: ▪ Jerry has eaten more of his granola bar. ▪ $\frac{3}{6} > \frac{3}{8}$. ▪ $\frac{3}{6}$ is greater than $\frac{3}{8}$ because the units are larger.

A Progression Toward Mastery				
4 **3.NF.2a, b** **3.NF.3a–d** **3.NF.1**	The student does not demonstrate understanding of the meaning of the question and does not produce meaningful work.	The student completes part of the problem correctly but fails to draw accurate models or explain reasoning.	The student completes Parts (a), (b), and (c) correctly; the explanation includes some reasoning.	The student correctly: ▪ Shows all of the fractions from $\frac{0}{3}$ up to $\frac{12}{3}$ numerically, including renaming the wholes. ▪ Explains $\frac{1}{4}$ or $\frac{3}{12}$ of the whole roll was eaten with an accurate model in Part (b). ▪ Uses words, pictures, and numbers to explain that $\frac{1}{3}$ is equal to $\frac{2}{6}$ in Part (c).

Name Gina Date

1. Jerry put 7 equally spaced hooks on a straight wire so students could hang up their coats. The whole length is from the first hook to the last hook.

   a. On the picture below, label the fraction of the wire's length where each hook is located.

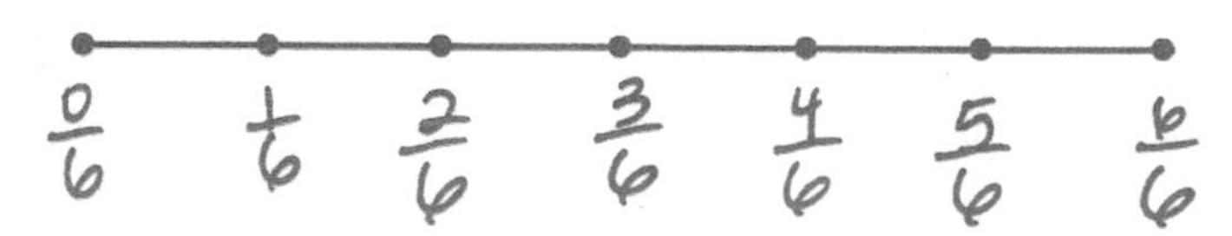

$\frac{0}{6}$ $\frac{1}{6}$ $\frac{2}{6}$ $\frac{3}{6}$ $\frac{4}{6}$ $\frac{5}{6}$ $\frac{6}{6}$

   b. At what fraction is Betsy's coat if she hangs it at the halfway point? $\frac{3}{6}$

   c. Write a fraction that is equivalent to your answer for Part (b). $\frac{1}{2}$

2. Jerry used the picture below to show his son how to find a fraction equal to $\frac{2}{3}$. Explain what Jerry might have said and done using words, pictures, and numbers.

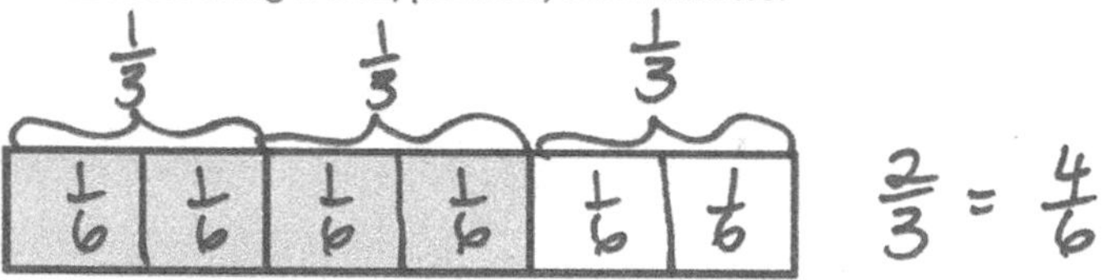

$\frac{2}{3} = \frac{4}{6}$

I made each $\frac{1}{3}$ into 2 smaller, equal parts. So then it wasn't just thirds anymore, it was sixths too! I can see from the shading that $\frac{2}{3}$ is the same as $\frac{4}{6}$.

3. Jerry and his son have the exact same granola bars. Jerry has eaten $\frac{3}{6}$ of his granola bar. His son has eaten $\frac{3}{8}$ of his. Who has eaten more? Explain your answer using words, pictures, and numbers.

Jerry $\frac{3}{6}$ $\frac{3}{6} > \frac{3}{8}$

Son $\frac{3}{8}$

Jerry ate more because his pieces are bigger than his son's pieces and they ate the same number of pieces.

4. Jerry has a fruit roll that is 4 feet long.

 a. Label the number line to show how Jerry might cut his fruit roll into pieces $\frac{1}{3}$ of a foot long. Label every fraction on the number line, including renaming the wholes.

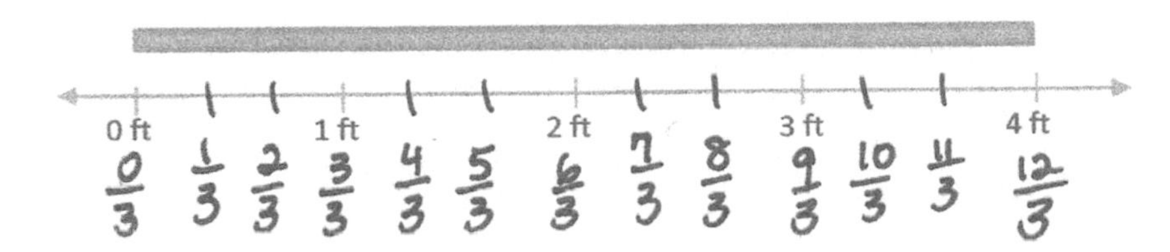

 b. Jerry cut his fruit roll into pieces that are $\frac{1}{3}$ of a foot long. Jerry and his 2 sons each eat one piece. What fraction of the whole fruit roll is eaten? Explain your answer using words, pictures, and numbers.

0 ft 1 ft 2 ft 3 ft 4 ft

$\frac{1}{4}$ of the whole roll was eaten because together they ate 1 of the 4 feet. Or, you can say $\frac{3}{12}$ was eaten because there are 12 pieces and they ate 3 pieces.

 c. Jerry's son says that 1 third is the same as 2 sixths. Do you agree? Why or why not? Use words, pictures, and numbers to explain your answer.

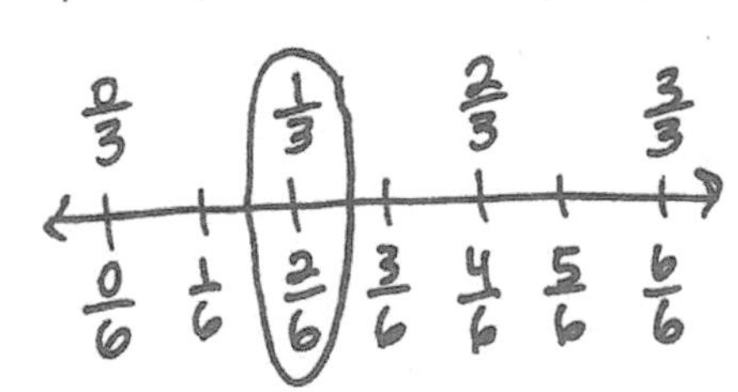

Yes, I agree. When I draw a number line with thirds and sixths, $\frac{1}{3}$ and $\frac{2}{6}$ are at the same point. That means they're equal!

This page intentionally left blank

Answer Key

Eureka Math Grade 3 Module 5

Special thanks go to the Gordon A. Cain Center and to the Department of Mathematics at Louisiana State University for their support in the development of *Eureka Math*.

For a free *Eureka Math* Teacher Resource Pack, Parent Tip Sheets, and more please visit www.Eureka.tools

Published by the non-profit Great Minds

Printed in the U.S.A.
This book may be purchased from the publisher at eureka-math.org
10 9 8 7 6 5 4 3 2

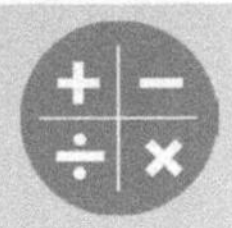

Answer Key

GRADE 3 • MODULE 5

Fractions as Numbers on the Number Line

Lesson 1

Problem Set

1. Answer provided; 1 fourth shaded; 1 third shaded
2. 1 third; 1 sixth; 1 fourth
3. a. Rectangle drawn; 1 line, 1 half
   b. Rectangle drawn; 2 lines, 1 third
   c. Rectangle drawn; 3 lines, 1 fourth
4. a. Sevenths are shown; ninths are shown
   b. Answers will vary; 19 lines
5. 1 half; picture drawn to show 2 halves

Exit Ticket

1. 1 fourth
2. Rectangle partitioned into thirds
3. 1 fourth

Homework

1. Answer provided; 1 fifth shaded; 1 sixth shaded
2. 1 third; 1 fourth; 1 seventh
3. Halves are shown; thirds are shown; sixths are shown
4. Lines drawn to show halves; lines drawn to show fourths; lines drawn to show eighths
5. Lines drawn to show sixths; lines drawn to show thirds
6. 1 sixth; picture drawn to show 6 equal parts
7. 6 grams

Lesson 2

Problem Set

1. First and last strips circled
2. a. 4, 2
   b. 6, 5
   c. 7, 3
   d. 7, 0
3. 1 third; bar drawn and labeled appropriately; 1 third labeled
4. a. 1 fourth; fraction strip drawn and labeled correctly
   b. Sixths; fraction strip drawn and labeled correctly

Exit Ticket

1. Second model
2. 10, 7
3. Answers will vary, showing 4 equal parts.

Homework

1. Second and third strips circled
2. a. 2, 1
   b. 3, 1
   c. 5, 1
   d. 14, 7
3. Answers will vary.
4. a. 1 eighth
   b. 5 eighths

Lesson 3

Sprint

Side A

1.	6	12.	42	23.	60	34.	54
2.	6	13.	42	24.	54	35.	36
3.	12	14.	48	25.	24	36.	18
4.	12	15.	48	26.	48	37.	12
5.	18	16.	54	27.	18	38.	42
6.	18	17.	54	28.	42	39.	48
7.	24	18.	60	29.	36	40.	66
8.	24	19.	60	30.	60	41.	66
9.	30	20.	18	31.	30	42.	72
10.	30	21.	6	32.	24	43.	72
11.	36	22.	12	33.	6	44.	78

Side B

1.	6	12.	42	23.	54	34.	24
2.	6	13.	42	24.	18	35.	54
3.	12	14.	48	25.	48	36.	12
4.	12	15.	48	26.	24	37.	42
5.	18	16.	54	27.	42	38.	18
6.	18	17.	54	28.	30	39.	48
7.	24	18.	60	29.	36	40.	66
8.	24	19.	60	30.	30	41.	66
9.	30	20.	6	31.	60	42.	72
10.	30	21.	60	32.	6	43.	72
11.	36	22.	12	33.	36	44.	78

Problem Set

1. Eighths; 5 eighths
   Thirds; 3 thirds
   Halves; 1 half
2. First, third, and fifth shapes circled; sentences will vary.
3. Shapes divided into 4 equal parts; fourths
4. Shapes divided and shaded appropriately
5. Answers will vary.
6. Candy bar drawn and divided into 5 equal parts; 1 fifth

Exit Ticket

1. 4
2. First and third shapes circled
3. 1 fourth

Homework

1. Fifths; 4 fifths
   Sixths; 3 sixths
   Halves; 0 halves
2. Answers will vary.
3. Calendar drawn and divided into 12 equal parts; 1 twelfth

Lesson 4

Sprint

Side A

1. 12
2. 18
3. 24
4. 30
5. 6
6. 2
7. 3
8. 5
9. 1
10. 4
11. 36
12. 42
13. 48
14. 54
15. 60
16. 8
17. 7
18. 9
19. 6
20. 10
21. 5
22. 1
23. 10
24. 2
25. 3
26. 10
27. 5
28. 1
29. 2
30. 3
31. 6
32. 7
33. 9
34. 8
35. 7
36. 9
37. 6
38. 8
39. 66
40. 11
41. 72
42. 12
43. 84
44. 14

Side B

1. 6
2. 12
3. 18
4. 24
5. 30
6. 3
7. 2
8. 4
9. 1
10. 5
11. 60
12. 36
13. 42
14. 48
15. 54
16. 7
17. 6
18. 8
19. 10
20. 9
21. 1
22. 5
23. 2
24. 10
25. 3
26. 2
27. 1
28. 10
29. 5
30. 3
31. 3
32. 4
33. 9
34. 7
35. 8
36. 9
37. 6
38. 7
39. 66
40. 11
41. 72
42. 12
43. 78
44. 13

Problem Set

1. Answers will vary.
2. Answers will vary.
3. Answers will vary.
4. Answers will vary.
5. Answers will vary.
6. Answers will vary.

Exit Ticket

1. Lines drawn to show 1 fourth
2. Lines drawn to show 1 fifth
3. 1 sixth

Homework

1. Lines drawn to show halves for each figure; each figure shaded to show 1 half
2. Lines drawn to show fourths for each figure; each figure shaded to show 1 fourth
3. Lines drawn to show thirds for each figure; each figure shaded to show 1 third
4. Fractions matched to equivalent shape

Lesson 5

Problem Set

1. a. 2, 1, 1 half, $\frac{1}{2}$
   b. 3, 1, 1 third, $\frac{1}{3}$
   c. 4, 1, 1 fourth, $\frac{1}{4}$
   d. 5, 1, 1 fifth, $\frac{1}{5}$
   e. 6, 1, 1 sixth, $\frac{1}{6}$
   f. 8, 1, 1 eighth, $\frac{1}{8}$
2. No; explanations will vary.
3. Lines drawn to show tenths; $\frac{1}{10}$
4. Rectangles are drawn and labeled to show $\frac{1}{10}$ and $\frac{1}{8}$; $\frac{1}{8}$ is bigger than $\frac{1}{10}$.

Exit Ticket

1. 6, 1, 1 sixth, $\frac{1}{6}$
2. $\frac{1}{7}$; $\frac{1}{2}$; $\frac{1}{9}$
3. Rectangles are drawn and labeled to show $\frac{1}{5}$ and $\frac{1}{8}$; $\frac{1}{5}$ is bigger than $\frac{1}{8}$.

Homework

1. a. 2, 1, 1 half, $\frac{1}{2}$
   b. 3, 1, 1 third, $\frac{1}{3}$
   c. 10, 1, 1 tenth, $\frac{1}{10}$
   d. 5, 1, 1 fifth, $\frac{1}{5}$
   e. 4, 1, 1 fourth, $\frac{1}{4}$
2. No; explanations will vary.
3. Lines drawn to show fourths
4. Rectangles drawn and labeled to show $\frac{1}{7}$ and $\frac{1}{10}$; explanations will vary.

Lesson 6

Sprint

Side A

1. 7	12. 42	23. 70	34. 63
2. 7	13. 49	24. 63	35. 28
3. 14	14. 56	25. 28	36. 21
4. 14	15. 56	26. 56	37. 14
5. 21	16. 63	27. 21	38. 49
6. 21	17. 63	28. 49	39. 56
7. 28	18. 70	29. 42	40. 77
8. 28	19. 70	30. 70	41. 77
9. 35	20. 21	31. 35	42. 84
10. 35	21. 7	32. 42	43. 84
11. 42	22. 14	33. 7	44. 91

Side B

1. 7	12. 42	23. 63	34. 28
2. 7	13. 49	24. 21	35. 63
3. 14	14. 56	25. 56	36. 14
4. 14	15. 56	26. 28	37. 49
5. 21	16. 63	27. 49	38. 21
6. 21	17. 63	28. 35	39. 56
7. 28	18. 70	29. 42	40. 77
8. 28	19. 70	30. 35	41. 77
9. 35	20. 7	31. 70	42. 84
10. 35	21. 70	32. 7	43. 84
11. 42	22. 14	33. 42	44. 91

Problem Set

1. Each shape partitioned, labeled, and shaded correctly
   a. $\frac{3}{4}$
   b. $\frac{3}{7}$
   c. $\frac{4}{5}$
   d. $\frac{2}{6}$
2. a. $\frac{1}{8}$
   b. $\frac{7}{8}$
3. a. 9, 5, $\frac{1}{9}$, $\frac{5}{9}$
   b. 7, 3, $\frac{1}{7}$, $\frac{3}{7}$
   c. 5, 4, $\frac{1}{5}$, $\frac{4}{5}$
   d. 6, 2, $\frac{1}{6}$, $\frac{2}{6}$
   e. 8, 8, $\frac{1}{8}$, $\frac{8}{8}$

Exit Ticket

1. $\frac{2}{5}$; fraction strip partitioned, labeled, and shaded correctly
2. a. $\frac{1}{8}$
   b. $\frac{7}{8}$
3. 4, 2, $\frac{1}{4}$, $\frac{2}{4}$

Homework

1. Each shape partitioned, labeled, and shaded correctly
   a. $\frac{2}{3}$
   b. $\frac{5}{7}$
   c. $\frac{3}{5}$
   d. $\frac{2}{8}$
2. a. $\frac{1}{6}$
   b. $\frac{5}{6}$
3. a. 4, 3, $\frac{1}{4}$, $\frac{3}{4}$
   b. 9, 6, $\frac{1}{9}$, $\frac{6}{9}$
   c. 7, 4, $\frac{1}{7}$, $\frac{4}{7}$
   d. 6, 3, $\frac{1}{6}$, $\frac{3}{6}$

Lesson 7

Sprint

Side A

1. 14	12. 49	23. 10	34. 8
2. 21	13. 56	24. 2	35. 7
3. 28	14. 63	25. 3	36. 9
4. 35	15. 70	26. 10	37. 6
5. 7	16. 8	27. 5	38. 8
6. 2	17. 7	28. 1	39. 77
7. 3	18. 9	29. 2	40. 11
8. 5	19. 6	30. 3	41. 84
9. 1	20. 10	31. 6	42. 12
10. 4	21. 5	32. 7	43. 98
11. 42	22. 1	33. 9	44. 14

Side B

1. 7	12. 42	23. 2	34. 7
2. 14	13. 49	24. 10	35. 8
3. 21	14. 56	25. 3	36. 9
4. 28	15. 63	26. 2	37. 6
5. 35	16. 7	27. 1	38. 7
6. 3	17. 6	28. 10	39. 77
7. 2	18. 8	29. 5	40. 11
8. 4	19. 10	30. 3	41. 84
9. 1	20. 9	31. 3	42. 12
10. 5	21. 1	32. 4	43. 91
11. 70	22. 5	33. 9	44. 13

Problem Set

1. 1 half
2. 3 fourths
3. 8 ninths
4. 5 sixths
5. 4 fifths
6. 2 thirds
7. 6 sevenths
8. 7 eighths
9. a. 8
   b. 9
   c. 12
10. $\frac{1}{5}, \frac{4}{5}; \frac{1}{7}, \frac{6}{7}; \frac{1}{11}, \frac{10}{11}$
11. 5 sixths

Exit Ticket

1. $\frac{7}{8}$
2. 6
3. $\frac{3}{4}, \frac{1}{4}$
4. $\frac{1}{10}$

Homework

1. 3 fourths
2. 9 tenths
3. 1 half
4. 2 thirds
5. 6 sevenths
6. 4 fifths
7. 10 elevenths
8. 5 sixths
9. $\frac{4}{5}, \frac{1}{5}; \frac{1}{12}, \frac{11}{12}$
10. 3 fourths; picture drawn and labeled to show $\frac{1}{4}$ finished and $\frac{3}{4}$ unfinished
11. 1 eighth; picture drawn and labeled to show $\frac{1}{8}$ uneaten and $\frac{7}{8}$ eaten

Lesson 8

Sprint

Side A

1.	1/2	12.	1/3	23.	3/4	34.	4/5
2.	1/3	13.	1/3	24.	3/4	35.	4/5
3.	1/5	14.	1/3	25.	3/4	36.	1/10
4.	1/2	15.	2/3	26.	1/2	37.	2/10
5.	1/3	16.	2/3	27.	1/2	38.	3/10
6.	1/4	17.	2/3	28.	1/5	39.	8/10
7.	1/2	18.	1/4	29.	1/5	40.	5/10
8.	1/3	19.	1/4	30.	2/5	41.	7/10
9.	1/4	20.	1/4	31.	2/5	42.	6/10
10.	1/2	21.	1/4	32.	3/5	43.	5/6
11.	1/4	22.	3/4	33.	3/5	44.	1/6

Side B

1.	1/2	12.	1/3	23.	3/4	34.	4/5
2.	1/3	13.	1/3	24.	3/4	35.	4/5
3.	1/4	14.	1/3	25.	3/4	36.	5/10
4.	1/2	15.	2/3	26.	1/2	37.	1/10
5.	1/3	16.	2/3	27.	1/2	38.	2/10
6.	1/4	17.	2/3	28.	1/5	39.	3/10
7.	1/2	18.	1/4	29.	1/5	40.	8/10
8.	1/3	19.	1/4	30.	2/5	41.	6/10
9.	1/4	20.	1/4	31.	2/5	42.	7/10
10.	1/2	21.	1/4	32.	3/5	43.	1/6
11.	1/4	22.	3/4	33.	3/5	44.	5/6

Problem Set

1. Number bond showing $\frac{3}{5}$ and $\frac{2}{5}$ equals 1 whole; second visual model drawn
2. Number bond showing $\frac{3}{4}$ and $\frac{1}{4}$ equals 1 whole; second visual model drawn
3. Number bond showing $\frac{3}{6}$ and $\frac{3}{6}$ equals 1 whole; second visual model drawn
4. Number bond showing $\frac{2}{9}$ and $\frac{7}{9}$ equals 1 whole; second visual model drawn
5. a. Number bond showing $\frac{3}{4}$ and $\frac{1}{4}$ equals 1 whole; $\frac{3}{4}$ decomposed showing 3 units of $\frac{1}{4}$
   b. Number bond showing $\frac{2}{3}$ and $\frac{1}{3}$ equals 1 whole; $\frac{2}{3}$ decomposed showing 2 units of $\frac{1}{3}$
   c. Number bond showing $\frac{2}{4}$ and $\frac{2}{4}$ equals 1 whole; both of $\frac{2}{4}$ bonds decomposed showing 2 units of $\frac{1}{4}$
   d. Number bond showing $\frac{2}{5}$ and $\frac{3}{5}$ equals 1 whole; $\frac{2}{5}$ decomposed showing 2 units of $\frac{1}{5}$; $\frac{3}{5}$ decomposed showing 3 units of $\frac{1}{5}$
6. a. $\frac{3}{4}$
   b. 3
   c. Number bond showing $\frac{1}{4}$ and $\frac{3}{4}$ equals 1 whole; $\frac{3}{4}$ decomposed showing 3 units of $\frac{1}{4}$; second visual model drawn

Exit Ticket

1. Number bond showing $\frac{2}{5}$ and $\frac{3}{5}$ equals 1 whole; $\frac{2}{5}$ decomposed showing 2 units of $\frac{1}{5}$; $\frac{3}{5}$ decomposed showing 3 units of $\frac{1}{5}$
2. $\frac{5}{7}$; shape drawn and shaded to match the completed number bond

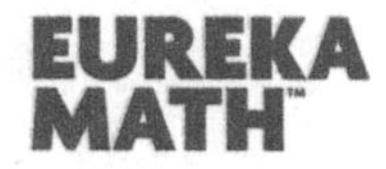

Homework

1. Number bond showing $\frac{2}{4}$ and $\frac{2}{4}$ equals 1 whole; second visual model drawn
2. Number bond showing $\frac{5}{7}$ and $\frac{2}{7}$ equals 1 whole; second visual model drawn
3. Number bond showing $\frac{4}{5}$ and $\frac{1}{5}$ equals 1 whole; second visual model drawn
4. Number bond showing $\frac{0}{8}$ and $\frac{8}{8}$ equals 1 whole; second visual model drawn
5. a. Number bond showing $\frac{2}{3}$ and $\frac{1}{3}$ equals 1 whole; $\frac{2}{3}$ decomposed showing 2 units of $\frac{1}{3}$
   b. Number bond showing $\frac{4}{5}$ and $\frac{1}{5}$ equals 1 whole; $\frac{4}{5}$ decomposed showing 4 units of $\frac{1}{5}$
   c. Number bond showing $\frac{3}{5}$ and $\frac{2}{5}$ equals 1 whole; $\frac{3}{5}$ decomposed showing 3 units of $\frac{1}{5}$; $\frac{2}{5}$ decomposed showing 2 units of $\frac{1}{5}$
6. Drawing showing 3 equal parts; 2 thirds shaded; number bond showing $\frac{2}{3}$ and $\frac{1}{3}$ equals 1 whole; $\frac{2}{3}$

Lesson 9

Sprint

Side A

1. 8
2. 8
3. 16
4. 16
5. 24
6. 24
7. 32
8. 32
9. 40
10. 40
11. 48
12. 48
13. 56
14. 56
15. 64
16. 72
17. 72
18. 80
19. 80
20. 8
21. 80
22. 16
23. 72
24. 24
25. 64
26. 32
27. 56
28. 40
29. 48
30. 40
31. 80
32. 8
33. 48
34. 32
35. 72
36. 16
37. 56
38. 24
39. 64
40. 88
41. 88
42. 96
43. 96
44. 104

Side B

1. 8
2. 8
3. 16
4. 16
5. 24
6. 24
7. 32
8. 32
9. 40
10. 40
11. 48
12. 48
13. 56
14. 56
15. 64
16. 72
17. 72
18. 80
19. 80
20. 24
21. 8
22. 16
23. 80
24. 72
25. 32
26. 64
27. 24
28. 56
29. 48
30. 80
31. 40
32. 48
33. 8
34. 72
35. 32
36. 24
37. 16
38. 56
39. 64
40. 88
41. 88
42. 96
43. 96
44. 104

Problem Set

1. a. Answer provided
   b. $\frac{1}{8}$, 15, $\frac{15}{8}$
   c. $\frac{1}{6}$, 14, $\frac{14}{6}$
   d. $\frac{1}{5}$, 8, $\frac{8}{5}$
   e. $\frac{1}{4}$, 9, $\frac{9}{4}$
   f. $\frac{1}{3}$, 7, $\frac{7}{3}$
2. a. Each whole partitioned into sixths; 8 sixths shaded; $\frac{8}{6}$
   b. Each whole partitioned into fourth; 7 fourths shaded; $\frac{7}{4}$
   c. Each whole partitioned into fifths; 6 fifths shaded; 6 fifths
   d. Each whole partitioned into halves; 5 halves shaded; 5 halves
3. a. 2 equivalent wholes drawn; each whole partitioned into 8 equal pieces; 10 pieces shaded
   b. $\frac{10}{8}$

Exit Ticket

1. $\frac{1}{3}$, 11, $\frac{11}{3}$
2. a. Each whole partitioned into thirds; 4 thirds shaded; $\frac{4}{3}$
   b. Each whole partitioned into fourths; 10 fourths shaded; 10 fourths

Homework

1 a. Answer provided
   b. $\frac{1}{6}$, 9, $\frac{9}{6}$
   c. $\frac{1}{4}$, 15, $\frac{15}{4}$
   d. $\frac{1}{2}$, 6, $\frac{6}{2}$
   e. $\frac{1}{3}$, 4, $\frac{4}{3}$
   f. $\frac{1}{3}$, 4, $\frac{4}{3}$
2. a. Each whole partitioned into thirds; 5 thirds shaded; $\frac{5}{3}$
   b. Each whole partitioned into thirds; 9 thirds shaded; 9 thirds
3. a. 2 equivalent wholes drawn; each whole partitioned into 4 equal pieces; 5 pieces shaded
   b. $\frac{5}{4}$

Lesson 10

Sprint

Side A

1. 16
2. 24
3. 32
4. 40
5. 8
6. 2
7. 3
8. 5
9. 1
10. 4
11. 48
12. 56
13. 64
14. 72
15. 80
16. 8
17. 7
18. 9
19. 6
20. 10
21. 5
22. 1
23. 10
24. 2
25. 3
26. 10
27. 5
28. 1
29. 2
30. 3
31. 6
32. 7
33. 9
34. 8
35. 7
36. 9
37. 6
38. 8
39. 88
40. 11
41. 96
42. 12
43. 112
44. 14

Side B

1. 8
2. 16
3. 24
4. 32
5. 40
6. 3
7. 2
8. 4
9. 1
10. 5
11. 80
12. 48
13. 56
14. 64
15. 72
16. 7
17. 6
18. 8
19. 10
20. 9
21. 1
22. 5
23. 2
24. 10
25. 3
26. 2
27. 1
28. 10
29. 5
30. 3
31. 3
32. 4
33. 9
34. 7
35. 8
36. 9
37. 6
38. 7
39. 88
40. 11
41. 96
42. 12
43. 104
44. 13

Problem Set

1. Specified fractional unit shaded in each strip
2. a. Greater than
   b. Less than
   c. Less than
   d. Greater than
   e. Less than
   f. Less than
   g. Greater than
   h. Greater than
3. More oil; explanations will vary.
4. a. >
   b. <
   c. =
   d. >
   e. <
   f. =
   g. =, <, <, <, =
5. No; explanations will vary.

Exit Ticket

1. Specified fractional unit shaded in each strip; circled $\frac{1}{2}$; shaded star drawn next to $\frac{1}{4}$
2. a. >
   b. =
   c. <

Homework

1. Specified fractional unit shaded in each strip
2. a. Greater than
   b. Less than
   c. Less than
   d. Greater than
   e. Less than
   f. Less than
   g. Greater than
   h. Greater than
3. More water; explanations will vary.
4. a. >
   b. <
   c. =
   d. <
   e. <
   f. =
   g. >
5. Answers will vary.

Lesson 11

Problem Set

1. $\frac{1}{3}$; answers will vary.
2. $\frac{1}{5}$; answers will vary.
3. $\frac{1}{10}$; answers will vary.
4. $\frac{1}{12}$; answers will vary.
5. Answers will vary; $\frac{1}{8}$
6. Answers will vary; $\frac{1}{9}$
7. Answers will vary; $\frac{1}{12}$
8. Answers will vary.
9. No, explanations will vary.
10. No, explanations will vary.

Exit Ticket

1. Answers will vary.
2. Answers will vary.

Homework

1. $\frac{1}{8}$; answers will vary.
2. $\frac{1}{4}$; answers will vary.
3. $\frac{1}{10}$; answers will vary.
4. $\frac{1}{9}$; answers will vary.
5. Answers will vary; $\frac{1}{2}$
6. Answers will vary; $\frac{1}{4}$
7. Answers will vary; $\frac{1}{12}$
8. Answers will vary.
9. a. Explanations will vary.
   b. Explanations will vary.

Lesson 12

Sprint

Side A

1. 9
2. 9
3. 18
4. 18
5. 27
6. 27
7. 36
8. 36
9. 45
10. 45
11. 54
12. 54
13. 63
14. 63
15. 72
16. 72
17. 81
18. 90
19. 90
20. 9
21. 90
22. 18
23. 81
24. 27
25. 72
26. 36
27. 63
28. 45
29. 54
30. 45
31. 90
32. 9
33. 54
34. 36
35. 81
36. 18
37. 63
38. 27
39. 72
40. 99
41. 99
42. 108
43. 108
44. 117

Side B

1. 9
2. 9
3. 18
4. 18
5. 27
6. 27
7. 36
8. 36
9. 45
10. 45
11. 54
12. 54
13. 63
14. 63
15. 72
16. 72
17. 81
18. 90
19. 90
20. 27
21. 9
22. 18
23. 90
24. 81
25. 36
26. 72
27. 27
28. 63
29. 54
30. 90
31. 45
32. 54
33. 9
34. 81
35. 36
36. 27
37. 18
38. 63
39. 72
40. 99
41. 99
42. 108
43. 108
44. 117

Problem Set

1. Answers will vary.
2. Answers will vary.
3. Answers will vary.
4. Answers will vary.
5. Answers will vary.
6. Answers will vary.

Exit Ticket

1. Picture representing 1 whole
2. Picture representing 1 whole
3. Both, explanations will vary.

Homework

1. Picture representing 2 halves
2. Picture representing 6 sixths
3. Picture representing 3 thirds
4. Picture representing 4 fourths
5. Answer provided
6. Picture representing 2 halves; number bond showing 2 units of $\frac{1}{2}$ equals 1 whole
7. Picture representing 5 fifths; number bond showing 5 units of $\frac{1}{5}$ equals 1 whole
8. Picture representing 7 sevenths; number bond showing 7 units of $\frac{1}{7}$ equals 1 whole
9. No, explanations will vary.

Lesson 13

Problem Set

1. a. $\frac{1}{2}$
   b. Shaded part divided to show $\frac{1}{2}$
2. a. $\frac{1}{4}$
   b. Shaded part divided to show $\frac{1}{4}$
3. a. $\frac{1}{3}$
   b. Shaded part divided to show $\frac{1}{3}$
4. a. $\frac{1}{5}$
   b. Shaded part divided to show $\frac{1}{5}$
5. a. $\frac{1}{6}$
   b. Shaded part divided to show $\frac{1}{6}$
6. a. C
   b. B
   c. A
   d. 2; $\frac{1}{2}$
   e. $\frac{1}{2}$; $\frac{1}{4}$
7. Answers will vary.

Exit Ticket

Both; explanations will vary.

Homework

1. a. $\frac{1}{2}$
   b. Shaded part divided to show $\frac{1}{2}$
2. a. $\frac{1}{3}$
   b. Shaded part divided to show $\frac{1}{3}$
3. a. $\frac{1}{4}$
   b. Shaded part divided to show $\frac{1}{4}$
4. a. $\frac{1}{5}$
   b. Shaded part divided to show $\frac{1}{5}$
5. a. B
   b. A
   c. 3; 2
   d. 3; number bond showing 3 units of $\frac{1}{3}$ equals 1 whole
   e. 2; number bond showing 2 units of $\frac{1}{2}$ equals 1 whole
6. Strings drawn correctly

Lesson 14

Problem Set

1. a. Answer provided; fraction strip partitioned and labeled to show halves; number line partitioned and labeled correctly from $\frac{0}{2}$ to $\frac{2}{2}$
   b. Number bond showing 3 units of $\frac{1}{3}$; fraction strip partitioned and labeled to show thirds; number line partitioned and labeled correctly from $\frac{0}{3}$ to $\frac{3}{3}$
   c. Number bond showing 4 units of $\frac{1}{4}$; fraction strip partitioned and labeled to show fourths; number line partitioned and labeled correctly from $\frac{0}{4}$ to $\frac{4}{4}$
   d. Number bond showing 5 units of $\frac{1}{5}$; fraction strip partitioned and labeled to show fifths; number line partitioned and labeled correctly from $\frac{0}{5}$ to $\frac{5}{5}$
2. Number line showing fourths; each quarter (fourth) hour from $\frac{0}{4}$ to $\frac{4}{4}$ correctly labeled correctly, including 0 hours and 1 hour
3. Number line showing fifths; each fifth meter from $\frac{0}{5}$ to $\frac{5}{5}$ correctly labeled, including 0 meters and 1 meter

Exit Ticket

1. Number bond showing 6 units of $\frac{1}{6}$; fraction strip partitioned and labeled to show sixths; number line partitioned and labeled correctly from $\frac{0}{6}$ to $\frac{6}{6}$
2. Number bond showing 5 units of $\frac{1}{5}$; number line partitioned and labeled correctly from $\frac{0}{5}$ to $\frac{5}{5}$
   a. $\frac{1}{5}$
   b. 20 cents or $0.20

Homework

1. a. Answer provided; fraction strip is partitioned and labeled correctly to show halves; number line partitioned and labeled correctly from $\frac{0}{2}$ to $\frac{2}{2}$
   b. Number bond is drawn correctly to show 8 units of $\frac{1}{8}$; fraction strip is partitioned and labeled correctly to show eighths; number line partitioned and labeled correctly from $\frac{0}{8}$ to $\frac{8}{8}$
   c. Number bond is drawn correctly to show 5 units of $\frac{1}{5}$; fraction strip is partitioned and labeled correctly to show fifths; number line partitioned and labeled correctly from $\frac{0}{5}$ to $\frac{5}{5}$
2. Yes
3. Fraction strip drawn and labeled
   a. 9 seeds
   b. 36 seeds
   c. Number line is drawn and partitioned correctly to show ninths

Lesson 15

Problem Set

1. a. Number line partitioned into thirds and labeled correctly with $\frac{0}{3}$, $\frac{2}{3}$, $\frac{3}{3}$; answer provided

   b. Number line partitioned into fourths and labeled correctly with $\frac{0}{4}$, $\frac{3}{4}$, $\frac{4}{4}$; number bond showing $\frac{3}{4}$ and $\frac{1}{4}$ equals 1 whole

   c. Number line partitioned into fifths and labeled correctly with $\frac{0}{5}$, $\frac{3}{5}$, $\frac{5}{5}$; number bond showing $\frac{3}{5}$ and $\frac{2}{5}$ equals 1 whole

   d. Number line partitioned into sixths and labeled correctly with $\frac{0}{6}$, $\frac{5}{6}$, $\frac{6}{6}$; number bond showing $\frac{5}{6}$ and $\frac{1}{6}$ equals 1 whole

   e. Number line partitioned into tenths and labeled correctly with $\frac{0}{10}$, $\frac{3}{10}$, $\frac{10}{10}$; number bond showing $\frac{3}{10}$ and $\frac{7}{10}$ equals 1 whole

2. Number line drawn with 0 and 1 labeled correctly; fraction strip used appropriately to partition and label a number line to show eighths; number line labeled correctly from $\frac{0}{8}$ to $\frac{8}{8}$

3. a. 4 equal parts; rope labeled correctly from $\frac{0}{4}$ to $\frac{4}{4}$

   b. $\frac{2}{4}$

   c. $\frac{1}{5}$

Exit Ticket

1. Number line partitioned into fifths and labeled correctly with $\frac{0}{5}$, $\frac{3}{5}$, $\frac{5}{5}$; number bond showing $\frac{3}{5}$ and $\frac{2}{5}$ equals 1 whole

2. Number line partitioned into sixths; $\frac{1}{6}$, $\frac{3}{6}$, $\frac{5}{6}$ placed correctly on the number line

Homework

1. a. Answer provided

   b. Number line partitioned into sixths and labeled correctly with $\frac{0}{6}, \frac{3}{6}, \frac{6}{6}$; number bond showing $\frac{3}{6}$ and $\frac{3}{6}$ equals 1 whole

   c. Number line partitioned into fifths and labeled correctly with $\frac{0}{5}, \frac{2}{5}, \frac{5}{5}$; number bond showing $\frac{2}{5}$ and $\frac{3}{5}$ equals 1 whole

   d. Number line partitioned into tenths and labeled correctly with $\frac{0}{10}, \frac{7}{10}, \frac{10}{10}$; number bond showing $\frac{7}{10}$ and $\frac{3}{10}$ equals 1 whole

   e. Number line partitioned into sevenths and labeled correctly with $\frac{0}{7}, \frac{3}{7}, \frac{7}{7}$; number bond showing $\frac{3}{7}$ and $\frac{4}{7}$ equals 1 whole

2. a. Henry: $\frac{5}{10}$; Ben: $\frac{9}{10}$; Tina: $\frac{2}{10}$

   b. Number line partitioned into tenths; $\frac{2}{10}, \frac{5}{10}, \frac{9}{10}$ placed correctly on the number line

3. a. Number line drawn with 0 and 1 labeled correctly; fraction strip used appropriately to partition and label number line to show eighths; number line labeled correctly from $\frac{0}{8}$ to $\frac{8}{8}$

   b. Touched and counted each fraction from 0 eighths to 8 eighths

Lesson 16

Sprint

Side A

1. 18
2. 27
3. 36
4. 45
5. 9
6. 2
7. 3
8. 5
9. 1
10. 4
11. 54
12. 63
13. 72
14. 81
15. 90
16. 8
17. 7
18. 9
19. 6
20. 10
21. 5
22. 1
23. 10
24. 2
25. 3
26. 10
27. 5
28. 1
29. 2
30. 3
31. 6
32. 7
33. 9
34. 8
35. 7
36. 9
37. 6
38. 8
39. 99
40. 11
41. 108
42. 12
43. 126
44. 14

Side B

1. 9
2. 18
3. 27
4. 36
5. 45
6. 3
7. 2
8. 4
9. 1
10. 5
11. 90
12. 54
13. 63
14. 72
15. 81
16. 7
17. 6
18. 8
19. 10
20. 9
21. 1
22. 5
23. 2
24. 10
25. 3
26. 2
27. 1
28. 10
29. 5
30. 3
31. 3
32. 4
33. 9
34. 7
35. 8
36. 9
37. 6
38. 7
39. 99
40. 11
41. 108
42. 12
43. 117
44. 13

Problem Set

1. a. Answer provided
   b. Number line partitioned into thirds and labeled; $\frac{3}{3}$, $\frac{6}{3}$ boxed
   c. Number line partitioned into halves and labeled; $\frac{4}{2}$, $\frac{6}{2}$, $\frac{8}{2}$ boxed
   d. Number line partitioned into fourths and labeled; $\frac{12}{4}$, $\frac{16}{4}$, $\frac{20}{4}$ boxed; 4 labeled below $\frac{16}{4}$
   e. Number line partitioned into thirds and labeled; $\frac{18}{3}$, $\frac{21}{3}$, $\frac{24}{3}$, $\frac{27}{3}$ boxed; 7 labeled below $\frac{21}{3}$, 8 labeled below $\frac{24}{3}$
2. Number line partitioned into fifths and labeled; $\frac{0}{5}$, $\frac{5}{5}$, $\frac{10}{5}$ boxed
3. Number line partitioned into thirds and labeled; $\frac{3}{3}$, $\frac{6}{3}$, $\frac{9}{3}$, $\frac{12}{3}$ boxed
4. Number line drawn with endpoints 0 and 3; wholes labeled; number line partitioned and labeled

Exit Ticket

1. Number line partitioned into fifths and labeled; $\frac{10}{5}$, $\frac{15}{5}$ boxed
2. Number line drawn with endpoints 0 and 2, wholes labeled; number line partitioned and labeled; $\frac{0}{6}$, $\frac{6}{6}$, $\frac{12}{6}$ boxed

Homework

1. a. Answer provided
   b. Number line partitioned into eighths and labeled; $\frac{16}{8}$, $\frac{24}{8}$ boxed
   c. Number line partitioned into fourths and labeled; $\frac{8}{4}$, $\frac{12}{4}$, $\frac{16}{4}$ boxed
   d. Number line partitioned into halves and labeled; $\frac{6}{2}$, $\frac{8}{2}$, $\frac{10}{2}$ boxed; 4 labeled below $\frac{8}{2}$
   e. Number line partitioned into fifths and labeled; $\frac{30}{5}$, $\frac{35}{5}$, $\frac{40}{5}$, $\frac{45}{5}$ boxed; 7 labeled below $\frac{35}{5}$, 8 labeled below $\frac{40}{5}$
2. Number line partitioned into sixths and labeled; $\frac{18}{6}$, $\frac{24}{6}$, $\frac{30}{6}$ boxed
3. Number line partitioned into halves and labeled; $\frac{8}{2}$, $\frac{10}{2}$, $\frac{12}{2}$, $\frac{14}{2}$ boxed
4. Number line with endpoints 0 and 3; wholes labeled; number line partitioned and labeled

Lesson 17

Sprint

Side A

1. 1
2. 1
3. 1
4. 1
5. 0
6. 0
7. 0
8. 0
9. 2
10. 3
11. 4
12. 5
13. 2
14. 3
15. 4
16. 5
17. 6
18. 6
19. 7
20. 7
21. 2
22. 10
23. 8
24. 8
25. 3
26. 10
27. 9
28. 9
29. 4
30. 10
31. 5
32. 4
33. 6
34. 6
35. 7
36. 8
37. 8
38. 9
39. 11
40. 9
41. 8
42. 9
43. 8
44. 8

Side B

1. 1
2. 1
3. 1
4. 1
5. 0
6. 0
7. 0
8. 0
9. 2
10. 3
11. 4
12. 5
13. 2
14. 3
15. 4
16. 5
17. 6
18. 6
19. 7
20. 7
21. 10
22. 2
23. 8
24. 8
25. 10
26. 3
27. 9
28. 9
29. 10
30. 4
31. 4
32. 5
33. 6
34. 6
35. 8
36. 7
37. 9
38. 8
39. 11
40. 9
41. 9
42. 8
43. 9
44. 7

Problem Set

1. Number line partitioned into sixths; given fractions located and labeled
2. Number line partitioned into fourths; given fractions located and labeled
3. Number line partitioned into thirds; given fractions located and labeled
4. Alex; number line partitioned into fourths; 2 inches and $\frac{7}{4}$ inches located and labeled; number line showing 2 inches is longer than $\frac{7}{4}$ inches
5. Number line with endpoints 0 km to 4 km, partitioned into fifths; $\frac{0}{5}$ (0) km, $\frac{20}{5}$ (4) km, $\frac{7}{5}$ km, $\frac{12}{5}$ km located and labeled

Exit Ticket

1. Number line partitioned into thirds; given fractions located and labeled
2. $\frac{1}{2}$; number line drawn

Homework

1. Number line partitioned into halves; given fractions located and labeled
2. Number line partitioned into thirds; given fractions located and labeled
3. Number line partitioned into fourths; given fractions located and labeled
4. Number line drawn with endpoints 0 km to 4 km, partitioned correctly into thirds; $\frac{0}{3}$ (0) km, $\frac{12}{3}$ (4) km, $\frac{4}{3}$ km, $\frac{10}{3}$ km located and labeled
5. Yes; number line partitioned into fourths; $\frac{19}{4}$ ft. located and labeled; number line showing $\frac{19}{4}$ ft. is smaller than 5 ft.

Lesson 18

Problem Set

1. Answer provided
2. Number line partitioned into sixths; $\frac{2}{6}$ and $\frac{3}{6}$ placed; $\frac{2}{6}$ circled; $<$
3. Number line partitioned into halves and fourths; $\frac{1}{2}$ and $\frac{1}{4}$ placed; $\frac{1}{4}$ circled; $>$
4. Number line partitioned into thirds and sixths; $\frac{2}{3}$ and $\frac{2}{6}$ placed; $\frac{2}{6}$ circled; $>$
5. Number line partitioned into eighths and fourths; $\frac{11}{8}$ and $\frac{7}{4}$ placed; $\frac{11}{8}$ circled; $<$
6. JoAnn, explanations will vary.
7. Red thread, explanations will vary.
8. Number line partitioned into eighths, fourths, and halves; $\frac{7}{8}$, $\frac{7}{4}$, and $\frac{4}{2}$ placed; $\frac{7}{8} < \frac{7}{4} < \frac{4}{2}$; explanations will vary.

Exit Ticket

1. Number line partitioned into fifths; $\frac{3}{5}$ and $\frac{1}{5}$ placed; $\frac{1}{5}$ circled; $>$
2. Number line partitioned into halves and fourths; $\frac{1}{2}$ and $\frac{3}{4}$ placed; $\frac{1}{2}$ circled; $<$
3. No; $\frac{3}{2}$; number line drawn and partitioned into thirds and halves; $\frac{2}{3}$ and $\frac{3}{2}$ placed correctly on the number line; number line showing $\frac{2}{3}$ is closer to 0 than $\frac{3}{2}$

Homework

1. Number line partitioned into thirds; $\frac{1}{3}$ and $\frac{2}{3}$ placed; $\frac{1}{3}$ circled; <
2. Number line partitioned into sixths; $\frac{4}{6}$ and $\frac{1}{6}$ placed; $\frac{1}{6}$ circled; >
3. Number line partitioned into fourths and eighths; $\frac{1}{4}$ and $\frac{1}{8}$ placed; $\frac{1}{8}$ circled; >
4. Number line partitioned into fifths and tenths; $\frac{4}{5}$ and $\frac{4}{10}$ placed; $\frac{4}{10}$ circled; >
5. Number line partitioned into sixths and thirds; $\frac{8}{6}$ and $\frac{5}{3}$ placed; $\frac{8}{6}$ circled; <
6. Jay, explanations will vary.
7. Wendy, explanations will vary.
8. Number line partitioned into sixths, thirds, and halves; $\frac{5}{6}$, $\frac{5}{3}$, and $\frac{3}{2}$ placed; $\frac{5}{6} < \frac{3}{2} < \frac{5}{3}$; explanations will vary.

Lesson 19

Sprint

Side A

1. 2	12. 4	23. 2	34. 9
2. 1	13. 1	24. 1	35. 6
3. 2	14. 5	25. 3	36. 7
4. 3	15. 3	26. 3	37. 6
5. 5	16. 10	27. 4	38. 9
6. 4	17. 2	28. 5	39. 8
7. 5	18. 4	29. 4	40. 8
8. 1	19. 2	30. 5	41. 8
9. 2	20. 1	31. 7	42. 9
10. 3	21. 4	32. 6	43. 9
11. 5	22. 3	33. 7	44. 8

Side B

1. 5	12. 4	23. 2	34. 7
2. 1	13. 10	24. 1	35. 6
3. 2	14. 1	25. 4	36. 7
4. 3	15. 5	26. 3	37. 8
5. 5	16. 3	27. 4	38. 9
6. 4	17. 2	28. 5	39. 8
7. 2	18. 4	29. 4	40. 6
8. 1	19. 2	30. 5	41. 9
9. 2	20. 1	31. 9	42. 8
10. 3	21. 3	32. 6	43. 9
11. 5	22. 3	33. 7	44. 8

Problem Set

1. a. Number line divided into halves; given fractions placed; each whole written correctly as a fraction
   b. Number line divided into fourths; given fractions placed; each whole written correctly as a fraction
   c. Number line divided into eighths; given fractions placed; each whole written correctly as a fraction
2. Row 1: <, <, >
   Row 2: >, <, =
   Row 3: <, >, >
3. Answers will vary.
4. Answers will vary.
5. Answers will vary.

Exit Ticket

1. Number line divided into fourths; given fractions placed; each whole written correctly as a fraction
2. From left to right: <, >, >
3. $\frac{9}{4}$; number line showing 2 wholes, or $\frac{8}{4}$, is closer to 0

Homework

1. a. Number line divided into thirds; given fractions placed; each whole written correctly as a fraction
   b. Number line divided into sixths; given fractions placed; each whole written correctly as a fraction
   c. Number line divided into fifths; given fractions placed; each whole written correctly as a fraction
2. Row 1: >, <, >
   Row 2: =, <, >
   Row 3: >, =, >
3. Answers will vary.
4. Answers will vary.
5. Answers will vary.

Lesson 20

Pattern Sheet

7	14	21	28
35	7	14	7
21	7	28	7
35	7	14	21
14	28	14	35
14	7	14	21
7	21	14	21
28	21	35	21
28	7	28	14
28	21	28	35
28	35	7	35
14	35	21	35
28	14	28	21
35	21	14	28
21	35	14	28

Problem Set

1. a. $\frac{4}{8}, \frac{4}{8}, \frac{3}{8}, \frac{4}{8}$; first, second, and last shapes circled
   b. $\frac{2}{5}, \frac{1}{5}, \frac{2}{5}, \frac{2}{5}$; first, third, and last shapes circled
   c. $\frac{2}{6}, \frac{2}{6}, \frac{4}{6}, \frac{3}{6}$; first and second shapes circled
2. a. $\frac{1}{4}$; two different representations of $\frac{1}{4}$ drawn
   b. $\frac{1}{7}$; two different representations of $\frac{1}{7}$ drawn
3. a. Triangles, squares
   b. 4 triangles, 4 squares
   c. At least two different representations of Ann's set of shapes drawn with no overlaps; $\frac{2}{6}$
4. Cristina, explanations will vary.

Exit Ticket

1. $\frac{3}{5}, \frac{3}{5}, \frac{2}{5}, \frac{3}{5}$; first, second, and last shapes circled
2. a. $\frac{5}{11}$; two different representations of $\frac{5}{11}$ drawn
   b. $\frac{2}{10}$; two different representations of $\frac{2}{10}$ drawn

Homework

1. $\frac{3}{7}$; two different representations of $\frac{3}{7}$ drawn
2. a. No, these shapes are not equivalent. Although both shapes show $\frac{4}{5}$, the units are not the same size.
   b. Two different representations of $\frac{4}{5}$ that are equivalent are drawn correctly.
3. Neither, explanations will vary.

Lesson 21

Problem Set

1. halves: $\frac{0}{2}, \frac{3}{2}$; fourths: $\frac{1}{4}, \frac{2}{4}, \frac{4}{4}, \frac{6}{4}, \frac{8}{4}$

   halves: $\frac{0}{2}, \frac{2}{2}, \frac{3}{2}$; sixths: $\frac{1}{6}, \frac{3}{6}, \frac{6}{6}, \frac{9}{6}, \frac{12}{6}$

2. Shaded blue: $\frac{1}{2}, \frac{2}{4}; \frac{1}{2}, \frac{3}{6}$

   Shaded yellow: $\frac{2}{2}, \frac{4}{4}; \frac{2}{2}, \frac{6}{6}$

   Shaded green: $\frac{3}{2}, \frac{6}{4}; \frac{3}{2}, \frac{9}{6}$

   Shaded red: $\frac{4}{2}, \frac{8}{4}; \frac{4}{2}, \frac{12}{6}$

3. 3; 2, $\frac{4}{4}$; 9, 4

4. $\frac{4}{8}$ inch; number line drawn

5. Yes; $\frac{1}{2} = \frac{2}{4} = \frac{4}{8}$; explanations will vary.

Exit Ticket

Number line drawn; explanations will vary.

Homework

1. fourths: $\frac{0}{4}, \frac{3}{4}, \frac{5}{4}, \frac{6}{4}$; eighths: $\frac{2}{8}, \frac{4}{8}, \frac{7}{8}, \frac{8}{8}, \frac{12}{8}, \frac{16}{8}$

   thirds: $\frac{1}{3}, \frac{3}{3}, \frac{5}{3}$; sixths: $\frac{1}{6}, \frac{4}{6}, \frac{6}{6}, \frac{8}{6}, \frac{10}{6}, \frac{12}{6}$

2. Shaded purple: $\frac{4}{4}$, 1, $\frac{8}{8}$; $\frac{3}{3}$, 1, $\frac{6}{6}$

   Shaded yellow: $\frac{2}{4}, \frac{4}{8}$

   Shaded blue: $\frac{8}{4}$, 2, $\frac{16}{8}$; $\frac{6}{3}$, 2, $\frac{12}{6}$

   Shaded green: $\frac{5}{3}, \frac{10}{6}$

   Answers will vary.

3. Row 1: 2; 8; 4

   Row 2: 6; 6; 16

4. a. Group B; explanations will vary.

   b. $\frac{4}{6}$ (or $\frac{2}{3}$); explanations will vary.

Lesson 22

Problem Set

1. $\frac{1}{2}$ matched to $\frac{2}{4}$
   $\frac{4}{6}$ matched to $\frac{2}{3}$
   $\frac{3}{4}$ matched to $\frac{6}{8}$
   $\frac{3}{9}$ matched to $\frac{1}{3}$
2. 2; 8; 16
3. Explanations will vary.
4. 2 sixths, explanations will vary.
5. Explanations will vary.

Exit Ticket

1. Answers will vary.
2. Answers will vary.

Homework

1. $\frac{1}{2}$ matched to $\frac{3}{6}$
   $\frac{2}{5}$ matched to $\frac{4}{10}$
   $\frac{8}{10}$ matched to $\frac{4}{5}$
   $\frac{2}{8}$ matched to $\frac{1}{4}$
2. 8; 6; 18
3. Explanations will vary.
4. 3 ninths, explanations will vary.
5. 6; explanations will vary.

Lesson 23

Sprint

Side A

1. 6	12. 31	23. 13	34. 84
2. 7	13. 41	24. 23	35. 15
3. 8	14. 51	25. 33	36. 25
4. 9	15. 61	26. 43	37. 35
5. 10	16. 91	27. 53	38. 45
6. 10	17. 12	28. 83	39. 95
7. 9	18. 22	29. 14	40. 81
8. 8	19. 32	30. 24	41. 62
9. 7	20. 42	31. 34	42. 83
10. 6	21. 52	32. 44	43. 94
11. 21	22. 82	33. 54	44. 105

Side B

1. 6	12. 21	23. 13	34. 94
2. 7	13. 31	24. 23	35. 15
3. 8	14. 41	25. 33	36. 25
4. 9	15. 51	26. 43	37. 35
5. 10	16. 81	27. 53	38. 45
6. 10	17. 12	28. 73	39. 85
7. 9	18. 22	29. 14	40. 61
8. 8	19. 32	30. 24	41. 82
9. 7	20. 42	31. 34	42. 63
10. 6	21. 52	32. 44	43. 104
11. 11	22. 92	33. 54	44. 95

Problem Set

1. Number line divided into fourths and labeled correctly in red pencil
2. Number line divided into eighths and labeled correctly in blue pencil
3. $\frac{0}{4} = \frac{0}{8}$, $\frac{1}{4} = \frac{2}{8}$, $\frac{2}{4} = \frac{4}{8}$, $\frac{3}{4} = \frac{6}{8}$, $\frac{4}{4} = \frac{8}{8}$, $\frac{5}{4} = \frac{10}{8}$, $\frac{6}{4} = \frac{12}{8}$, $\frac{7}{4} = \frac{14}{8}$, $\frac{8}{4} = \frac{16}{8}$, $\frac{9}{4} = \frac{18}{8}$, $\frac{10}{4} = \frac{20}{8}$, $\frac{11}{4} = \frac{22}{8}$, $\frac{12}{4} = \frac{24}{8}$
4. $\frac{7}{2} = \frac{14}{4} = \frac{28}{8}$; number line drawn, divided, and labeled correctly with these fractions
5. $\frac{1}{3}$, $\frac{2}{6}$; $\frac{2}{4}$, $\frac{1}{2}$ or $\frac{4}{8}$; $\frac{5}{4}$, $\frac{10}{8}$; $\frac{10}{5}$, answers will vary.
6. No; explanations will vary.

Exit Ticket

$\frac{2}{3}$; number line drawn, divided, and labeled correctly to explain the answer

Homework

1. Number line divided into thirds and labeled correctly with a colored pencil
2. Number line divided into sixths and labeled correctly with another colored pencil
3. $\frac{0}{3} = \frac{0}{6}$, $\frac{1}{3} = \frac{2}{6}$, $\frac{2}{3} = \frac{4}{6}$, $\frac{3}{3} = \frac{6}{6}$, $\frac{4}{3} = \frac{8}{6}$, $\frac{5}{3} = \frac{10}{6}$, $\frac{6}{3} = \frac{12}{6}$, $\frac{7}{3} = \frac{14}{6}$, $\frac{8}{3} = \frac{16}{6}$, $\frac{9}{3} = \frac{18}{6}$
4. 10; 24; number line drawn and labeled with appropriate fractions
5. $\frac{2}{3} = \frac{4}{6}$; $\frac{1}{4} = \frac{2}{8}$; $\frac{7}{4} = \frac{14}{8}$; $\frac{7}{5} = \frac{14}{10}$
6. $\frac{4}{12}$; number line drawn to explain the answer

Lesson 24

Sprint

Side A

1. 7	12. 41	23. 13	34. 15
2. 8	13. 51	24. 23	35. 25
3. 9	14. 91	25. 33	36. 35
4. 10	15. 71	26. 43	37. 45
5. 10	16. 12	27. 53	38. 85
6. 9	17. 22	28. 73	39. 16
7. 8	18. 32	29. 14	40. 26
8. 7	19. 42	30. 24	41. 36
9. 11	20. 52	31. 34	42. 46
10. 21	21. 82	32. 44	43. 56
11. 31	22. 62	33. 94	44. 86

Side B

1. 7	12. 41	23. 13	34. 15
2. 8	13. 51	24. 23	35. 25
3. 9	14. 81	25. 33	36. 35
4. 10	15. 61	26. 43	37. 45
5. 10	16. 12	27. 53	38. 95
6. 9	17. 22	28. 83	39. 16
7. 8	18. 32	29. 14	40. 26
8. 7	19. 42	30. 24	41. 36
9. 11	20. 52	31. 34	42. 46
10. 21	21. 92	32. 44	43. 56
11. 31	22. 72	33. 74	44. 96

Problem Set

1. Halves: Answer provided

   Thirds: Number bond showing 3 units of $\frac{1}{3}$; number line partitioned and labeled from 0 to 1

   Fourths: Number bond showing 4 units of $\frac{1}{4}$; number line partitioned and labeled from 0 to 1

   Fifths: Number bond showing 5 units of $\frac{1}{5}$; number line partitioned and labeled from 0 to 1

2. Fractions equal to 1 circled; $\frac{3}{3}, \frac{4}{4}, \frac{5}{5}$

3. Answers will vary.

4. No, explanations will vary.

Exit Ticket

1. Fourths: Number bond showing 4 units of $\frac{1}{4}$; number line partitioned and labeled from 0 to 1

2. 4 copies; $\frac{4}{4}$

Homework

1. Fifths: Number bond showing 5 units of $\frac{1}{5}$; number line partitioned and labeled from 0 to 1

   Sixths: Number bond showing 6 units of $\frac{1}{6}$; number line partitioned labeled from 0 to 1

   Sevenths: Number bond showing 7 units of $\frac{1}{7}$; number line partitioned and labeled from 0 to 1

   Eighths: Number bond showing 8 units of $\frac{1}{8}$; number line partitioned and labeled from 0 to 1

2. Fractions equal to 1 circled; $\frac{6}{6}, \frac{7}{7}, \frac{8}{8}$

3. Answers will vary, $\frac{9}{9}$

4. No, explanations will vary.

Lesson 25

Sprint

Side A

1.	10	12.	53	23.	17	34.	29
2.	0	13.	4	24.	27	35.	79
3.	20	14.	14	25.	57	36.	59
4.	1	15.	64	26.	77	37.	84
5.	11	16.	5	27.	8	38.	47
6.	31	17.	15	28.	18	39.	36
7.	2	18.	75	29.	28	40.	65
8.	12	19.	6	30.	68	41.	68
9.	42	20.	16	31.	48	42.	89
10.	3	21.	76	32.	9	43.	45
11.	13	22.	7	33.	19	44.	86

Side B

1.	0	12.	83	23.	17	34.	29
2.	10	13.	4	24.	27	35.	69
3.	20	14.	14	25.	47	36.	49
4.	1	15.	84	26.	67	37.	64
5.	11	16.	5	27.	8	38.	57
6.	61	17.	15	28.	18	39.	46
7.	2	18.	35	29.	28	40.	75
8.	12	19.	6	30.	58	41.	58
9.	72	20.	16	31.	38	42.	79
10.	3	21.	36	32.	9	43.	85
11.	13	22.	7	33.	19	44.	46

Problem Set

1. Answer provided; $\frac{3}{2}$; $\frac{3}{1}$; $\frac{4}{4}$; $\frac{4}{2}$; $\frac{4}{1}$; $\frac{6}{6}$; $\frac{6}{3}$; $\frac{6}{1}$
2. $\frac{0}{1}, \frac{1}{1}, \frac{2}{1}, \frac{3}{1}, \frac{4}{1}, \frac{5}{1}, \frac{6}{1}$

   10, 11, $\frac{12}{1}$, $\frac{13}{1}$, 14, $\frac{15}{1}$, $\frac{16}{1}$
3. Explanations will vary.

Exit Ticket

1. $\frac{5}{1}$
2. Number line partitioned into thirds; fraction for 3 wholes renamed as $\frac{9}{3}$

Homework

1. $\frac{4}{4}$; $\frac{4}{2}$; $\frac{4}{1}$; $\frac{8}{8}$; $\frac{8}{4}$; $\frac{8}{1}$
2. $\frac{0}{1}, \frac{4}{1}, \frac{6}{1}, \frac{12}{1}$

   15, 16, $\frac{17}{1}$, $\frac{18}{1}$, 19, $\frac{20}{1}$, $\frac{21}{1}$
3. Explanations will vary.

Lesson 26

Sprint

Side A

1. 8
2. 9
3. 10
4. 10
5. 9
6. 8
7. 11
8. 21
9. 31
10. 41
11. 51
12. 91
13. 12
14. 22
15. 32
16. 42
17. 52
18. 82
19. 13
20. 23
21. 33
22. 43
23. 73
24. 14
25. 24
26. 34
27. 44
28. 94
29. 54
30. 15
31. 25
32. 35
33. 45
34. 85
35. 16
36. 26
37. 36
38. 46
39. 76
40. 17
41. 27
42. 37
43. 47
44. 97

Side B

1. 8
2. 9
3. 10
4. 10
5. 9
6. 8
7. 11
8. 21
9. 31
10. 41
11. 51
12. 81
13. 12
14. 22
15. 32
16. 42
17. 52
18. 92
19. 13
20. 23
21. 33
22. 43
23. 63
24. 14
25. 24
26. 34
27. 44
28. 74
29. 64
30. 15
31. 25
32. 35
33. 45
34. 75
35. 16
36. 26
37. 36
38. 46
39. 86
40. 17
41. 27
42. 37
43. 47
44. 97

Problem Set

1. Halves: 0, 0; 2, 2; 4; number bonds completed

   Thirds: 6, 6; 9, 9; 12, 12; number bonds completed

2. Halves: Answer provided

   Thirds: $\frac{6}{3}, \frac{9}{3}, \frac{12}{3}$

   Fourths: $\frac{8}{4}, \frac{12}{4}, \frac{16}{4}$

   Sixths: $\frac{12}{6}, \frac{18}{6}, \frac{24}{6}$

3. a. Number line representing 1 meter of wire; partitioned correctly into fourths; 4

   b. 12 days

4. a. Number line representing 1 pound of food; partitioned correctly into thirds

   b. Second number line representing 4 pounds of food; partitioned correctly into thirds; 1

   c. 2

Exit Ticket

a. Number line drawn to represent 2 yards of fabric

b. Number line partitioned and labeled to show fifths

c. 10; number bond completed

Homework

1. Sixths: 0, 0; 6, 6; 12; number bond completed

   Fifths: 10, 10; 15, 15; 20, 20; number bonds completed

2. Thirds: Answer provided

   Sevenths: $\frac{14}{7}, \frac{21}{7}, \frac{28}{7}$

   Eighths: $\frac{16}{8}, \frac{24}{8}, \frac{32}{8}$

   Tenths: $\frac{20}{10}, \frac{30}{10}, \frac{40}{10}$

3. Number line drawn to represent the basketball court, partitioned into thirds, and labeled correctly;

   Day 1: $\frac{1}{3}$, Day 2: $\frac{2}{3}$, Day 3: $\frac{3}{3}$; $\frac{3}{3}$

Lesson 27

Sprint

Side A

1. 10
2. 0
3. 20
4. 1
5. 11
6. 31
7. 2
8. 12
9. 42
10. 3
11. 13
12. 53
13. 4
14. 14
15. 64
16. 5
17. 15
18. 75
19. 6
20. 16
21. 76
22. 7
23. 17
24. 27
25. 57
26. 77
27. 8
28. 18
29. 28
30. 68
31. 48
32. 9
33. 19
34. 29
35. 79
36. 59
37. 83
38. 46
39. 35
40. 64
41. 67
42. 49
43. 88
44. 85

Side B

1. 0
2. 10
3. 20
4. 1
5. 11
6. 61
7. 2
8. 12
9. 72
10. 3
11. 13
12. 83
13. 4
14. 14
15. 84
16. 5
17. 15
18. 35
19. 6
20. 16
21. 36
22. 7
23. 17
24. 27
25. 47
26. 67
27. 8
28. 18
29. 28
30. 58
31. 38
32. 9
33. 19
34. 29
35. 69
36. 49
37. 63
38. 56
39. 45
40. 74
41. 67
42. 59
43. 78
44. 45

Problem Set

1. 2, 2, bigger, less
   4, 4, smaller, more
2. $\frac{1}{2}$ of a candy bar, $\frac{1}{2} = \frac{2}{4} = \frac{3}{6}$
3. Explanations will vary.
4. 2 sixths; model drawn
5. Answers will vary.

Exit Ticket

1. 8; 8
2. Two models showing fractions equivalent to those in Problem 1 drawn and labeled
3. Answers will vary.

Homework

1. 1, 1, bigger
   3, 3, smaller
2. $\frac{1}{4}$ of a pizza, $\frac{1}{4} = \frac{2}{8}$
3. Explanations will vary.
4. 2 eighths; model drawn to support the answer
5. 16 slices; explanations will vary.

Lesson 28

Sprint

Side A

1. 10
2. 0
3. 20
4. 1
5. 11
6. 31
7. 2
8. 12
9. 42
10. 3
11. 13
12. 63
13. 4
14. 14
15. 74
16. 5
17. 15
18. 75
19. 6
20. 16
21. 26
22. 46
23. 66
24. 7
25. 17
26. 27
27. 77
28. 57
29. 8
30. 18
31. 28
32. 88
33. 68
34. 9
35. 19
36. 29
37. 79
38. 59
39. 62
40. 54
41. 76
42. 58
43. 83
44. 67

Side B

1. 0
2. 10
3. 20
4. 1
5. 11
6. 61
7. 2
8. 12
9. 52
10. 3
11. 13
12. 73
13. 4
14. 14
15. 44
16. 5
17. 15
18. 85
19. 6
20. 16
21. 26
22. 66
23. 86
24. 7
25. 17
26. 27
27. 87
28. 67
29. 8
30. 18
31. 28
32. 58
33. 38
34. 9
35. 19
36. 29
37. 89
38. 69
39. 72
40. 63
41. 45
42. 37
43. 79
44. 46

Problem Set

1. Models shaded correctly; 2 thirds circled
2. Models shaded correctly; 2 eighths circled
3. Models shaded correctly; 3 fourths circled
4. Models shaded correctly; 4 sixths circled
5. Models shaded correctly; 3 thirds circled
6. Kelly; tape diagrams drawn correctly
7. Becky; tape diagrams drawn correctly
8. Doll B, Doll A, Doll C; picture drawn

Exit Ticket

1. 2 thirds; explanations will vary.
2. Models drawn correctly for each fraction; 3 sevenths circled

Homework

1. Models shaded correctly; 1 half circled
2. Models shaded correctly; 2 fourths circled
3. Models shaded correctly; 4 fifths circled
4. Models shaded correctly; 5 sevenths circled
5. Models are shaded correctly; 4 fourths circled
6. Saleem; tape diagrams drawn correctly
7. Lily; tape diagrams drawn correctly

Lesson 29

Pattern Sheet

8	16	24	32
40	48	56	64
72	80	40	48
40	56	40	64
40	72	40	80
48	40	48	56
48	64	48	72
48	56	48	56
64	56	72	56
64	48	64	56
64	72	72	48
72	56	72	64
72	64	48	72
56	72	48	64
72	56	48	64

Problem Set

1. Answer provided
2. $\frac{3}{4} > \frac{3}{8}$
3. $\frac{1}{4} < \frac{1}{2}$
4. $\frac{4}{4} > \frac{4}{6}$
5. a. <
   b. >
   c. >
6. Models drawn correctly; <
7. Models drawn correctly; >
8. Nicholas; models drawn correctly
9. Robbie; models drawn correctly

Exit Ticket

1. >
2. Number lines drawn; explanations will vary.

Homework

1. $\frac{5}{12} < \frac{5}{6}$
2. $\frac{6}{6} > \frac{6}{12}$
3. $\frac{2}{4} > \frac{2}{9}$
4. $\frac{1}{2} > \frac{1}{6}$
5. a. <

   b. <

   c. >
6. Models drawn correctly; <
7. Models drawn correctly; >
8. Michello; models drawn correctly
9. Jahsir; models drawn correctly

Lesson 30

Pattern Sheet

9	18	27	36
45	9	18	9
27	9	36	9
45	9	18	27
18	36	18	45
18	9	18	27
9	27	18	27
36	27	45	27
36	9	36	18
36	27	36	45
36	45	9	45
18	45	27	45
36	18	36	27
45	27	18	36
27	45	18	36

Homework

Answers will vary.